# GEOMETRY
## OF
# TIME AND SPACE

# CAMBRIDGE
## UNIVERSITY PRESS

University Printing House, Cambridge CB2 8BS, United Kingdom

Published in the United States of America by Cambridge University Press, New York

Cambridge University Press is part of the University of Cambridge.

It furthers the University's mission by disseminating knowledge in the pursuit of education, learning and research at the highest international levels of excellence.

www.cambridge.org
Information on this title: www.cambridge.org/9781107631809

© Cambridge University Press 1936

First published 1936
First paperback edition 2014

A catalogue record for this publication is available from the British Library

ISBN 978-1-107-63180-9 Paperback

Cambridge University Press has no responsibility for the persistence or accuracy of URLs for external or third-party internet websites referred to in this publication, and does not guarantee that any content on such websites is, or will remain, accurate or appropriate.

# GEOMETRY

## OF

# TIME AND SPACE

by

ALFRED A. ROBB

Sc.D., D.Sc., Ph.D., F.R.S.

CAMBRIDGE

AT THE UNIVERSITY PRESS

1936

"The Bird of Time has but a little way
To fly—and Lo! the Bird is on the Wing."

OMAR KHAYYÁM

"I could not have been in two places at once
unless I were a bird."  SIR BOYLE ROCHE

Contrary to the view so generally held; not
even "the Bird of Time" can be in two places
at once.  AUTHOR

# PREFACE

THE present volume is essentially a second edition of one which was published by the author in 1914 under the title: *A Theory of Time and Space*. An alteration of the title has been made, since it was considered that the word *geometry* conveyed a somewhat better idea of the nature of the contents of the book than did the word *theory*.

The first edition was going through the press at the time of the outbreak of the war, so that its publication took place under very unfavourable circumstances. The present volume differs from its predecessor in several respects. The Introduction has been re-written and extended; while the proofs of a number of theorems, which were rather lengthy, have been curtailed and simplified.

A considerable amount of new matter has also been introduced, making the book more self-contained and complete.

The demonstrations have all been carried out as deductions from certain postulates expressed in terms of the relations of *after* and *before*; so that the whole work may be regarded as a demonstration of the fundamental character of these relations in Time-Space theory.

So far as I am aware, the book, in its original form, was the first of its kind to be written, and a brief account of its origin may be of interest. At the meeting of the British Association held at Belfast in 1902, Lord Rayleigh gave a paper entitled: *Does Motion through the Ether cause double Refraction?* in which he described certain experiments which he had carried out with the object of testing this matter, and which seemed to indicate that the answer was in the negative.

I remember that he inquired of Professor Larmor, who was present on this occasion, whether, from his theory, he would expect double refraction to be produced in this way. Professor Larmor replied that he would not, and, in the discussion which followed considerable surprise was expressed that, in any attempt to detect motion through the aether, things seemed to conspire together so as to give null results. The impression which this discussion made upon me was, that, in order properly to understand the matter, it would be necessary to make some sort of analysis of one's ideas concerning equality of lengths, etc., and I decided that, at some future time, I should attempt to carry this out. I am not quite certain that I had not some idea of the sort prior to this meeting, but, in any case, the inspiration came from Professor Larmor, either then, or on some previous occasion while attending his lectures.

Some years later I attempted to carry out this scheme, and, while doing so, I heard for the first time of Einstein's work.

I may say that, from the first, I felt dissatisfied with his approach to the subject, and I decided to continue my own efforts to find a suitable basis for a theory.

The first work which I published on the subject was a pamphlet which appeared in 1911 entitled: *Optical Geometry of Motion: A New View of the Theory of Relativity.*

This pamphlet was of an exploratory character and did not profess to give a complete logical analysis of the subject; but nevertheless, although bearing a very different aspect, it contained some of the germs of my later work. It was, in fact, an attempt to describe Time-Space relations without making any assumption as to the simultaneity of events at different places. Later on, the idea of *Conical Order* occurred to me, in which instants at different places are regarded as definitely distinct; so that there is no such simultaneity.

As it was evident that a thorough working out of this idea would entail a great deal of labour, I published, in 1913, a short preliminary account of it under the title: *A Theory of Time and Space.*

In 1914, as above mentioned, I published a book bearing the same title, of which the present volume is a second edition.

The working out of a scientific theory in the form of a sequence of propositions, such as was done by Euclid, Newton and others, seems largely to have gone out of vogue in these latter days and I consider that this is rather regrettable.

No doubt, in doing exploratory work, other methods are permissible and necessary, but I think that the incorporation of the more fundamental parts of a theory in a sequence of propositions should always be kept in view, since, in this way, one is able to see much more readily what are our primary assumptions, and one is able to fall back upon these in cases of difficulty.

One can also test the effect on a theory of an alteration in one or more of these primary assumptions such, for instance, as that produced in ordinary geometry by the rejection of the Euclidean axiom of parallels and the substitution for it of some other primary assumption, such as that of Lobatschewski. It will be found that the theory developed in this work is dependent upon the rejection of one generally accepted postulate with regard to instants of time and the substitution of others.

In conclusion, I desire once more to express my indebtedness to Sir Joseph Larmor, without whom this book would never have been written; and to convey my best thanks to the officials and staff of the Cambridge University Press for the care and skill with which they have carried out the printing.

ALFRED A. ROBB

CAMBRIDGE
20 *November* 1935

# INTRODUCTION

IN beginning the study of Geometrical Science it is customary to start with a course of pure geometry and, when a foundation of this has been laid, to proceed to the introduction of coordinate methods.

Thus, before being introduced to Cartesian geometry, one is taught certain propositions concerning the congruence of triangles, the properties of parallels, the theorem of Pythagoras, the theory of proportion, etc. To a large extent the methods of pure and of coordinate geometry are then carried on side by side, and it is customary, in proving a proposition, to make use of whichever method appears to be more convenient for the particular purpose in hand.

Speaking generally, no doubt, this is the course of procedure by which progress is most rapidly made, but I do not think that anyone would have the temerity to suggest that coordinate methods should be taken up without some prior grounding in pure geometry.

When one goes on to the study of other types of geometry than the Euclidean, the importance of logical sequence should become apparent, but I am sorry to say that it does not always seem to do so.

In many discussions of Time-Space theory we find ideas of ordinary Euclidean geometry carried forward into a domain in which they no longer apply, with occasional disastrous results.

The extension of Cartesian coordinates from three to four or more dimensions does not offer any very serious difficulties, since the formula for the square of the distance between two points, which, in three dimensions, has the form

$$s^2 = (x_1 - x_2)^2 + (y_1 - y_2)^2 + (z_1 - z_2)^2,$$

becomes simply

$$s^2 = (x_1 - x_2)^2 + (y_1 - y_2)^2 + (z_1 - z_2)^2 + (w_1 - w_2)^2$$

in four dimensions; with a similar extension for any larger number.

It was found, however, by Minkowski that many of the facts connected with Time-Space theory could conveniently be represented by a four-dimensional coordinate geometry in which a formula

$$s^2 = (x_1 - x_2)^2 + (y_1 - y_2)^2 + (z_1 - z_2)^2 - (t_1 - t_2)^2$$

held; that is to say, a formula in which one square is affected with the negative sign.

This negative sign makes an enormous difference in the subject and renders invalid a great part of what holds in ordinary Euclidean geometry.

Some idea of the extent of the modifications required may perhaps be obtained when I state that the construction of the very first proposition of Euclid becomes impossible except in a certain type of plane, and that two other types of plane occur in which an equilateral triangle cannot exist.

Numerous other features of this Time-Space geometry are so curious as to seem at first quite paradoxical, and some consideration of a few of these from the coordinate standpoint may perhaps emphasise the importance of laying a proper foundation for the subject on the purely geometrical side.

It is to be observed in the first place that whereas the expression

$$(x_1 - x_2)^2 + (y_1 - y_2)^2 + (z_1 - z_2)^2$$

(which may briefly be written in the form

$$\delta x^2 + \delta y^2 + \delta z^2)$$

is always positive for real values of $\delta x$, $\delta y$, $\delta z$ which are not all zero; the expression

$$(x_1 - x_2)^2 + (y_1 - y_2)^2 + (z_1 - z_2)^2 - (t_1 - t_2)^2$$

(or

$$\delta x^2 + \delta y^2 + \delta z^2 - \delta t^2)$$

may be either positive, zero or negative for real values of $\delta x$, $\delta y$, $\delta z$, $\delta t$ differing from zero.

Three types of line joining the points $(x_1, y_1, z_1, t_1)$, $(x_2, y_2, z_2, t_2)$ exist corresponding to these three cases.

When the expression is positive the square of the distance between the points is given by the formula

$$\delta s^2 = \delta x^2 + \delta y^2 + \delta z^2 - \delta t^2.$$

When the expression is negative, then, analytically, $\delta s$ becomes a pure imaginary; but if we write

$$\delta \bar{s}^2 = -\delta s^2$$

and recollect that we are now dealing with a line of a different type, we get the square of the distance in these new units given by

$$\delta \bar{s}^2 = \delta t^2 - \delta x^2 - \delta y^2 - \delta z^2.$$

When the expression is zero one is tempted to think that the distance between the points must be zero; but this is a misleading interpretation. The real interpretation of the equation

$$(x_1 - x_2)^2 + (y_1 - y_2)^2 + (z_1 - z_2)^2 - (t_1 - t_2)^2 = 0$$

is that the points $(x_1, y_1, z_1, t_1)$, $(x_2, y_2, z_2, t_2)$ lie in a particular type of line.

For this type of line the conception of length partially, but not entirely, breaks down.

We may compare lengths along a given line of this kind, or along two such parallel lines, but not along two which are not parallel.

Consider the case of lines for which the expression

$$\delta x^2 + \delta y^2 + \delta z^2 - \delta t^2$$

is positive.

It is obvious that the axes of $x$, $y$ and $z$ (but *not* the axis of $t$), are lines of this character.

Now let $O$ be the origin of coordinates and let $P$ be any point in the positive axis of $x$ and let $OP = 2l$.

Let $A_1$, $A_2$ and $A_3$ be three points whose coordinates are given by the following table:

|     | $A_1$ | $A_2$ | $A_3$ |
| --- | --- | --- | --- |
| $x$ | $l$ | $l$ | $l$ |
| $y$ | $b$ | $c$ | $0$ |
| $z$ | $0$ | $0$ | $0$ |
| $t$ | $0$ | $c$ | $Kl$ |

(where $K^2 < 1$).

Then
$$OA_1^2 = l^2 + b^2 \qquad \therefore OA_1 > l,$$
$$OA_2^2 = l^2 \qquad \therefore OA_2 = l,$$
$$OA_3^2 = (1 - K^2)\, l^2 \quad \therefore OA_3 < l.$$

Similarly
$$PA_1 > l,$$
$$PA_2 = l,$$
$$PA_3 < l.$$

Thus
$$OA_1 + PA_1 > OP,$$
$$OA_2 + PA_2 = OP,$$
$$OA_3 + PA_3 < OP,$$

so that, in this geometry, we have two sides of a triangle together greater than, equal to, or less than the third. But the side $OP$ is common to all three triangles, so that its length is neither a minimum nor a maximum.

The question naturally arises: If such a line be neither a minimum nor a maximum, what is it? Our ordinary idea of a straight line breaks down.

The case is rather different if we take a triangle one of whose sides is a part of the axis of $t$. Thus let $Q$ be any point on the positive axis of $t$ such that $OQ = \lambda$ and let $A$ be a point whose coordinates are $(a, b, c, d)$, where $\lambda > d > 0$.

In order that the three sides of our triangle may be all lines of the same kind we shall suppose that $a$, $b$ and $c$ are so small that

$$a^2 + b^2 + c^2 < d^2,$$

and also

$$a^2 + b^2 + c^2 < (\lambda - d)^2.$$

Then

$$OA = \sqrt{d^2 - a^2 - b^2 - c^2} < d,$$

and

$$AQ = \sqrt{(\lambda - d)^2 - a^2 - b^2 - c^2} < \lambda - d.$$

Thus

$$OA + AQ < OQ$$

and we have two sides of the triangle together less than the third.

This will be the case for all values of $a$, $b$ and $c$ provided that these are sufficiently small, and it is obvious that a similar property will hold for any part of the interval $OQ$: so that here $OQ$ is a line of *maximum length* in the mathematical sense.

This again is something quite different from what we have in Euclidean geometry and once more our ordinary idea of a straight line breaks down.

The *normality* of lines, etc., exhibits some very curious features in this geometry.

The equations of a line may be put in the form

$$\frac{x - x_1}{l} = \frac{y - y_1}{m} = \frac{z - z_1}{n} = \frac{t - t_1}{p}.$$

If

$$\frac{x - x_2}{l'} = \frac{y - y_2}{m'} = \frac{z - z_2}{n'} = \frac{t - t_2}{p'}$$

be a second line, the analytic condition of normality is found to be

$$ll' + mm' + nn' - pp' = 0.$$

If a line be such that

$$l^2 + m^2 + n^2 - p^2 = 0,$$

then analytically, it must be regarded as being *normal to itself*, and this is the type of line for which, as we have already remarked, the conception of length partially breaks down.

Proceeding from the purely analytical standpoint it is easily shown that a line

$$\frac{x}{l} = \frac{y}{m} = \frac{z}{n} = \frac{t}{p}$$

will be normal to a threefold whose equation is

$$lx + my + nz - pt = 0.$$

Now any other line through the origin whose equations are

$$\frac{x}{l'} = \frac{y}{m'} = \frac{z}{n'} = \frac{t}{p'}$$

will lie in this threefold provided that

$$ll' + mm' + nn' - pp' = 0.$$

If the line $(l, m, n, p)$ be such that

$$l^2 + m^2 + n^2 - p^2 = 0,$$

then it must itself lie in this threefold to which it is normal.

Thus all lines in such a threefold will be normal to this particular line and, of course, to its parallels, and, if we take $z = 0$, we get a plane such that all its lines are normal to a particular set of parallel lines lying in the plane.

Here again is something quite different from what we get in ordinary Euclidean geometry.

It will be found that there are three types of plane and three types of threefold, just as there are three types of line, and the geometrical characters of these are quite distinct from one another.

From the analytical examples which we have given it is evident that this geometry differs in some of the most fundamental respects from that of Euclid and it is clear that, from the pure geometrical standpoint, it must be built up in an entirely different way from that which he employed.

It will be found however that it not only contains Euclidean geometry as an essential part, but that it supplies also what is perhaps the most satisfactory theoretical basis upon which to construct the Euclidean system.

Now we have seen that in this geometry we cannot take a "straight" line as being a *shortest* line and it will be necessary to define it in some other way.

Further, since the coordinate axes in Minkowski's analytical work are supposed to be "straight", we are faced with a serious difficulty even before we are in a position to set up a system of axes.

Moreover the Minkowski axes are supposed to be "normal" to one another and we have seen that there are some rather curious features connected with normality.

We must accordingly build up the subject from the very beginning and must look about us for suitable postulates.

Now in the first place: what is geometry in the general abstract sense?

Geometry has been defined by Whitehead as the "science of cross classification".

The fundamental elements classified are usually called "points" but any entities which satisfy certain postulates may serve the purpose.

Using this definition we may have "geometries" with only a finite number of fundamental elements; but, though interesting as logical curiosities, such systems have no special application in the present state of Science.

The types of geometry with which we are specially concerned when we attempt to map out time and space involve an infinite set of elements forming what is called a "continuum".

The classes of these elements, such as lines, planes, etc., with which we are concerned, are defined by means of certain relations among the elements involved.

In order that the system should be of any use for mapping purposes it is necessary that these relations should have their counterparts in physical space or time.

As to whether these physical counterparts exist or not, the geometry, as a branch of pure mathematics, need not concern itself; but, since the interest of the subject to many persons depends mainly upon the application, we shall devote a little time to a consideration of these matters.

Now in considering the subject of time as it presents itself to our experience there is one very important respect in which it appears to differ from our spacial experience.

Of any two instants which one experiences *in one's own mind* one is *after* the other.

This relation of *after* is what is called an *asymmetrical relation*; by which is meant a relation $R$ such that if $B$ bears the relation $R$ to $A$ then $A$ does not bear the relation $R$ to $B$.

Thus, in the particular case considered, if $B$ is *after* $A$, then $A$ is not *after* $B$.

There are however relations which are *symmetrical*; such, for example, as the relation of *equality*, where if $B$ is *equal to* $A$ then $A$ is *equal to* $B$.

Now the relation of two points or two particles in space is a *symmetrical relation* and, if $A$ and $B$ be taken as two distinct points, there is no reason why we should say that $B$ is *after* $A$ rather than that $A$ is *after* $B$.

If we consider points in a straight line it would, of course, be possible to set up some convention according to which we might regard one point as being *after* another; but such convention would be perfectly arbitrary and would not correspond to any natural distinction, as in the case of instants of time in our own consciousness.

Let us consider what actually does hold with regard to the latter.

It is hardly possible to describe what we mean when we use the word *Now*. *Now* singles itself out in the mind and is, as the Germans say, "ausgezeichnet" in some way or other.

Though we speak of *Now* as an *instant*, yet there are innumerable instants, each of which is in its turn a *Now*.

These instants which one experiences in one's own mind have, as already pointed out, an asymmetrical relation one to another; and our very thoughts themselves have a time order, so that we recognize one thought as following *after* another, even if we close our eyes and other channels of sense as far as possible.

We shall not therefore attempt to make any unreal distinction between what is physical and what is mental in respect of the perceptions of a single individual.

These perceptions form a complex picture which is continually changing and, if one splits it up into component parts, one is able to say (at least approximately) that certain events occur at the same instant, while others occur at different instants.

This simultaneity, or lack of it, is an ultimate fact and must be regarded as *absolute*; but we must carefully note what things we are asserting to be simultaneous or otherwise. We are making the assertion about *certain perceptions of a single individual*.

A normal individual who is not a solipsist (and a solipsist could hardly be regarded as a normal individual) believes in the existence of more than his own self and his own perceptions, and one is accustomed to regard these perceptions, under normal circumstances, as representing things as real as one's self but in some sense *external*.

One naturally thinks of these assumed external events as having a time order, and the first standpoint which one is accustomed to adopt, and which, as a matter of fact, serves for most of the purposes of our daily life, is that these external events occur at the instants at which one perceives them.

More careful observation however convinces one that this cannot be strictly correct, at any rate for all our perceptions, since the perception of an event by one of our senses may be *after* the perception of the same event by another sense.

Thus the visual and auditory perceptions of a blow being struck by a hammer are practically simultaneous when the occurrence is close at hand; but the auditory perception is appreciably *after* the visual perception when the occurrence is at a distance from the observer.

Thus the auditory perception, at any rate, cannot be simultaneous with the distant event and the question naturally arises whether the visual perception is so or not; and, once more, the answer is in the negative.

The first indication that this is the case was obtained by Römer in 1675–6, through observations of the eclipses of Jupiter's satellites; and, though there was a possibility of some other explanation of these observations, such possibility practically vanished when Fizeau, in 1849, was able to test the matter by direct experiment.

Fizeau found that when a flash of light was sent out from the neighbourhood of an observer to a distant mirror which reflected it back to him, the return of the flash occurred at an instant appreciably *after* the instant of its departure.

Thus the instant of one's visual perception of a distant event cannot be identical with the instant at which the event occurs, and we perceive near and distant events simultaneously which certainly do not occur simultaneously.

This fact cannot be ignored if we attempt to correlate astronomical events with one another or with terrestrial ones; although in the ordinary affairs of daily life we can and do ignore it with impunity.

If now we attempt to identify the instant at which a distant event occurs with that of some event near at hand, we find ourselves confronted with very serious difficulties, since this question is intimately bound up with the question of the identification of one and the same point of space (or the aether of space, if there be such a thing) at different instants of time.

If this latter were possible one would be able to tell when a particle was at "absolute rest" that is to say it would be possible to state that it remained at the same point of space (or of the aether).

If we had an apparatus such as that which Fizeau employed and we could be assured that it remained at "absolute rest" in this sense, and if, for the moment, we neglect any difficulties which there may be in connexion with measurement of space intervals, time intervals or velocity, it would be reasonable to assume that light travels through space (or the aether) with uniform velocity and would take equal intervals of time on its outward and return journeys; so that the instant at the observing station which was midway between the instants of departure and return of the light flash would be identical with the instant of its reflection at the distant mirror.

If, on the other hand, we suppose that observer and apparatus are both in uniform motion, say in the direction of the outward going light, then the mirror would retire in front of the outward going flash, while the observer would advance to meet the returning one, so that the light would have further to travel on its outward than on its return journey, and the instant at the observing station midway between the instants of departure and return would no longer be identical with the instant of reflection.

Now according to the classical mechanics a system of bodies whose centre of inertia is in uniform motion in a straight line is indistinguishable, so far as mechanical effects are concerned, from a similar system whose centre of inertia is at rest.

It is conceivable that some difference might be detected by some optical or electrical device, and many attempts have been made with the object of detecting the motion of the earth through the aether; but none of these attempts has been successful.

Of these attempts, the best known is the celebrated experiment of Michelson and Morley, which consisted in dividing a beam of light into two portions which travelled, the one in one direction and the other in a transverse direction and were reflected back again by mirrors.

If we adopt ordinary ideas for the moment and suppose the light to be propagated with a velocity $v$ through a medium and that the apparatus moves through the medium with velocity $u$; then it is

easy to calculate the time of the double journey for the two portions of
the beam.

For the case of a part of the beam which travels in the direction of
motion of the apparatus the time occupied by the double journey is
found to be

$$t_1 = \frac{2va_1}{v^2 - u^2},$$

where $a_1$ is the distance between the point of the apparatus where the
beam divides and the corresponding mirror.

If $a_2$ be the corresponding distance for the case of the transverse
portion of the beam, then we can easily show that the time of the
double journey should be

$$t_2 = \frac{2a_2}{\sqrt{v^2 - u^2}}.$$

Now if the distances $a_1$ and $a_2$ be adjusted so that the times occupied
by the two portions of the beam on their journeys are equal, we have

$$\frac{2va_1}{v^2 - u^2} = \frac{2a_2}{\sqrt{v^2 - u^2}}.$$

From this it follows that:

$$a_1 = a_2 \sqrt{1 - \left(\frac{u}{v}\right)^2};$$

so that $a_1$ should be somewhat less than $a_2$.

Now the necessary adjustment can be made with extreme accuracy
by means of the optical interference bands which are produced and the
remarkable fact is observed that, when the whole apparatus is caused
to revolve at a uniform slow rate, the one adjustment holds for all
positions.

Thus the apparatus gives no evidence of the motion of the earth,
although it might be expected to do so.

In order to explain this result the hypothesis was put forward by
FitzGerald and Lorentz that the material of the apparatus contracts
along the direction of its motion through the aether in the ratio

$$1 : \sqrt{1 - \left(\frac{u}{v}\right)^2}.$$

If however this FitzGerald-Lorentz contraction occurs and bodies
change their dimensions in this manner when they move, and if we are
unable to detect this motion, what do we really mean by a body remaining of constant length, or of being equal in length to another body?

If such be the case, the distance between the graduations of the most rigid measuring rod will change as the rod is turned in different directions with respect to the earth's motion, and similarly, the shape of the most rigid material triangle will change when we try to superpose it on a material triangle of different orientation.

Admittedly such changes would be very minute under ordinary circumstances, and could generally be neglected; but then, on the other hand, if they are non-existent, how can one explain the null result of the Michelson-Morley experiment, especially in view of the results of some experiments by Lodge which seemed to show that the aether was not carried along by matter moving in the neighbourhood?

Thus we appear to be confronted with formidable difficulties, since, not only can we give no criterion by which to decide that a distant event is simultaneous with one near at hand, but even those physical properties of solid bodies of which use is made in the ordinary measurements of length appear open to question.

The first great steps towards reducing this matter to order were taken by Larmor and Lorentz. These writers showed that the electromagnetic equations could be reduced to the same form for a system moving through an assumed aether as they had for a system "at rest"; and, on the question being raised by Lord Rayleigh in 1902, as to whether rotatory polarisation would be influenced by the earth's motion, and whether such motion would cause double refraction, Larmor was able, from his theory, to predict that no such effects would occur; and this was confirmed by Lord Rayleigh's experiments.

The transformation of the electromagnetic equations involved the introduction of a so-called "local time" and this raises the question as to what is the philosophical significance of this conception.

The view which was put forward by Einstein was that events could be simultaneous for one observer but not simultaneous for another moving with respect to the first.

This view, in my opinion, gives an air of unreality to the external world which cannot be justified; since the events might be the impacts of particles moving with respect to one another, and therefore associated with different "local times", although an impact necessarily involves both particles which impinge and cannot be described without mention of both.

We also think of a definite instant of impact which can be referred to without any mention of "local times" in this sense.

As has already been pointed out, the only simultaneity with which one is directly acquainted, namely, that of perceptions or ideas in one's own mind, is of an *absolute* character and my contention is that any real simultaneity of external events is also absolute in a similar way.

Let us now examine Einstein's standpoint in order to show in what respect he departs from actually observed or observable facts.

If a flash of light goes out from the neighbourhood of an observer to a distant mirror and is there reflected back to him, then, according to Einstein, the reflection at the distant mirror is simultaneous with an event at the observing station which takes place at the instant midway between the instants of departure and return of the flash of light.

Einstein supposes the instant midway between the instants of departure and return to be determined by means of a clock. Ignoring for the moment the difficulty involved in obtaining an accurate clock; let us consider what this implies.

Let us suppose that to-day I were to observe the outburst of a new star which, in astronomical language, was at a distance of 100 light years, then according to Einstein's view this outburst was simultaneous with terrestrial events which occurred before I was born.

It is evident that this could not be a fact of observation, so far as I am concerned: so that it would be incorrect to speak of such events as *simultaneous for one observer*.

It is frequently asserted that Einstein's theory keeps strictly to observed or observable facts; but here would be a palpable departure from the facts of observation.

The actual observed fact in such a case would be that my *perception* of the outburst would be simultaneous with other experiences of mine occurring to-day.

These are the sort of events which are simultaneous to one observer, and not the occurrence of a distant and a near event, and such simultaneity is *absolute*.

In case it be contended that the above is Einstein's definition of the simultaneity of distant and near events, then our reply is: that if this be so the word simultaneity is being used in two utterly distinct senses in a manner which may lead to great confusion of thought.

In the one case the word is employed correctly to describe something *absolute* while in the other it would be used to describe a mere convention which has not even the merit of being definite without the limitation that the observing station is *unaccelerated* in the interval between the departure and return of the flash of light (or else is accelerated in certain restricted ways).

It is perhaps desirable to point out that it is Einstein's philosophy which I am here attacking and not his mathematics.

This is all the more justifiable, in that the conception of "local time" as a mathematical quantity was introduced, not by Einstein, but by others who did not hold his views.

Even if it were the case that near and distant events were simultaneous, we have, as already pointed out, no means at our disposal of testing this by observation.

A much more notable advance was that made by Minkowski, when he developed a type of four-dimensional analytical geometry in which the change from one set of Time-Space variables to another corresponded to a change of coordinate axes.

The idea of time as a fourth dimension is however much older than Minkowski and dates at least as far back as the time of Lagrange.

The work of Minkowski is purely analytical and does not touch on the difficulties which lie in the application of measurement to time and space intervals and the introduction of a coordinate system.

As regards such measurement; one cannot regard either clocks or measuring rods as satisfactory bases on which to build up a theoretical structure such as we require in this subject.

One knows only too well the difficulty there is in getting clocks to agree with one another; while measuring rods expand or contract in a greater or lesser degree as compared with others.

The existence of a substance such as india-rubber should be sufficient to upset ones trust in measuring rods as ultimate standards; when one considers that it only possesses in an exaggerated degree a property of extensibility common to all solid bodies.

It is not sufficient to say that Einstein's clocks and measuring rods are *ideal* ones: for, before we are in a position to speak of them as being ideal, it is necessary to have some clear conception as to how one could, at least theoretically, recognise ideal clocks or measuring rods in case one were ever sufficiently fortunate as to come across such things; and in case we have this clear conception, it is quite unnecessary, in our theoretical investigations, to introduce clocks or measuring rods at all.

We have in fact a problem to consider regarding the measurement of time and space intervals which may be compared to that which Lord Kelvin set himself in connexion with the measurement of temperature, and which he solved by the invention of the thermodynamic scale.

Now we have seen that in Minkowski's analytical geometry the length of an interval of a line such as the axis of $x$, $y$ or $z$ is neither a minimum nor a maximum, while the length of an interval of a line such as the axis of $t$ is actually a mathematical maximum.

Further there are certain lines (which I have called "optical lines") for which the conception of length partially, but not entirely, breaks down.

It thus appears that the conception of length is not at all so simple as is generally supposed and, as a matter of fact, *it is not a fundamental concept at all in Time-Space theory*.

If the measurement of time and space intervals is not fundamental, it may be asked: what is to take its place?

I say that ideas of *order* must take the place of measurement as a theoretical basis; and conceptions of measurement constructed from them.

The process by which this is done is somewhat lengthy, but will be found to shed an important light on the seeming paradoxes above mentioned.

In constructing this system it is necessary to modify certain currently accepted notions, but the modifications required all appear to be capable of justification and the structure when completed will be found closely to resemble our ordinary conceptions.

We shall regard an instant as a fundamental concept which, for present purposes, it is unnecessary further to analyse, and shall consider the relations of order among the instants of which I am directly conscious.

Thus for such instants we find the following properties:

(1) If an instant $B$ be *after* an instant $A$, then the instant $A$ is not *after* the instant $B$, and is said to be *before* it.

(2) If $A$ be any instant, there is at least one instant which is *after* $A$ and also at least one instant which is *before* $A$.

(3) If an instant $B$ be *after* an instant $A$, there is at least one instant which is both *after* $A$ and *before* $B$.

(4) If an instant $B$ be *after* an instant $A$ and an instant $C$ be *after* the instant $B$, then the instant $C$ is *after* the instant $A$.

($5_0$) If an instant $A$ be neither *before* nor *after* an instant $B$, the instant $A$ is identical with the instant $B$.

Now it appears to have been too hastily assumed, because the set of instants of which a single individual is directly conscious possess all these properties, that therefore they must hold in respect of all instants throughout the universe.

It would appear that people in general have been making a somewhat similar blunder to the one ascribed to Sir Boyle Roche; who is alleged to have asserted in a speech, that he could not have been in two places at once unless he were a bird.

They have assumed that an instant, like Sir Boyle Roche's bird, could be in two places at once, and, in consequence, they have found extreme difficulty in identifying it as one and the same instant in the two places.

Had Sir Boyle Roche pursued the subject further, he might perhaps have arrived at a form of relativity theory whereby a bird might be simultaneously in two places to one observer, but not to another. However he got sufficiently far to achieve immortal fame.

I, however, venture to dissent from the generally accepted view that an instant can be in two places at once and, while still regarding postulate ($5_0$) as holding for the set of instants of which any one individual is directly conscious, or which any single particle occupies, I propose to reject it for the universe in general and to substitute for it the following:

(5) If $A$ be any instant, there is at least one instant distinct from $A$ which is neither *before* nor *after* $A$.

If I am directly conscious of the instant $A$ then any instant such as that here postulated will be one of which I am not directly conscious, but only indirectly apprehend, and which is, as I say, *elsewhere*.

The other four postulates are however to be regarded as holding in general and not merely for a single individual or a single particle.

Since we are able to distinguish an instant elsewhere in terms of *before* and *after* relations, it is unnecessary to have any separate concept of *space*; since the geometrical properties of space can be expressed in terms of these relations; although, of course, this involves an elaborate logical analysis.

While the set of instants of which any single individual is conscious or which any single particle of matter occupies have a linear order, the set of instants for the universe in general appear to have what I have called a *Conical Order*.

I have given it this name because it may be illustrated by means of ordinary geometric cones: while it contains within itself the possibility of defining particular sets of instants having a simple linear order such as that with which each of us is familiar.

It is to be recollected that this illustration is given merely as a mental aid to enable us to grasp a certain set of abstract relations, just as figures are an aid in doing geometry; but, as in the latter case, everything which we may introduce incidentally and which cannot be described in terms of the abstract relations is to be ignored.

Without some such picture it would be rather difficult for most people to retain all the relations in mind in complicated cases, and moreover, as in ordinary geometry, a diagram will often suggest that certain theorems may hold and may also suggest methods of proof.

Let us suppose that we have a system of right circular cones of equal angle and with their axes all parallel (or identical). We shall suppose each cone to terminate at the vertex, which however is to be regarded as a point of the cone.

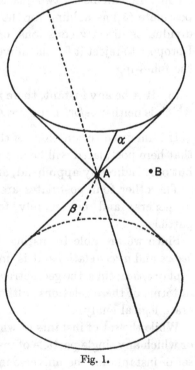

We shall call such a cone having its opening pointed in one direction (say upwards) an α cone and one having the opening pointed in the opposite direction a β cone.

Corresponding to any point of space we shall have an α cone and a β cone having the point as a common vertex.

Now it is possible by using such cones and making a convention with respect to the use of the words *before* and *after* to set up a type of order of the points of space.

For the purposes of this illustration we shall make the convention that if *A* be the common vertex of such a pair of α and β cones, then any point will be said to be *after A* provided that it is distinct from *A* and lies

Fig. 1.

either on or inside the α cone of *A* and will be said to be *before A* provided that it is distinct from *A* and lies either on or inside the β cone of *A*, but not otherwise.

Thus any point which is either identical with $A$ or else lies outside both the cones $\alpha$ and $\beta$ will be neither *before* nor *after* $A$.

It is easy to see that postulates (1), (2), (3), (4) and (5) hold generally in this illustration substituting points for instants, but that $(5_0)$ only holds for certain sets of points forming lines straight or curved.

We may, by a study of such models, ascertain various other *before* and *after* relations which hold among the points and we can then reverse the process by taking the *before* and *after* relations as starting point instead of the cones, and propositions expressed in terms of these relations as postulates, and can, in this way, build up a system of geometrical relations very closely analogous to, but not quite identical with, those from which we started out and which involve nothing except what can be expressed in terms of the *before* and *after* relations.

In this way we are able to define what I call $\alpha$ and $\beta$ sub-sets, which have many, although not all the properties which we assigned to the $\alpha$ and $\beta$ cones, and can gradually, step by step, build up a system of geometry which is equivalent to the analytical system of Minkowski.

Our model is only three-dimensional, while the geometry of Minkowski is four-dimensional; but, in spite of this, most of our postulates may be represented in three dimensions, and, in fact, there are only two which cannot. One of these introduces a fourth dimension, while the other limits the number of dimensions to four.

We could extend the system to a larger number of dimensions if required, but we do not propose to do so in this work.

If we consider straight lines in our model passing through the point $A$, we observe that such a line may be of three distinct types. The first type falls within the cones $\alpha$ and $\beta$; the second type forms a generator of these cones; while the third type falls outside the cones.

The first and second types have this in common, that, if we consider two distinct points lying in either type of line, one is *after* the other; while if we consider any two distinct points in the third type of line the one is neither *before* nor *after* the other.

Again, if we consider planes through the point $A$ we see that they too may be of three distinct types. The first type intersects the cones $\alpha$ and $\beta$ in two generators; the second type touches the cones along a generator; while the third type has no point in common with the cones except the point $A$.

Similarly there are three types of threefold, but in order to represent them we should require a four-dimensional model.

These different types of line, plane and threefold may all be defined in terms of *before* and *after* relations.

In one important respect however our model differs from our logical constructions. Equal lengths in the model do not, in general, represent equal lengths in our geometry: the latter being defined by a certain analysable similarity of *before* and *after* relations.

The reason why there is this difference between the model and the system of geometry which we build up, is that the model has already got a system of measurement imposed upon it, owing to the fact that it is constructed in ordinary three-dimensional space, and so involves more than the mere *before* and *after* relations which it was designed to illustrate.

Finally we are able to introduce coordinates and the system is then seen to be equivalent to the analytical geometry of Minkowski.

In such a system as he employed, one coordinate is measured along a line corresponding to one lying within the cones in the model (and which we shall call an inertia line), and represents what clocks purport to measure. The other three coordinates are measured along lines corresponding to those lying outside the cones (or separation lines) and these represent what we call spacial distances.

The four coordinate axes in this system are all normal to one another (normality being also capable of definition in terms of *before* and *after*), but, if we do not insist on normality, it is possible to introduce a symmetrical system of coordinates in which all four are measured along lines of the same type.

Now, as the *before* and *after* relations from which our whole theory is built up have a temporal significance, we appear to have absorbed the theory of space in a theory of time, in which instants have a conical order instead of the purely linear order which they are generally regarded as having.

An instant for the universe in general is identified by four coordinates in this theory instead of merely one coordinate as is generally assumed.

An instant is localised and does not range all over the universe like Sir Boyle Roche's bird: so that *the present instant does not extend beyond here, and the only really simultaneous events are events which occur at the same place.*

In Minkowski's system of coordinates the so-called "local time" is merely the value of that particular coordinate which is measured along an inertia line.

If we take a second normal coordinate system in which the inertia axis is not parallel to the former, we have one which is appropriate to a material system which is moving uniformly with respect to the first, and we have a different "local time".

An inertia line is the time path of an unaccelerated particle, and, since it is defined in terms of *before* and *after* relations, we are able to say in terms of these relations what we mean by a particle being *unaccelerated*.

We can however assign no meaning to a particle being at "absolute rest" since, in this geometry, any inertia line is exactly on a par with any other one.

Thus instead of regarding ourselves as, so to speak, swimming along in an ocean of space (as we usually do), we are to think of ourselves rather as somehow pursuing a course in an ocean of time; while *spacial relations are to be regarded as the manifestation of the fact that the elements of time form a system in conical order : a conception which may be analysed in terms of the relations of after and before.*

It should be noted that the fundamental relation of *after* which serves as a basis for Time-Space theory is simpler than the relation which geometers are accustomed to make use of in building up ordinary three-dimensional geometry.

The basic relation which they employ is generally the relation of *between*: one point being linearly between two others. This is a relation involving three terms, whereas the *after* relation is one involving only two.

It will appear in the course of this work that a relation of *linearly between* may be defined in terms of *before* and *after* relations for the case of three elements in a separation line; although no one of these three elements is either *before* or *after* either of the other two.

One could scarcely hope to do the converse of this, that is to say, to define an asymmetrical relation of two elements in terms of one like *linearly between* involving three.

It is true that spacial models involving cones may be used to illustrate graphically various postulates employed in our geometry, but this can only be done by means of an arbitrary convention as to what should represent *after* and what *before*.

This convention might have been reversed without affecting the usefulness of the representation; but, by no stretch of the imagination can one (so far as I can see) reverse the time relations of *before* and *after* which one perceives directly in one's own mind. *The thought process is essentially an irreversible one.*

Another interesting point to note is that whereas, on the one hand, if ordinary geometry is built up from the *between* relation, the theory of *congruence* appears as something extraneous grafted on to an otherwise complete scheme; on the other hand, if the *before* and *after* relations are used as a basis, *congruence* appears as an intrinsic part of the subject.

Let us now consider what is the physical peculiarity of the time relations of *before* and *after* which gives them their asymmetrical character.

One thing seems clear: If I at the instant $A$ can produce any effect, however slight, at a distinct instant $B$, then this is sufficient to imply that $B$ is *after* $A$.

A present action of mine may produce some effect to-morrow, but nothing which I may do now can have any effect on what occurred yesterday.

It appears to me that we have here the essential feature of what we mean when we use the word *after*, and that the abstract power of a person or living being at the instant $A$ to produce an effect at a distinct instant $B$ is not merely a *sufficient* but also a *necessary* condition that $B$ is *after* $A$.

If however a person at the instant $A$ cannot produce an effect at the instant $B$, it does not follow that $B$ is *before* $A$.

In order that this should be so it would be necessary that a person at $B$ should be able to produce an effect at $A$; since *before* and *after* are converse relations.

Thus the significance of an instant $A$ being neither *before* nor *after* a distinct instant $B$, is that a person at $A$ should be unable to produce any effect at $B$ and a person at $B$ should be unable to produce any effect at $A$.

We shall have to give some further consideration to this idea of *possibility of producing an effect*; but, before doing so, we shall first consider the physical circumstances under which one instant is neither *before* nor *after* another.

In the first place it is to be observed that, regarded from the standpoint of pure mathematics, the system of geometry which we are about to develop only presupposes that there should be a set of elements which are related in a certain way which can be analysed in terms of a certain asymmetrical relation.

In our attempt to apply this, we identify an element with an *instant*, and the asymmetrical relation with the physical relation of *after*.

The suitability or otherwise of this abstract geometry for describing actual time and space relations is dependent upon the degree of accuracy with which the various postulates of the geometry correspond with various physical facts.

Now it appears to be possible to establish a very close, although perhaps not an exact, correspondence of this sort by means of the physical properties of light.

Let $P$ and $Q$ be two separate and distinct particles and let a flash of light be sent out from $P$ at the instant $A$ so as to arrive directly at $Q$ at the instant $B$, then, according to our interpretation of *after*, $B$ is *after* $A$.

Further, there are strong physical grounds for believing, at any rate in the absence of appreciable quantities of matter, that light supplies a criterion which, with the meaning we have above ascribed to *after*, enables us to say that $B$ is the first instant at $Q$ which is *after* $A$ and that $A$ is the last instant at $P$ which is *before* $B$.

It will be observed that no mention of *velocity* is made in this statement but merely the *before* and *after* relations.

The conception of velocity involves the conception of *measurement* of space and time intervals and these are supposed to be not yet defined.

Let us suppose next that the light flash is reflected directly back from $Q$ to $P$ and that it arrives there at the instant $C$, then, if the view we have mentioned be correct, any instant at $P$ which is *after* $A$ and *before* $C$ will be neither *before* nor *after* $B$.

Now Fizeau's apparatus is an arrangement in which this is practically carried out: so that we can say that *any instant at the sending apparatus which is* after *the instant of departure of a flash of light and* before *the instant of its return is neither* before *nor* after *the instant of reflection at the distant mirror*.

It is possible that the analytical geometry of Minkowski, with these optical interpretations of our postulates, gives only an approximate, although under ordinary circumstances a very closely approximate representation of time and space relations, and this is the view now held by Einstein and others; but even so, it does not follow that with some slightly different interpretation it may not be exact.

But, as we shall see, the *before* and *after* relations enable us to define equality of intervals in Minkowski's geometry, and, however the Time-Space universe may be constituted, these relations certainly have some physical significance; so that there can be little doubt that they must play just as important a rôle in the foundations of any generalised theory as they do in the simple one.

In fact the Minkowski theory might perhaps be regarded as giving the constitution of Time-Space provided that we do not consider too large a portion of it, while the so-called generalised theory would be the sort of thing we should get provided, in our model, the cones, instead of being all similar and similarly situated, varied from one point to another.

I ought perhaps to remark that any proper quadric cone would serve equally well to illustrate all our postulates and it is only for the sake of simplicity that I supposed the cones to be right circular ones.

Before one is in a position to set up any type of coordinates it is fairly evident that one must, either tacitly or explicitly, make use of considerations of order, if these coordinates are to have any sort of system about them, and the *before* and *after* relations appear to have the requisite fundamental character to supply this.

The view that time relations are fundamental appears to have an important bearing on what Professor William James called the theory of a "block universe": by which name he referred to the theory that the universe is something like a cinematograph film in which the photographs have already been taken and which is merely in process of being exhibited to us.

Most writers on this subject treat time as if it were merely a fourth dimension of space: an attitude which encourages one to favour the "block universe" idea.

When instead, we regard *before* and *after* relations as fundamental, and analyse spacial relations up in terms of these, the whole subject appears in a very different light and the "block universe" theory does not commend itself so strongly.

If the universe were in this way like a cinematograph film which is merely being displayed before us, then its innumerable details must have been fixed through all eternity and there would be complete determinism as to the future.

But have we really any grounds for thinking that the universe is of this nature: or, reverting to the cinematograph analogy, is it any simpler to suppose that the film has already been taken than to suppose that the play is in process of being acted?

If the *after* relation has the significance which I suggested and if what we call time and space may be analysed in terms of *before* and *after* then it would seem that instead of having grounds for belief in a "block universe" we have actually got grounds for an opposite view.

It seems therefore that the question turns on the significance of the *after* relation and its asymmetric character.

It is interesting to note that recently, on quite different grounds, some physicists are coming round to the view that the universe is not strictly deterministic.

Scientific predictions as to future events are made on the assumption that certain uniformities will continue.

If they do continue the prediction may be a logical consequence of their doing so, but, if the uniformities do not continue, the conclusion may be unwarranted.

The continuance of the uniformities is only an assumption for which we have no absolute guarantee, and, should they cease, no promise is broken, since none was ever made. A departure from uniformity initiated at an instant $A$ may extend to an instant $B$ which is *after* $A$; and this would be an *effect* at $B$ of the departure from uniformity initiated at $A$.

All applied mathematics becomes pure mathematics when we get away from our fundamental assumptions and begin to draw logical conclusions from them.

Now I have ascribed certain characteristics to instants and to *before* and *after* relations which may or may not be strictly correct, but which serve as the basis by means of which one may apply a certain type of pure geometry to map out time and spacial relations.

The geometry, as I have already pointed out, is a logical structure built up from certain postulates which I shall formulate.

As a logical structure a geometry may have more than one application, as for instance, ordinary Euclidean plane geometry might be taken primarily as applying to figures on what we call a plane and again to geodesic lines drawn on a developable surface.

For the purposes of physical science, however, it is not sufficient merely that we should say, for instance, that *there are* such things as "straight lines" or that *there are* lengths which are equal, but it is necessary to have criteria by which we can say (at least approximately) "*here are* points which lie in a straight line" and "*here is* a length which is equal to yonder length".

In other words we must have more or less clear ideas of the physical things to which we apply our abstract theory.

The abstract theory itself does not require this, but the physical application does; and for this reason, I have tried to make clear the sort of physical meaning which I ascribe to the notions of an instant, the *before* and *after* relations and the criteria given by light flashes.

If we should discover, for instance, that the formal properties which we provisionally ascribe to light actually hold for some other influence; then the geometry which I propose to develop would apply with this new interpretation of its postulates.

Now I have made use of ordinary geometric cones in order to enable us to form a concrete picture of what I mean by *"conical order"*, but the idea of conical order is not at all dependent upon this graphic representation, but is built up by a rather lengthy piece of reasoning from the asymmetrical relations which I denote by the words *before* and *after*.

The representation by means of cones may be compared to the rough scaffolding used in the erection of a building which is removed when the building is complete and its component parts in position.

We must, however, be certain that the building is not supported by the scaffolding, or it will not be able to stand alone.

In order to make sure of this in our theory, great care must be taken not to take things for granted because they hold in our models.

In the first place we are not at liberty to introduce coordinates except for scaffolding purposes until we have defined them. Neither are we at liberty to speak of "velocity" except for scaffolding purposes till its meaning is defined.

Moreover in the actual proof of theorems we must not employ the ideas of equality of lengths or angles until these ideas are seen to be definable in terms of *before* and *after* relations.

We may however, and actually do, make use of such non-permissible ideas in our graphic representation.

Thus in the models we supposed the cones to have their axes parallel (or identical) and to have equal vertical angles, and neither the idea of *cone*, of *parallel*, of *axis*, of *angle*, nor of *equal* has been analysed in terms of *before* and *after* and therefore must be excluded in defining the α and β sub-sets, which are the names which I shall hereafter apply to the entities corresponding to the α and β cones.

The *before* and *after* relations are converse asymmetrical relations and either may be defined in terms of the other; so that it is a matter of indifference which of them we take as fundamental.

I actually take the relation of *after* as fundamental and define *before* in terms of it.

As regards the postulates which are expressed in terms of these relations, they generally consist of two parts (marked *a* and *b*) in which the *before* and *after* relations are interchanged.

In some of the postulates, however, the one part follows from the other on account of the mutual relations of *after* and *before*: while in some others the *before* and *after* relations are involved symmetrically.

We shall now proceed with the formal development of the subject.

# CONICAL ORDER

WE shall suppose that we have a set of elements and that certain of these elements stand in a relation to certain other elements of the set which we denote by saying that one element is *after* another.

We shall further assume the following conditions:

POSTULATE I. **If an element B be after an element A, then the element A is not after the element B.**

This is merely the condition that *after* should be an asymmetrical relation. If an element $B$ be *after* an element $A$, it follows directly from Post. I that $A$ and $B$ must be distinct elements, for, if we substitute $A$ for $B$ in the postulate, it becomes self-contradictory.

*Definition.* If an element $B$ be *after* an element $A$, then the element $A$ will be said to be *before* the element $B$.

POSTULATE II. (*a*) **If A be any element, there is at least one element which is after A.**

(*b*) **If A be any element, there is at least one element which is before A.**

POSTULATE III. **If an element B be after an element A, and if an element C be after the element B, the element C is after the element A.**

POSTULATE IV. **If an element B be after an element A, there is at least one element which is both after A and before B.**

POSTULATE V. **If A be any element, there is at least one other element distinct from A, which is neither before nor after A.**

POSTULATE VI. (*a*) **If A and B be two distinct elements, one of which is neither before nor after the other, there is at least one element which is after both A and B, but is not after any other element which is after both A and B.**

(*b*) **If A and B be two distinct elements, one of which is neither after nor before the other, there is at least one element which is before both A and B, but is not before any other element which is before both A and B.**

*Definition.* (*a*) If $A$ be any element of the set, then an element $X$ will be said to be a member of the $\alpha$ sub-set of $A$ provided $X$ is either identical with $A$, or else provided there exists at least one element $Y$ distinct from $A$ and neither *before* nor *after* $A$ and such that $X$ is *after* both $A$ and $Y$ but is not *after* any other element which is *after* both $A$ and $Y$.

(b) If $A$ be any element of the set, then an element $X$ will be said to be a member of the $\beta$ sub-set of $A$ provided $X$ is either identical with $A$, or else provided there exists at least one element $Y$ distinct from $A$ and neither *after* nor *before* $A$ and such that $X$ is *before* both $A$ and $Y$ but is not *before* any other element which is *before* both $A$ and $Y$.

If $A$ be any element then, by Post. V, there is at least one other element distinct from $A$ which is neither *before* nor *after* $A$ and so it follows directly by Post. VI (a) that there is at least one other element besides $A$ which is a member of the $\alpha$ sub-set of $A$.

Similarly, by Post. VI (b), there is at least one other element besides $A$ which is a member of the $\beta$ sub-set of $A$.

*Notation.* We shall denote by $\alpha_1$ and $\beta_1$ the sub-sets corresponding to an element $A_1$, and by $\alpha_2$ and $\beta_2$ those corresponding to an element $A_2$, etc.

**POSTULATE VII.** (a) **If $A_1$ and $A_2$ be elements and if $A_2$ be a member of $\alpha_1$, then $A_1$ is a member of $\beta_2$.**

(b) **If $A_1$ and $A_2$ be elements and if $A_2$ be a member of $\beta_1$, then $A_1$ is a member of $\alpha_2$.**

**POSTULATE VIII.** (a) **If $A_1$ be any element and $A_2$ be any other element in $\alpha_1$, there is at least one other element distinct from $A_2$ which is a member both of $\alpha_1$ and of $\alpha_2$.**

(b) **If $A_1$ be any element and $A_2$ be any other element in $\beta_1$, there is at least one other element distinct from $A_2$ which is a member both of $\beta_1$ and of $\beta_2$.**

### THEOREM 1

*If $A_1$ be any element and $A_2$ be any other element in $\alpha_1$, then any element $A_3$ which is both after $A_1$ and before $A_2$, must be a member both of $\alpha_1$ and $\beta_2$.*

By the definition of a member of the sub-set $\alpha_1$ there exists at least one element, say $A_4$, distinct from $A_1$ and neither *before* nor *after* $A_1$ and such that $A_2$ is *after* both $A_1$ and $A_4$ but is not *after* any other element which is *after* both $A_1$ and $A_4$.

Then $A_4$ cannot be *after* $A_3$, for if it were then, by Post. III, $A_4$ would be *after* $A_1$ contrary to hypothesis.

Further $A_4$ cannot be identical with $A_3$, for then again we should have $A_4$ *after* $A_1$ contrary to hypothesis.

Again $A_4$ cannot be *before* $A_3$ for then we should have $A_2$ *after* the element $A_3$ which would be *after* both $A_1$ and $A_4$ contrary to the hypothesis that $A_2$ is *after* both $A_1$ and $A_4$ but not *after* any other element which is *after* both $A_1$ and $A_4$.

Thus $A_4$ is distinct from $A_3$ and is neither *before* nor *after* $A_3$.

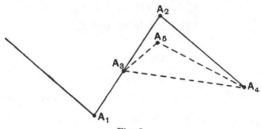

Fig. 2.

Now $A_2$ cannot be *after* any other element which is *after* both $A_3$ and $A_4$, for if $A_5$ were such an element it would follow by Post. III that since $A_3$ is *after* $A_1$ we should have $A_5$ *after* $A_1$.

Thus we should have $A_2$ *after* $A_5$ which would be *after* both $A_1$ and $A_4$ contrary to hypothesis.

Thus no such element as $A_5$ can exist and so $A_2$ satisfies the definition of being a member of $\alpha_3$.

Thus by Post. VII $(a)$ it follows that $A_3$ is a member of $\beta_2$.

Again by Post. VII $(a)$ since $A_2$ is a member of $\alpha_1$ it follows that $A_1$ is a member of $\beta_2$, and so by a similar method we may prove that $A_3$ is a member of $\alpha_1$. Thus the theorem is proved.

## THEOREM 2

$(a)$ *If $A_1$ be any element and $A_2$ be any other element in $\alpha_1$, there is at least one other element in $\alpha_1$ distinct from $A_2$ which is neither* before *nor* after $A_2$.

Since $A_2$ is a member of $\alpha_1$ it follows by Post. VII $(a)$ that $A_1$ is a member of $\beta_2$.

Thus there exists at least one other element, say $A_3$, distinct from $A_2$ and neither *before* nor *after* $A_2$ and such that $A_1$ is *before* $A_2$ and $A_3$, but is not *before* any other element which is *before* both $A_2$ and $A_3$.

Thus $A_1$ satisfies the definition of being a member both of $\beta_2$ and $\beta_3$ and so, by Post. VII $(b)$, $A_3$ is also a member of $\alpha_1$. Thus since $A_3$ is distinct from $A_2$ and neither *before* nor *after* $A_2$, the theorem is proved.

(b) *If $A_1$ be any element and $A_2$ be any other element in $\beta_1$, there is at least one other element in $\beta_1$ distinct from $A_2$ which is neither* after *nor* before *$A_2$.*

*Definition.* If $A_1$ be any element and $A_2$ be any other element in $\alpha_1$, the *optical line $A_1A_2$* is defined as the aggregate of all elements which lie either

|  | (1) both in $\alpha_1$ and $\alpha_2$, |
|---|---|
| or | (2) both in $\alpha_1$ and $\beta_2$, |
| or | (3) both in $\beta_1$ and $\beta_2$. |

### Theorem 3

(a) *If a be any optical line, there exists at least one element which is not an element of the optical line, but is* before *some element of it.*

If $A_1$ be any element and $A_2$ be any other element in $\alpha_1$ then, by Post. VII (a), $A_1$ is a member of $\beta_2$.

Thus by Theorem 2 (b) there is at least one other element in $\beta_2$ distinct from $A_1$ which is neither *after* nor *before* $A_1$.

Call such an element $A_3$.

Then since $A_3$ is in $\beta_2$ and distinct from $A_2$ it is *before* $A_2$.

But $A_3$ cannot lie in the optical line $A_1A_2$, for by the definition of the optical line $A_1A_2$, in order to lie in it $A_3$ would require to lie also either in $\alpha_1$ or $\beta_1$.

But if $A_3$ should lie in $\alpha_1$ it would be either *after* $A_1$ or identical with $A_1$, while if it should lie in $\beta_1$ it would be either *before* $A_1$ or identical with $A_1$.

But $A_3$ is distinct from $A_1$ and is neither *after* nor *before* $A_1$ and therefore does not lie in the optical line $A_1A_2$, although it is *before* $A_2$ an element of it.

(b) *If a be any optical line, there exists at least one element which is not an element of the optical line, but is* after *some element of it.*

**Postulate IX.** (a) **If $a$ be an optical line and if $A_1$ be any element which is not in the optical line but before some element of it, there is one single element which is an element both of the optical line $a$ and the sub-set $\alpha_1$.**

(b) **If $a$ be an optical line and if $A_1$ be any element which is not in the optical line but after some element of it, there is one single element which is an element both of the optical line $a$ and the sub-set $\beta_1$.**

## THEOREM 4

(a) *If $A_1$ be any element there is at least one other element which is after $A_1$ but is not a member of the sub-set $\alpha_1$.*

Let $A_2$ be any other member of the sub-set $\alpha_1$ distinct from $A_1$.

Then $A_2$ is *after* $A_1$ and so by Post. IV there is at least one element, say $A_3$, which is both *after* $A_1$ and *before* $A_2$.

By Theorem 1 $A_3$ is a member both of $\alpha_1$ and of $\beta_2$ and is therefore an element of the optical line $A_1 A_2$.

But since $A_3$ is a member of $\beta_2$ it follows that $A_2$ is a member of $\alpha_3$ and so by Theorem 2 there is at least one other element in $\alpha_3$ distinct from $A_2$ which is neither *before* nor *after* $A_2$.

Let $A_4$ be such an element.

Then since $A_4$ is neither *before* nor *after* $A_2$ it cannot be a member either of $\beta_2$ or $\alpha_2$ and so $A_4$ is not an element of the optical line $A_1 A_2$ although it is *after* $A_3$ an element of it.

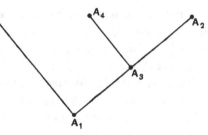

Fig. 3.

But since $A_4$ is a member of $\alpha_3$ it follows by Post. VII (a) that $A_3$ is a member of the sub-set $\beta_4$.

Thus $A_3$ is the *one single element* which by Post. IX (b) is an element both of the optical line and the sub-set $\beta_4$.

But $A_4$ cannot be a member of $\alpha_1$, for then $A_1$ would be a member of $\beta_4$ and so $A_1$ would be a second element common to the optical line $A_1 A_2$ and the sub-set $\beta_4$, which is impossible by Post. IX (b).

Further, $A_4$ is *after* $A_3$ and $A_3$ is *after* $A_1$ and therefore $A_4$ is *after* $A_1$.

Thus $A_4$ is *after* $A_1$ but is not a member of the sub-set $\alpha_1$.

(b) *If $A_1$ be any element there is at least one other element which is before $A_1$ but is not a member of the sub-set $\beta_1$.*

## THEOREM 5

*If $A_1$ be any element and $A_2$ be any other element which is after $A_1$, there is at least one other distinct element which is a member of both $\alpha_1$ and $\beta_2$.*

Two cases arise: (1) $A_2$ may be a member of $\alpha_1$ or (2) $A_2$ may not be a member of $\alpha_1$.

If $A_2$ is a member of $\alpha_1$ then by Post. IV there is at least one element which is both *after* $A_1$ and *before* $A_2$, and by Theorem 1 such an element is a member both of $\alpha_1$ and $\beta_2$.

Thus case (1) is proved.

Suppose next that $A_2$ is not a member of $\alpha_1$ and let $A_3$ be any element of $\alpha_2$ distinct from $A_2$.

Then the optical line $A_2 A_3$ which for brevity we may call $a$, consists of the aggregate of all elements which lie either

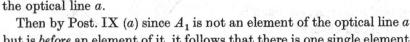

(1)  both in $\alpha_2$ and $\alpha_3$,

or       (2)  both in $\alpha_2$ and $\beta_3$,

or       (3)  both in $\beta_2$ and $\beta_3$.

Fig. 4.

Since $A_2$ is not a member of $\alpha_1$ it follows that $A_1$ is not a member of $\beta_2$ and so, since $A_1$ is *before* $A_2$ it follows that $A_1$ is not an element of the optical line $a$.

Then by Post. IX ($a$) since $A_1$ is not an element of the optical line $a$ but is *before* an element of it, it follows that there is one single element which is an element both of the optical line $a$ and the sub-set $\alpha_1$.

Let $A_4$ be this element.

Then since we have supposed that $A_2$ is not a member of $\alpha_1$ it follows that $A_4$ is not identical with $A_2$.

Further, $A_4$ cannot be *after* $A_2$ for then we should have $A_2$ *after* $A_1$ and *before* $A_4$ and so by Theorem 1 we should have $A_2$ a member of $\alpha_1$ contrary to hypothesis.

Thus $A_4$ cannot be a member of $\alpha_2$ and therefore since it is an element of the optical line $a$ it must be a member of $\beta_2$ and $\beta_3$.

Thus the element $A_4$ is a member of both $\alpha_1$ and $\beta_2$ and so the theorem is proved.

## THEOREM 6

(a) *If $A_1$ be any element and $A_2$ be any other element in $\alpha_1$, while $A_3$ is an element distinct from $A_2$, which is a member both of $\alpha_1$ and of $\alpha_2$, then there is at least one other element which is a member of $\alpha_1$, of $\alpha_2$ and of $\alpha_3$.*

By Post. VIII ($a$) since $A_3$ is an element of $\alpha_2$ distinct from $A_2$ there is at least one other element distinct from $A_3$ which is a member both of $\alpha_2$ and of $\alpha_3$. Call such an element $A_4$. Then since $A_4$ is in $\alpha_3$ and distinct from $A_3$ it is *after* $A_3$.

Thus $A_4$ is *after* an element of the optical line $A_1 A_2$.

But $A_4$ is a member of $\alpha_2$ and also of $\alpha_3$ and so by Post. VII (*a*) $A_2$ and $A_3$ are each members of $\beta_4$.

Now if $A_4$ were not in the optical line $A_1 A_2$ it would follow by Post. IX (*b*) that there was *one single element* which was an element both of the optical line and the sub-set $\beta_4$.

There are however *at least two elements* $A_2$ and $A_3$ with this property and so $A_4$ must be in the optical line $A_1 A_2$.

Also since $A_4$ is in $\alpha_2$ it must also be in $\alpha_1$ from the definition of the optical line.

Thus $A_4$ is a member of $\alpha_1$, of $\alpha_2$ and of $\alpha_3$.

(*b*) *If $A_1$ be any element and $A_2$ be any other element in $\beta_1$, while $A_3$ is an element distinct from $A_2$, which is a member both of $\beta_1$ and of $\beta_2$, then there is at least one other element which is a member of $\beta_1$, of $\beta_2$ and of $\beta_3$.*

### THEOREM 7

(*a*) *If $X$ be any element of an optical line there is at least one element of the optical line which is* after *$X$.*

Let the optical line be defined by any element $A_1$ and another element $A_2$ in $\alpha_1$. Then $X$ may lie either

|      | (1) both in $\alpha_1$ and $\alpha_2$, |
|------|----------------------------------------|
| or   | (2) both in $\alpha_1$ and $\beta_2$,  |
| or   | (3) both in $\beta_1$ and $\beta_2$.   |

If $X$ be not identical with $A_2$, then in cases (2) and (3) since $X$ lies in $\beta_2$, the element $A_2$ is *after $X$*.

If $X$ be identical with $A_2$, then by Post. VIII (*a*) there is at least one other element distinct from $A_2$ which is a member both of $\alpha_1$ and of $\alpha_2$ and is therefore an element of the optical line.

Since such an element is not identical with $A_2$ it must be *after $A_2$*; that is to say it must be *after $X$*.

Next suppose $X$ is in both $\alpha_1$ and $\alpha_2$ and is distinct from $A_2$.

It follows by Theorem 6 (*a*) that there is at least one *other* element which is a member of $\alpha_1$ and $\alpha_2$ and of the $\alpha$ sub-set of $X$.

Since such an element is not identical with $X$ and lies in the $\alpha$ sub-set of $X$ it must be *after $X$*.

Further since it is an element both of $\alpha_1$ and of $\alpha_2$ it lies in the optical

R

line. Thus in all cases there is at least one element of the optical line which is *after* $X$.

(*b*) *If $X$ be any element of an optical line there is at least one element of the optical line which is* before $X$.

THEOREM 8

(*a*) *If $A_1$ be any element and $A_2$ be any other element in $\alpha_1$, and if $A_3$ and $A_4$ be other distinct elements which are members of both $\alpha_1$ and $\alpha_2$, one of the two elements $A_3$ and $A_4$ is in the $\alpha$ sub-set of the other.*

Since $A_3$ is in $\alpha_2$ and distinct from $A_2$ therefore $A_2$ and $A_3$ define an optical line. Further since $A_2$ and $A_3$ both lie in $\alpha_1$ therefore $A_1$ lies in both $\beta_2$ and $\beta_3$.

Thus $A_1$ is an element of the optical line $A_2A_3$.

But $A_4$, since it is a member of $\alpha_1$ and not identical with $A_1$, is *after* $A_1$.

That is to say, it is *after* an element of the optical line $A_2A_3$.

If then $A_4$ were not an element of the optical line $A_2A_3$ there would, by Post. IX (*b*), be *one single element* which would be an element both of the optical line $A_2A_3$ and the sub-set $\beta_4$.

But $A_4$ is a member both of $\alpha_1$ and of $\alpha_2$ and so both $A_1$ and $A_2$ are members of $\beta_4$.

Thus since $A_1$ and $A_2$ are two distinct elements of the optical line $A_2A_3$ it follows that $A_4$ must be an element of the same optical line.

But $A_4$ is a member of $\alpha_2$ and therefore by the definition of the optical line $A_4$ must be either a member of $\alpha_3$ or of $\beta_3$.

If $A_4$ be a member of $\beta_3$, then we should have $A_3$ a member of $\alpha_4$.

Thus one of the two elements $A_3$ and $A_4$ lies in the $\alpha$ sub-set of the other.

It also follows since $A_3$ and $A_4$ are supposed to be distinct, that the one is *after* the other.

(*b*) *If $A_1$ be any element and $A_2$ be any other element in $\beta_1$, and if $A_3$ and $A_4$ be other distinct elements which are members of both $\beta_1$ and $\beta_2$, one of the two elements $A_3$ and $A_4$ is in the $\beta$ sub-set of the other.*

It also follows since $A_3$ and $A_4$ are supposed to be distinct that the one is *before* the other.

THEOREM 9

*If a pair of elements be in an optical line defined by another pair of elements, then one of the first pair is in the $\alpha$ sub-set of the other.*

Consider the optical line defined by the element $A_1$ and another element $A_2$ in $\alpha_1$. Suppose now in the first place that we have an element $A_3$ distinct from $A_1$ and $A_2$ and lying in the optical line.

Then by the definition of an optical line $A_3$ may be

$\qquad\qquad\qquad$ (1) both in $\alpha_1$ and $\alpha_2$,

or $\qquad\qquad\quad$ (2) both in $\alpha_1$ and $\beta_2$,

or $\qquad\qquad\quad$ (3) both in $\beta_1$ and $\beta_2$.

Thus if $A_1$ and $A_3$ be taken as a pair of elements in the optical line defined by $A_1$ and $A_2$, we have in the first and second cases $A_3$ is in $\alpha_1$, while in the third we have $A_3$ in $\beta_1$ and consequently $A_1$ in $\alpha_3$. Thus one of the pair $A_1$, $A_3$ is in the $\alpha$ sub-set of the other.

Again if $A_2$ and $A_3$ be taken as a pair of elements in the optical line defined by $A_1$ and $A_2$, we have in the first case $A_3$ is in $\alpha_2$, while in the second and third we have $A_3$ in $\beta_2$ and consequently $A_2$ in $\alpha_3$. Thus one of the pair $A_2$, $A_3$ is in the $\alpha$ sub-set of the other.

Next suppose that we have another element $A_4$ lying in the optical line and distinct from $A_1$, $A_2$ and $A_3$.

Then there are the following possibilities:

$$A_3 \text{ both in } \alpha_1 \text{ and } \alpha_2 \text{ with } \begin{cases} A_4 \text{ both in } \alpha_1 \text{ and } \alpha_2 \dots\dots(1), \\ \text{or } A_4 \text{ both in } \alpha_1 \text{ and } \beta_2 \dots\dots(2), \\ \text{or } A_4 \text{ both in } \beta_1 \text{ and } \beta_2 \dots\dots(3). \end{cases}$$

$$A_3 \text{ both in } \alpha_1 \text{ and } \beta_2 \text{ with } \begin{cases} A_4 \text{ both in } \alpha_1 \text{ and } \alpha_2 \dots\dots(4), \\ \text{or } A_4 \text{ both in } \alpha_1 \text{ and } \beta_2 \dots\dots(5), \\ \text{or } A_4 \text{ both in } \beta_1 \text{ and } \beta_2 \dots\dots(6). \end{cases}$$

$$A_3 \text{ both in } \beta_1 \text{ and } \beta_2 \text{ with } \begin{cases} A_4 \text{ both in } \alpha_1 \text{ and } \alpha_2 \dots\dots(7), \\ \text{or } A_4 \text{ both in } \alpha_1 \text{ and } \beta_2 \dots\dots(8), \\ \text{or } A_4 \text{ both in } \beta_1 \text{ and } \beta_2 \dots\dots(9). \end{cases}$$

In case (1) by Theorem 8 ($a$) one of the two elements $A_3$ and $A_4$ is in the $\alpha$ sub-set of the other. Similarly in case (9) by Theorem 8 ($b$) one of the two elements $A_3$ and $A_4$ is in the $\beta$ sub-set of the other, and therefore by Post. VII ($b$) one of them is in the $\alpha$ sub-set of the other.

Consider next case (2).

Since $A_4$ is in $\alpha_1$ and distinct from $A_1$ it follows that $A_4$ is *after* $A_1$.

Further, since $A_4$ is in $\beta_2$ and distinct from $A_2$ we have $A_2$ *after* $A_4$, and since $A_3$ is in $\alpha_2$ and distinct from $A_2$ we have $A_3$ *after* $A_2$.

Thus by Post. III $A_3$ is *after* $A_4$.

But, since $A_3$ is in $\alpha_1$, it follows by Theorem 1 that $A_4$ is in $\beta_3$ and consequently $A_3$ lies in $\alpha_4$.

Similarly in case (4) we may prove that $A_4$ must lie in $\alpha_3$.

In an analogous manner in case (8) since $A_4$ is in $\beta_2$ and distinct from $A_2$ we have $A_4$ is *before* $A_2$.

Further, since $A_4$ is in $\alpha_1$ and distinct from $A_1$, we have $A_4$ is *after* $A_1$, and since $A_3$ is in $\beta_1$ and distinct from $A_1$ we have $A_1$ is *after* $A_3$ and so, by Post. III, $A_4$ is *after* $A_3$.

But since $A_3$ lies in $\beta_2$ therefore $A_2$ lies in $\alpha_3$ and so, by Theorem 1, $A_4$ must lie in $\alpha_3$.

Similarly in case (6) we may prove that $A_3$ must lie in $\alpha_4$.

Consider next case (3).

We have $A_4$ in $\beta_2$ and therefore $A_2$ in $\alpha_4$.

Also we have $A_2$ in $\alpha_1$, and so $A_1$ in $\beta_2$.

Further we have $A_4$ in $\beta_1$, and so $A_1$ in $\alpha_4$.

Thus $A_4$ and $A_2$ determine an optical line which contains $A_1$.

But $A_3$ is in $\alpha_2$, and being distinct from $A_2$ it must be *after* $A_2$ an element of the optical line determined by $A_4$ and $A_2$.

Also since $A_3$ is in both $\alpha_1$ and $\alpha_2$ it follows that both $A_1$ and $A_2$ lie in $\beta_3$.

But by Post. IX (*b*) if $A_3$ were not in the optical line determined by $A_4$ and $A_2$ there would be *one single element* which would be an element both of the optical line and the sub-set $\beta_3$.

Thus since there are at least two distinct elements $A_1$ and $A_2$ common to the optical line and the sub-set $\beta_3$ it follows that $A_3$ must be an element of the optical line $A_4 A_2$. Further, since $A_3$ lies in $\alpha_2$ it must, by the definition of the optical line, lie also in $\alpha_4$.

We may in a similar manner show in case (7) that $A_4$ must lie in $\alpha_3$.

We are thus left with only case (5) to prove.

Now since $A_2$ is an element distinct from $A_1$ and lying in $\alpha_1$, therefore, by Post. VIII (*a*), there is at least one other element distinct from $A_2$ which is a member both of $\alpha_1$ and of $\alpha_2$.

Call such an element $A_5$. Then $A_2$ is *before* $A_5$.

But $A_3$ is distinct from $A_2$ and lies in $\beta_2$ and so $A_3$ is *before* $A_2$. Thus $A_3$ is *before* $A_5$.

Also $A_3$ is distinct from $A_1$ and lies in $\alpha_1$ and so $A_3$ is *after* $A_1$.

Thus, by Theorem 1, $A_3$ must be an element of the sub-set $\beta_5$.

Similarly $A_4$ must be an element of the sub-set $\beta_5$.

Also both $A_3$ and $A_4$ are elements of $\beta_2$ and so by Theorem 8 (*b*) one of the two elements $A_3$ and $A_4$ is in the $\beta$ sub-set of the other, and therefore by Post. VII (*b*) one is in the $\alpha$ sub-set of the other.

Thus the theorem is true in all cases.

*It follows directly from this theorem that of any two distinct elements in an optical line one is after the other.*

### THEOREM 10

*Any two elements of an optical line determine that optical line.*

Let $A_1$ be any element and $A_2$ any other element in $\alpha_1$, then the optical line $A_1 A_2$ is defined as the aggregate of all elements which lie either

|     | (1) both in $\alpha_1$ and $\alpha_2$, |
| --- | --- |
| or  | (2) both in $\alpha_1$ and $\beta_2$, |
| or  | (3) both in $\beta_1$ and $\beta_2$. |

Suppose $A_3$ and $A_4$ to be any pair of elements in the optical line $A_1 A_2$; then by Theorem 9 one of the pair $A_3$, $A_4$ is in the $\alpha$ sub-set of the other.

We may suppose without loss of generality that it is $A_4$ which is in the sub-set $\alpha_3$.

Consider now any element $A_5$ of the optical line $A_1 A_2$ such that $A_5$ is distinct from $A_3$ and $A_4$.

Then by Theorem 9 there are the following possibilities:

$$A_4 \text{ in } \alpha_5 \text{ and also } A_3 \text{ in } \alpha_5 \qquad \ldots\ldots(1),$$
$$A_4 \text{ in } \alpha_5 \text{ and also } A_5 \text{ in } \alpha_3 \qquad \ldots\ldots(2),$$
$$A_5 \text{ in } \alpha_4 \text{ and also } A_5 \text{ in } \alpha_3 \qquad \ldots\ldots(3),$$
$$A_5 \text{ in } \alpha_4 \text{ and also } A_3 \text{ in } \alpha_5 \qquad \ldots\ldots(4).$$

Case (4) must however be excluded, for since $A_3$, $A_4$ and $A_5$ are supposed distinct we should have $A_5$ *after* $A_4$ and $A_3$ *after* $A_5$ and therefore, by Post. III, $A_3$ *after* $A_4$.

We however supposed $A_4$ to be *after* $A_3$ and by Post. I we cannot have also $A_3$ *after* $A_4$. Thus case (4) is impossible.

The three permissible cases may be expressed thus:

$$A_5 \text{ both in } \beta_3 \text{ and } \beta_4 \qquad \ldots\ldots(1),$$
$$A_5 \text{ both in } \alpha_3 \text{ and } \beta_4 \qquad \ldots\ldots(2),$$
$$A_5 \text{ both in } \alpha_3 \text{ and } \alpha_4 \qquad \ldots\ldots(3).$$

Thus in all cases $A_5$ lies in the optical line defined by $A_3$ and $A_4$.

Similarly it may be shown that every element in the optical line defined by $A_3$ and $A_4$ lies in the optical line defined by $A_1$ and $A_2$.

Thus the optical lines $A_1 A_2$ and $A_3 A_4$ are identical.

## Theorem 11

*If $A_3$ and $A_4$ be any two elements of an optical line $A_1 A_2$ there is at least one element of the optical line which is after the one and before the other.*

Since $A_3$ and $A_4$ are both elements of the same optical line the one must be in the $\alpha$ sub-set of the other by Theorem 9.

We shall suppose that $A_4$ lies in $\alpha_3$.

Then since $A_3$ and $A_4$ are distinct, $A_4$ will be *after* $A_3$, and so by Theorem 5 there is at least one other distinct element which is a member both of $\alpha_3$ and of $\beta_4$.

Call such an element $A_5$.

Then $A_5$ is in the optical line $A_3 A_4$, and therefore by Theorem 10 in the optical line $A_1 A_2$.

Further since $A_5$ is distinct from $A_3$ and $A_4$ it must be *after* $A_3$ and *before* $A_4$.

*From the preceding results it follows that an optical line contains an infinite number of elements.*

## Theorem 12

*If an element $A_1$ be before an element of an optical line $a$, and be also after an element of $a$, then $A_1$ must be itself an element of the optical line $a$.*

Suppose that $A_1$ is *before* the element $A_2$ of $a$ and also *after* the element $A_3$ of $a$.

Then by Post. I $A_3$ cannot be identical with $A_2$, and by Theorem 9 one of the elements $A_2$ and $A_3$ must be in the $\alpha$ sub-set of the other.

Since $A_1$ is *after* $A_3$ and $A_2$ is *after* $A_1$ it follows that $A_2$ is *after* $A_3$ and so it must be $A_2$ which is in the $\alpha$ sub-set of $A_3$.

But, by Theorem 1, it follows that $A_1$ must lie in $\alpha_3$ and also in $\beta_2$, and accordingly $A_1$ lies in the optical line $A_3 A_2$.

Thus since, by Theorem 10, any two elements of an optical line determine that optical line, it follows that $A_1$ lies in the optical line $a$.

## Theorem 13

(a) *If $A_1$ be any element and $A_2$ be any other element in $\alpha_1$ and if $A_3$ be any element in $\alpha_1$ which is either before or after $A_2$, then $A_3$ lies in the optical line $A_1 A_2$.*

(1) Suppose $A_3$ is *before* $A_2$.

Then since $A_3$ lies in $\alpha_1$ it must be either identical with $A_1$, in which case it lies in the optical line $A_1 A_2$; or else $A_3$ is *after* $A_1$, in which case by Theorem 1 $A_3$ must lie both in $\alpha_1$ and $\beta_2$ and therefore must lie in the optical line $A_1 A_2$.

(2) Suppose $A_3$ is *after* $A_2$.

Then $A_3$ lies in $\alpha_1$ and $A_2$ is *after* $A_1$ and *before* $A_3$ and therefore, by Theorem 1, $A_2$ must lie both in $\alpha_1$ and $\beta_3$.

But if $A_2$ lies in $\beta_3$, it follows by Post. VII (*b*) that $A_3$ lies in $\alpha_2$.

Thus $A_3$ lies both in $\alpha_1$ and $\alpha_2$ and therefore lies in the optical line $A_1 A_2$.

(*b*) *If $A_1$ be any element and $A_2$ be any other element in $\beta_1$ and if $A_3$ be any element in $\beta_1$ which is either* after *or* before *$A_2$, then $A_3$ lies in the optical line $A_1 A_2$.*

## Theorem 14

*Three distinct elements cannot lie in pairs in three distinct optical lines.*

Let $A_1$, $A_2$ and $A_3$ be three distinct elements and let $A_1$ and $A_2$ lie in one optical line.

We may suppose that it is $A_2$ which lies in $\alpha_1$.

If then $A_1$ and $A_3$ lie in an optical line we may suppose either that $A_3$ lies in $\alpha_1$ or in $\beta_1$.

First suppose $A_3$ lies in $\alpha_1$.

Then if $A_3$ and $A_2$ lie in an optical line one of them must be *after* the other and so by Theorem 13 (*a*) $A_3$ must lie in the optical line $A_1 A_2$.

Next suppose that $A_3$ lies in $\beta_1$.

Then if $A_3$ and $A_2$ lie in one optical line, $A_1$ is *before* $A_2$ one element of it and *after* $A_3$ another element of it and so by Theorem 12 $A_1$ must lie in the optical line $A_3 A_2$. Thus the optical lines are not distinct and so the theorem is proved.

### Remarks

If $a$ and $b$ be two distinct optical lines having an element $E$ in common and if $O$ be any element of $a$ which is *before* $E$ while $D$ and $F$ are elements of $b$ which are respectively *before* and *after* $E$; then, $E$ being *after* $O$, we must have $F$ *after* $O$, but, by the last theorem, $F$ and $O$ cannot lie in an optical line.

Further, $D$ cannot be *before* $O$, for then we should have $O$ *after* one element of the optical line $b$ and *before* another element of it and yet not lie in the optical line which, by Theorem 12, is impossible.

Also $D$ cannot be *after* $O$, for then we should have $D$ *after* one element of the optical line $a$ and *before* another element of it and yet not lie in the optical line, which again is impossible.

Thus $D$ is neither *before* nor *after* $O$.

Again, if $b'$ be an optical line distinct from $a$ but having an element $E'$ in common with $a$ and such that the element $O$ of $a$ is *after* $E'$, while $D'$ and $F'$ are elements of $b'$ which are respectively *after* and *before* $E'$; we may show in a similar way that $F'$ is *before* $O$, but is not in an optical line with it; while $D'$ is neither *after* nor *before* $O$.

### THEOREM 15

(a) *If $A_1$ be any element and $A_2$ be any other element in $\alpha_1$ and $A_3$ be any element in $\alpha_1$ distinct from $A_2$ which is neither* before *nor* after *$A_2$, then $A_3$ is neither* before *nor* after *any element of the optical line $A_1A_2$ which is* after *$A_1$.*

The element $A_3$ cannot lie in the optical line $A_1A_2$, for then since it is distinct from $A_2$ it would be either *before* or *after* it, contrary to hypothesis.

Now any element of the optical line $A_1A_2$ which is *after* $A_1$ must lie in $\alpha_1$.

Let $A_4$ be any such element.

Then if $A_3$ were either *before* or *after* $A_4$ it would by Theorem 13 lie in the optical line $A_1A_4$, which by Theorem 10 is identical with the optical line $A_1A_2$, and this we have shown to be impossible.

Thus $A_3$ cannot be either *before* or *after* any element of the optical line $A_1A_2$ which is *after* $A_1$.

(b) *If $A_1$ be any element and $A_2$ be any other element in $\beta_1$ and $A_3$ be any element in $\beta_1$ distinct from $A_2$ which is neither* after *nor* before *$A_2$, then $A_3$ is neither* after *nor* before *any element of the optical line $A_2A_1$ which is* before *$A_1$.*

**POSTULATE X. (a) If $a$ be an optical line and if A be any element not in the optical line but before some element of it, there is one single optical line containing A and such that each element of it is before an element of $a$.**

**(b) If $a$ be an optical line and if A be any element not in the optical line but after some element of it, there is one single optical line containing A and such that each element of it is after an element of $a$.**

### THEOREM 16

(a) *If each element of one optical line be* before *an element of another optical line the two optical lines cannot have an element in common.*

Let $a$ and $b$ be two distinct optical lines such that each element of $b$ is *before* an element of $a$.

Suppose, if possible, that $a$ and $b$ have an element $A_1$ in common.

Let $A_2$ be any element of $b$ which is *after* and therefore distinct from $A_1$.

Then, by hypothesis, $A_2$ is *before* some element (say $A_3$) of $a$.

Thus we should have $A_2$ *after* one element $A_1$ and *before* another element $A_3$ of the optical line $a$ and therefore, by Theorem 12, it would follow that $A_2$ must be an element of the optical line $a$.

Thus $a$ and $b$ would have two elements in common and so could not be distinct optical lines, contrary to hypothesis.

Thus the supposition that $a$ and $b$ have an element in common leads to a contradiction and is therefore impossible.

(b) *If each element of one optical line be* after *an element of another optical line the two optical lines cannot have an element in common.*


## Theorem 17

(a) *If each element of an optical line $a$ be* before *an element of another optical line $b$, then through each element of $a$ there is* one single optical line *which contains also an element of $b$.*

By Theorem 16 an element of $a$ cannot also be an element of $b$.

Suppose then that $A_1$ be any element of $a$.

Then $A_1$ is not an element of $b$, but is *before* an element of $b$ and therefore by Post. IX (a) there is *one single element*, say $A_2$, which is an element both of the optical line $b$ and the sub-set $\alpha_1$. Since $A_2$ cannot be identical with $A_1$ it follows that $A_1$ and $A_2$ determine an optical line which contains an element of $a$ and also an element of $b$.

Further, there cannot be more than one optical line through $A_1$ which contains also an element of $b$; for such an element of $b$ must, by Theorem 9, lie either in $\alpha_1$ or $\beta_1$.

But by Post. IX (a) there is only *one single element* common to $b$ and the sub-set $\alpha_1$, and so if such an element of $b$ existed it would have to lie in $\beta_1$.

Call such a hypothetical element $A_3$.

Then since $A_3$ is supposed to lie in $\beta_1$, we should have $A_1$ in $\alpha_3$.

But $A_2$ lies in $\alpha_1$ and so $A_1$ lies in $\beta_2$, and thus if such an element as $A_3$ existed, $A_1$ would lie in the optical line $A_3 A_2$: that is, in the optical line $b$, which is impossible, and so there cannot be any such element as $A_3$.

Thus there is only one single optical line through $A_1$ which contains also an element of $b$.

(b) *If each element of an optical line a be* after *an element of another optical line b, then through each element of a there is* one single optical line *which contains also an element of b.*

*Definition.* If two distinct optical lines have an element in common they will be said to *intersect* one another in that element.

If $A_1$ and $A_2$ be two distinct elements one of which is neither *before* nor *after* the other, then we know by Post. VI that there is at least one element, say $X$, which is *after* both $A_1$ and $A_2$, but is not *after* any other element which is *after* both $A_1$ and $A_2$.

From the definition of $\alpha$ sub-sets it follows that $X$ lies both in $\alpha_1$ and $\alpha_2$, so that there is *at least one element* which is a member both of $\alpha_1$ and $\alpha_2$. Similarly there is *at least one element* which is a member both of $\beta_1$ and $\beta_2$.

These remarks prepare the way for Post. XI (a) and (b).

POSTULATE XI. (*a*) **If $A_1$ and $A_2$ be two distinct elements one of which is neither before nor after the other and X be an element which is a member both of $\alpha_1$ and $\alpha_2$, then there is at least one other element distinct from X which is a member both of $\alpha_1$ and $\alpha_2$.**

(*b*) **If $A_1$ and $A_2$ be two distinct elements one of which is neither after nor before the other and X be an element which is a member both of $\beta_1$ and $\beta_2$, then there is at least one other element distinct from X which is a member both of $\beta_1$ and $\beta_2$.**

The above is the first of our postulates which requires more than two dimensions for its representation.

It is to be noted that it may easily be combined with Post. VI as follows:

(*a*) *If A and B be two distinct elements one of which is neither* before *nor* after *the other, there are at least two distinct elements either of which is* after *both A and B but is not* after *any other element which is* after *both A and B.*

(*b*) *If A and B be two distinct elements one of which is neither* after *nor* before *the other, there are at least two distinct elements either of which is* before *both A and B but is not* before *any other element which is* before *both A and B.*

### THEOREM 18

(a) *If $A_1$ and $A_2$ be any two distinct elements one of which is neither* before *nor* after *the other, and if $A_3$ and $A_4$ be distinct elements which lie both in $\alpha_1$ and $\alpha_2$, then one of these latter two elements is neither* before *nor* after *the other.*

By the definition of $\alpha$ sub-sets $A_3$ is *after* both $A_1$ and $A_2$ but is not *after* any other element which is *after* both $A_1$ and $A_2$.

Similarly $A_4$ is *after* both $A_1$ and $A_2$ but is not *after* any other element which is *after* both $A_1$ and $A_2$.

Thus $A_3$ is not *after* $A_4$, and $A_4$ is not *after* $A_3$.

Thus $A_3$ is neither *before* nor *after* $A_4$.

(b) *If $A_1$ and $A_2$ be any two distinct elements, one of which is neither* after *nor* before *the other, and if $A_3$ and $A_4$ be distinct elements which lie both in $\beta_1$ and $\beta_2$, then one of these latter two elements is neither* after *nor* before *the other.*

### THEOREM 19

(a) *If $A_1$ be any element and $A_2$ and $A_3$ be two other distinct elements of $\alpha_1$, one of which is neither* before *nor* after *the other, there is at least one other distinct element in $\alpha_1$ which is neither* before *nor* after $A_2$ *and neither* before *nor* after $A_3$.

Since $A_2$ is a member of $\alpha_1$, therefore $A_1$ is a member of $\beta_2$.

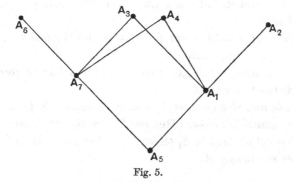

Fig. 5.

Thus by Post. VIII (b) there is at least one other element distinct from $A_1$ which is a member both of $\beta_2$ and of $\beta_1$.

Call such an element $A_5$.

Then $A_1$ and $A_2$ are both members of $\alpha_5$.

Thus by Theorem 2 (a) there is at least one other element in $\alpha_5$ distinct from $A_1$ which is neither *before* nor *after* $A_1$.

Call such an element $A_6$.

Now $A_3$ cannot lie in $\alpha_5$ for then, as it is an element of $\alpha_1$, it would lie in the optical line $A_5A_1$ along with $A_2$ and so $A_2$ and $A_3$ would either be identical or else $A_2$ would be either *before* or *after* $A_3$, contrary to hypothesis.

Now $A_3$ is *after* $A_1$ and $A_1$ is *after* $A_5$ and so by Post. III $A_3$ is *after* $A_5$, and since $A_3$ is not an element of $\alpha_5$ it cannot lie in the optical line $A_5A_6$.

Thus by Post. IX (*b*) there is *one single element* (say $A_7$) which is an element both of the optical line $A_5A_6$ and the sub-set $\beta_3$.

Now $A_5$ cannot be *after* $A_7$, for $A_3$ lies in $\alpha_7$ and so, by Theorem 1, $A_5$ would require to lie in $\beta_3$, which it cannot do since $A_3$ is not an element of $\alpha_5$.

Also $A_5$ cannot coincide with $A_7$ for then it would be in $\beta_3$.

Thus $A_7$ must be *after* $A_5$, and so by Theorem 15 $A_1$ is neither *before* nor *after* $A_7$.

Now $A_3$ lies both in $\alpha_1$ and in $\alpha_7$, and so by Post. XI (*a*) there is at least one other distinct element, say $A_4$, which lies both in $\alpha_1$ and in $\alpha_7$.

Then by Theorem 18 $A_4$ is neither *before* nor *after* $A_3$.

Further, $A_4$ cannot be either *before* or *after* $A_2$, for since $A_2$ and $A_4$ are both members of $\alpha_1$ it would follow by Theorem 13 that $A_4$ must lie in the optical line $A_1A_2$.

This would also be the case if $A_4$ coincided with $A_2$.

But then (since $A_4$ is *after* $A_1$ and therefore *after* $A_5$) we should have $A_4$ in $\alpha_5$ and $A_1$ and $A_7$ both in $\alpha_5$ and $\beta_4$, and thus $A_1$ and $A_7$ would lie in one optical line.

Thus $A_1$ and $A_7$ would either coincide or else the one would be *after* the other, which is impossible.

Thus $A_4$ is neither *before* nor *after* $A_2$ and is neither *before* nor *after* $A_3$ and is distinct from either.

(*b*) *If $A_1$ be any element and $A_2$ and $A_3$ be two other distinct elements of $\beta_1$, one of which is neither* after *nor* before *the other, there is at least one other distinct element in $\beta_1$ which is neither* after *nor* before $A_2$ *and neither* after *nor* before $A_3$.

## THEOREM 20

*If $A_1$ be any element there are at least three distinct optical lines containing $A_1$.*

Let $A_2$ be any element in $\alpha_1$ distinct from $A_1$.

Then by Theorem 2 (*a*) there is at least one other element in $\alpha_1$ distinct from $A_2$ which is neither *before* nor *after* $A_2$.

Call such an element $A_3$.

Further by Theorem 19 there is at least one other distinct element in $\alpha_1$ which is neither *before* nor *after* $A_2$ and neither *before* nor *after* $A_3$. Call such an element $A_4$.

Then $A_1$ and $A_2$ determine one optical line; $A_1$ and $A_3$ determine a second optical line; $A_1$ and $A_4$ determine a third optical line.

These are all distinct and all contain $A_1$.

If $a$ be an optical line and if $A$ be any element not in the optical line but *before* some element of it we have by Post. X ($a$) one single optical line containing $A$ and such that each element of it is *before* an element of $A$.

Further, we have seen in Theorem 17 that there is one single optical line containing $A$ and also intersecting $a$.

Also by Theorem 20 there are at least three optical lines containing $A$ and so there must be at least one optical line containing $A$ in addition to the two particular ones which we have already mentioned.

Similarly if $a$ be an optical line and if $A$ be any element not in the optical line but *after* some element of it, there is one single optical line containing $A$ and such that each element of it is *after* an element of $a$ and there is one single optical line containing $A$ and intersecting $a$.

In addition to these two particular optical lines Theorem 20 shows that there is at least one other optical line containing $A$.

These considerations prepare the way for Post. XII ($a$) and ($b$).

POSTULATE XII. ($a$) **If $a$ be an optical line and if A be any element not in the optical line but before some element of it, then each optical line through A, except the one which intersects $a$ and the one of which each element is before an element of $a$, has one single element which is neither before nor after any element of $a$.**

($b$) **If $a$ be an optical line and if A be any element not in the optical line but after some element of it, then each optical line through A, except the one which intersects $a$ and the one of which each element is after an element of $a$, has one single element which is neither after nor before any element of $a$.**

### THEOREM 21

($a$) *If each element of an optical line $a$ be after an element of a distinct optical line $b$, then each element of $b$ is before an element of $a$.*

Let $A_1$ be any element of $a$; then since $A_1$ is not in $b$ but *after* an element of $b$, there is one single element (say $A_2$) common to the optical line $b$ and the sub-set $\beta_1$ (Post. IX ($b$)).

Then $A_2$ is not an element of $a$ but is *before* the element $A_1$ of $a$ and so by Post. X (*a*) there is *one single optical line* (say $c$) containing $A_2$ and such that each element of it is *before* an element of $a$.

Now $b$ cannot be identical with the optical line $A_2A_1$, for then $a$ and $b$ would have the element $A_1$ in common, which is impossible by Theorem 16 (*b*).

Suppose now, if possible, that $b$ is not identical with $c$; then by Post. XII (*a*) there will be *one single element* in $b$ (say $A_3$) which will be neither *before* nor *after* any element of $a$.

Consider an element $A_4$ in $b$ and *after* $A_3$.

Since there can only be one element in $b$ which is neither *before* nor *after* any element of $a$, it would follow that $A_4$ must be either *before* or *after* some element of $a$.

Since $A_3$ is *before* $A_4$ it would follow, if $A_4$ were *before* an element of $a$, that $A_3$ was also *before* an element of $a$, contrary to hypothesis.

We should therefore require $A_4$ to be *after* some element (say $A_5$) of $a$: so that $A_5$ would be *before* $A_4$: an element of $b$.

But by hypothesis $A_5$ is *after* some element of $b$ and so, by Theorem 12, $A_5$ would require to lie in $b$.

Thus $a$ and $b$ would have an element in common, which is impossible by Theorem 16 (*b*).

Thus the supposition that $b$ is distinct from $c$ leads to a contradiction and therefore is not true.

Thus $b$ must be identical with $c$ and so each element of $b$ is *before* an element of $a$.

(*b*) *If each element of an optical line $a$ be* before *an element of a distinct optical line $b$, then each element of $b$ is* after *an element of $a$.*

## THEOREM 22

*If $a$ be an optical line and if $A_1$ be any element which is neither* before *nor* after *any element of $a$, there is* one single optical line *containing $A_1$ and such that no element of it is either* before *or* after *any element of $a$.*

Let $A_2$ be any selected element of $a$; then $A_1$ is neither *before* nor *after* $A_2$, and so by Post. VI (*b*) an element exists which is a member both of $\beta_1$ and of $\beta_2$.

Call such an element $A_3$.

Now $A_3$ is *before* $A_2$, an element of $a$, and does not lie in $a$ and therefore by Post. X (*a*) there is *one single optical line* (say $c$) containing $A_3$ and such that each element of $c$ is *before* an element of $a$.

Further, $A_1$ is *after* $A_3$, but is not *before* any element of $a$, and so does not lie in $c$.

Thus by Post. X (*b*) there is *one single optical line* (say $b$) containing $A_1$ and such that each element of $b$ is *after* an element of $c$.

Consider now any element $A_4$ other than $A_1$ in the optical line $b$; then $A_4$ cannot be an element of $a$ since otherwise $A_1$ would be either *before* or *after* an element of $a$, contrary to hypothesis.

Suppose now if possible that $A_4$ is *after* some element of $a$.

Then by Post. X (*b*) there is *one single optical line* (say $d$) containing $A_4$ and such that each element of $d$ is after an element of $a$.

But since each element of $a$ is *after* an element of $c$ therefore by Post. III each element of $d$ is *after* an element of $c$.

But by Post. X (*b*) there is only *one single optical line* containing $A_4$ which has this property and the optical line $b$ is such a one.

Thus the optical line $d$ must be identical with the optical line $b$.

Thus each element of $b$ would be *after* an element of $a$, contrary to the hypothesis that $A_1$ was neither *before* nor *after* any element of $a$.

Thus $A_4$ is not *after* any element of $a$.

Next suppose if possible that $A_4$ is *before* some element (say $A_5$) of $a$.

Then $A_5$ is not an element of $b$, but is *after* an element of $b$, and so by Post. X (*b*) there is *one single optical line* (say $e$) containing $A_5$ and such that each element of $e$ is *after* an element of $b$.

But each element of $b$ is *after* an element of $c$ and so by Post. III each element of $e$ is *after* an element of $c$.

There is however by Post. X (*b*) one *one single optical line* containing $A_5$ and having this property, and $a$ is such an optical line.

Thus $e$ must be identical with $a$, and so each element of $a$ must be *after* an element of $b$.

But if this were so then by Theorem 21 (*a*) each element of $b$ must be *before* an element of $a$, contrary to the hypothesis that $A_1$ is neither *before* nor *after* any element of $a$.

Thus $A_4$ is not *before* any element of $a$, and so no element of $b$ is either *before* or *after* any element of $a$.

We have thus shown that there is one optical line containing $A_1$ and having this property.

We have now to show that there is only one.

Consider any optical line containing $A_1$ other than the optical lines $b$ and $A_3 A_1$.

Call such an optical line $f$.

Then by Post. XII (*b*) there is *one single element* in $f$ (say $A_6$) such that $A_6$ is neither *before* nor *after* any element in $c$.

If then we take any element $A_7$ in $f$ and *before* $A_6$, such an element cannot be *after* any element in $c$, for then $A_6$ being *after* $A_7$ would be *after* an element of $c$, contrary to hypothesis.

Also, since there is only one element having the property of $A_6$ and lying in $f$, therefore $A_7$ must be *before* some element of $c$.

But this element is *before* some element of $a$, and so $A_7$ is *before* some element of $a$.

Thus there is only one optical line containing $A_1$ and such that no element of it is either *before* or *after* any element of $a$.

## THEOREM 23

*If a be an optical line and $A_1$ be any element which is neither* before *nor* after *any element of a while b is the one single optical line containing $A_1$ and such that no element of it is either* before *or* after *any element of a, then every optical line through $A_1$, with the exception of b, is divided by $A_1$ into elements which are* before *an element of a and elements which are* after *an element of a.*

We proved in Theorem 22 that there is only one optical line through $A_1$ having the property of $b$.

Thus if we take any other optical line $d$ through $A_1$ there must be at least one element of $d$ which is either *before* or *after* some element of $a$.

Suppose first that there is an element $A_3$ which is *before* some element of $a$.

Then $A_3$ cannot be *after* $A_1$, for since there is an element of $a$ *after* $A_3$ there would by Post. III be an element of $a$ *after* $A_1$, contrary to hypothesis.

Thus $A_3$ must be *before* $A_1$.

Further, $A_3$ cannot be an element of $a$, for then $A_1$ would be *after* an element of $a$, contrary to hypothesis.

Thus $A_3$ is not an element of $a$ but *before* an element of it, and so by Post. IX (*a*) there is *one single element* (say $A_2$) which is an element both of the optical line $a$ and the sub-set $\alpha_3$.

Further by Post. X (*a*) there is *one single optical line* (say $c$) containing $A_3$ and such that each element of it is *before* an element of $a$.

Then by Post. XII (*a*) since the optical line $d$ contains $A_3$ and is not identical with either of the optical lines $A_3 A_2$ or $c$ it follows that there is *one single element* of $d$ which is neither *before* nor *after* any element of $a$.

But by hypothesis $A_1$ has this property and so every other element of $d$ is either *before* or *after* an element of $a$.

However, as we have already seen, an element which is *after* $A_1$ in $d$ cannot be *before* an element of $a$ and so it must be *after* an element of $a$.

Similarly an element which is *before* $A_1$ in $d$ cannot be *after* an element of $a$, for then $A_1$ would be *after* an element of $a$ contrary to hypothesis, and so an element which is *before* $A_1$ in $d$ must be *before* an element of $a$.

We arrive at the same conclusion if we start off by supposing the existence in $d$ of an element $A_3'$ which is *after* some element of $a$. Thus the theorem is proved.

### Theorem 24

(a) *If each element of each of two distinct optical lines $a$ and $b$ be* after *elements of a third optical line $c$, and if one element $A_1$ of the optical line $b$ be* after *some element of the optical line $a$, then each element of $b$ is* after *an element of $a$.*

Let $b'$ be the *one single optical line* containing $A_1$ and such that each element of $b'$ is *after* an element of $a$.

Then since each element of $a$ is *after* an element of $c$ therefore by Post. III each element of $b'$ is *after* an element of $c$.

But by hypothesis each element of $b$ is *after* an element of $c$, and $b$ contains $A_1$ an element not in the optical line $c$ but *after* some element of it.

Thus by Post. X (b), since there is only one single optical line containing $A_1$ and having this property, it follows that $b'$ must be identical with $b$.

Thus each element of $b$ is *after* an element of $a$.

(b) *If each element of each of two distinct optical lines $a$ and $b$ be* before *elements of a third optical line $c$, and if one element $A_1$ of the optical line $b$ be* before *some element of the optical line $a$, then each element of $b$ is* before *an element of $a$.*

### Theorem 25

(a) *If each element of each of two distinct optical lines $a$ and $b$ be* after *elements of a third optical line $c$, and if one element $A_1$ of the optical line $b$ be neither* before *nor* after *any element of the optical line $a$, then no element of the optical line $b$ is either* before *or* after *any element of the optical line $a$.*

Since $A_1$ is not an element of $c$ but is *after* some element of it, therefore

R

by Post. IX $(b)$, there is one single element (say $A_3$) which is common to the optical line $c$ and the sub-set $\beta_1$.

Then since $A_3$ is not an element of $a$, but is *before* an element of $a$ (Theorem 21 $(a)$), therefore by Post. IX $(a)$ there is one single element (say $A_2$) which is common to the optical line $a$ and the sub-set $\alpha_3$.

The demonstration then follows as in Theorem 22.

($b$) *If each element of each of two distinct optical lines $a$ and $b$ be* before *elements of a third optical line $c$, and if one element $A_1$ of the optical line $b$ be neither* after *nor* before *any element of the optical line $a$, then no element of the optical line $b$ is either* after *or* before *any element of the optical line $a$.*

This may be demonstrated in an analogous manner.

### THEOREM 26

($a$) *If an optical line $a$ be such that no element of it is either* before *or* after *any element of the optical line $c$, and if another optical line $b$ be such that each element of it is* before *an element of $c$, then each element of $b$ is* before *an element of $a$.*

Since each element of $b$ is *before* an element of $c$, it follows by Theorem 21 $(b)$ that each element of $c$ is *after* an element of $b$.

Let $A_1$ be any element of $c$.

Then since $A_1$ is not an element of $b$ but is *after* an element of $b$, there is one single element common to the optical line $b$ and the sub-set $\beta_1$ (Post. IX $(b)$).

Let $A_2$ be this element.

Then $A_2$ and $A_1$ determine an optical line.

But by Theorem 23 every optical line containing $A_1$ except $c$ is divided by $A_1$ into elements which are *before* an element of $a$ and elements which are *after* an element of $a$, and since $A_2$ is *before* $A_1$ and lies in the optical line $A_1 A_2$, it follows that $A_2$ is also *before* an element of $a$ and is not an element of $a$.

Thus by Post. IX $(a)$ there is one single element (say $A_3$) common to the optical line $a$ and the sub-set $\alpha_2$.

Now $A_3$ is neither *before* nor *after* any element of $c$ and therefore if an optical line $a'$ be taken through $A_3$ such that each element of it is *after* an element of $b$, then by Theorem 25 $(a)$ no element of $a'$ is either *before* or *after* any element of $c$.

But by Theorem 22 there is only one optical line through $A_3$ having this property and $a$ is such an optical line.

Thus $a'$ is identical with $a$ and so each element of $a$ is *after* an element of $b$, and thus by Theorem 21 $(a)$ each element of $b$ is *before* an element of $a$.

$(b)$ *If an optical line $a$ be such that no element of it is either* after *or* before *any element of the optical line $c$, and if another optical line $b$ be such that each element of it is* after *an element of $c$, then each element of $b$ is* after *an element of $a$.*

### THEOREM 27

$(a)$ *If each element of an optical line $a$ be* after *an element of a distinct optical line $c$, and each element of another optical line $b$ be* before *an element of $c$, then each element of $a$ is* after *an element of $b$.*

By Theorem 21 $(b)$ each element of $c$ is *after* an element of $b$, and since each element of $a$ is *after* an element of $c$, therefore by Post. III each element of $a$ is *after* an element of $b$.

$(b)$ *If each element of an optical line $a$ be* before *an element of a distinct optical line $c$, and each element of another optical line $b$ be* after *an element of $c$, then each element of $a$ is* before *an element of $b$.*

### THEOREM 28

*If two distinct optical lines $a$ and $b$ be such that no element of either of them is either* before *or* after *any element of a third optical line $c$, then no element of $a$ is either* before *or* after *any element of $b$.*

For suppose, if possible, that some element $A_1$ of $a$ is *after* an element of $b$; then $A_1$ cannot lie in $b$ and by Post. IX $(b)$ there is one single element (say $A_2$) common to the optical line $b$ and the sub-set $\beta_1$.

But by Theorem 23 every optical line through $A_1$ except $a$ is divided by $A_1$ into elements which are *before* an element of $c$ and elements which are *after* an element of $c$.

Thus since $A_2$ and $A_1$ determine an optical line through $A_1$, and since $A_2$ is *before* $A_1$, therefore $A_2$ must be *before* an element of $c$, contrary to the hypothesis that no element of $b$ is either *before* or *after* any element of $c$.

Similarly if we suppose $A_1$ to be *before* an element of $b$ we are led to a conclusion contrary to hypothesis.

Thus no element of $a$ is either *before* or *after* any element of $b$.

*Definitions.* An optical line $a$ will be said to be parallel to a second distinct optical line $b$ when either:

(1) each element of $a$ is *after* an element of $b$,

or   (2) each element of $a$ is *before* an element of $b$,

or   (3) no element of $a$ is either *before* or *after* any element of $b$.

In case (1) $a$ will be said to be an *after-parallel* of $b$.

In case (2) $a$ will be said to be a *before-parallel* of $b$.

In case (3) $a$ will be said to be a *neutral-parallel* of $b$.

It follows from these definitions in conjunction with Theorem 21 that:

*If an optical line $a$ be parallel to an optical line $b$, then the optical line $b$ is parallel to the optical line $a$.*

Again, if $a$ be any optical line and $A$ be any element *not in the optical line*, $A$ may be *before* an element of $a$, or may be *after* an element of $a$, but by Theorem 12 $A$ cannot be *before* one element of $a$ and *after* another element of $a$.

By Post. XII $(a)$ and $(b)$ it follows that $A$ may be neither *before* nor *after* any element of $a$.

If $A$ be *before* an element of $a$, then by Post. X $(a)$, there is one single parallel to $a$ containing $A$.

If $A$ be *after* an element of $a$, then by Post. X $(b)$, there is one single parallel to $a$ containing $A$.

If $A$ be neither *before* nor *after* any element of $a$, then by Theorem 22 there is one single parallel to $a$ containing $A$.

Thus we can say in general:

*If $a$ be any optical line and $A$ be any element which is not in the optical line, then there is one single optical line parallel to $a$ and containing $A$.*

Further, combining Theorems 24 $(a)$, 24 $(b)$, 25 $(a)$, 25 $(b)$, 26 $(a)$, 26 $(b)$, 27 $(a)$, 27 $(b)$, 28, we have the general result that:

*If two distinct optical lines $a$ and $b$ are each parallel to a third optical line $c$, then the optical lines $a$ and $b$ are parallel one to another.*

*Definition.* If $a$ and $b$ be any pair of distinct optical lines one of which is an after-parallel of the other, then the aggregate of all elements of all optical lines which intersect both $a$ and $b$ will be called an *inertia plane.**

* In the first edition of this work the term *acceleration* plane was used instead of *inertia* plane. The change was made in order that the nomenclature might be more systematic.

## THEOREM 29

*If a be an optical line there are an infinite number of distinct inertia planes which all contain a.*

From Post. XII (*a*) and (*b*) it follows that there is at least one element, say $A_1$, which is neither *before* nor *after* any element of *a*.

If *b* be the one optical line through $A_1$ such that no element of it is either *before* or *after* any element of *a*, then by Theorem 23 every optical line through $A_1$ except *b* is divided by $A_1$ into elements which are *before* an element of *a* and elements which are *after* an element of *a*.

Let *f* be one particular optical line containing $A_1$ and distinct from *b*.

Let $A_2$ be any element in *f* other than $A_1$; then $A_2$ must be either *before* or *after* some element of *a* but is not itself an element of *a*.

Thus if an optical line *c* be taken through $A_2$ parallel to *a*, then *c* is either a before- or after-parallel of *a* and therefore along with *a* serves to define an inertia plane.

Let $A_3$ be another element of *f* distinct from $A_2$.

Then in order that $A_3$ should lie in the inertia plane defined by *a* and *c* it would have to lie in an optical line intersecting both *a* and *c*.

But since $A_3$ is distinct from $A_2$ and lies in the optical line *f* which also contains $A_2$ it must be either *before* or *after* $A_2$, and so by Post. IX (*a*) or Post. IX (*b*) there must be *one single element* which is an element both of the optical line *c* and the sub-set $\alpha_3$ or $\beta_3$ as the case may be.

But the element $A_2$ is such an element and therefore the optical line *f* containing $A_3$ and $A_2$ is the only optical line which intersects *c* and contains $A_3$.

Thus in order that $A_3$ should lie in the inertia plane defined by *a* and *c* it would be necessary for *f* to intersect *a* and this we know it does not do since if it did $A_1$ would be either *before* or *after* an element of *a*, contrary to hypothesis.

If then $A_3$ be distinct from $A_1$ it is either *before* or *after* an element of *a* and so if we take the optical line through $A_3$ parallel to *a*, it will be either a before- or after-parallel of *a*.

Call such an optical line *d*.

Then *d* and *a* define another inertia plane which is distinct from that defined by *c* and *a*, since the latter does not contain $A_3$.

If any other element $A_n$ in the optical line *f* be selected other than $A_2$ or $A_3$ and an optical line be taken through it parallel to *a*, then, provided $A_n$ is distinct from $A_1$, the parallel to *a* through $A_n$ will, along with *a*, define an inertia plane distinct from the others.

Thus each element of $f$ except $A_1$ corresponds to a distinct inertia plane and the number of elements in $f$ is infinite, while all the inertia planes contain $a$.

Thus there are an infinite number of distinct inertia planes all containing the optical line $a$.

From the last theorem it follows directly that it is permissible to speak of three or more inertia planes which have two elements in common.

This prepares the way for Post. XIII.

POSTULATE XIII. **If two distinct inertia planes have two elements in common, then any other inertia plane containing these two elements contains all elements common to the two first-mentioned inertia planes.**

### THEOREM 30

*If $a$ and $b$ be two distinct optical lines and if $a$ be an after-parallel of $b$, then if $c$ and $d$ be two other distinct optical lines intersecting both $a$ and $b$, one of these latter two optical lines is an after-parallel of the other.*

Let the optical line $c$ intersect $b$ in $A_1$ and $a$ in $A_2$ and let the other optical line $d$ intersect $b$ in $A_3$ and $a$ in $A_4$.

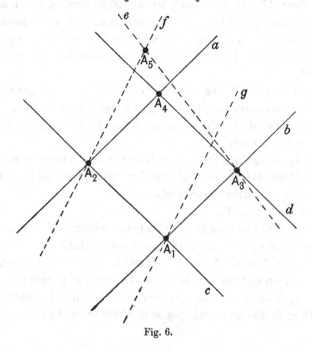

Fig. 6.

Then, by Theorem 17 $(a)$, it is not possible for $A_1$ and $A_3$ to be coincident while $A_2$ and $A_4$ are distinct; while, by Theorem 17 $(b)$, it is not possible for $A_2$ and $A_4$ to be coincident while $A_1$ and $A_3$ are distinct.

We may suppose without loss of generality that $A_3$ is *after* $A_1$.

Then since $a$ is an after-parallel of $b$ we must have $A_4$ *after* $A_3$ and therefore by Post. III $A_4$ is *after* $A_1$, or $A_1$ *before* $A_4$.

Further, since $a$ is an after-parallel of $b$, and since $A_1$ and $A_2$ lie in the optical line $c$, we must have $A_2$ *after* $A_1$ and therefore $A_2$ must lie in $\alpha_1$.

But now $A_4$ could not be *before* $A_2$, for then, by Theorem 1, $A_4$ would lie in $\alpha_1$ and, since it is distinct from $A_2$, we should have two elements common to the optical line $a$ and the sub-set $\alpha_1$; which is impossible.

Thus since $A_4$ and $A_2$ both lie in the optical line $a$ we must have $A_4$ *after* $A_2$ and so $A_4$ lies in $\alpha_2$.

Now let $e$ be the optical line through $A_3$ parallel to $c$; then $e$ is an after-parallel of $c$ since $A_3$ is *after* $A_1$.

Again there is one single optical line (say $f$) through $A_2$ intersecting $e$ in some element, say $A_5$ which lies in $\alpha_2$.

Now, since $A_2$ and $A_3$ are distinct elements both lying in $\alpha_1$, and since $A_2$ does not lie in the optical line $A_1 A_3$, it follows by Theorem 13 that $A_2$ is neither *before* nor *after* $A_3$ and therefore $A_5$ lies in $\alpha_3$.

Suppose now, if possible, that $A_5$ is distinct from $A_4$; then by Theorem 18 $(a)$ since $A_4$ and $A_5$ lie both in $\alpha_2$ and $\alpha_3$, the one is neither *before* nor *after* the other.

Thus $A_5$ could not lie either in $a$ or $d$ since then it would have to be either *before* or *after* $A_4$.

Neither can $A_5$ lie in $b$, for since $A_2$ is *after* $A_1$ and $A_5$ is *after* $A_2$, and $A_1$ is an element of $b$ it would then follow by Theorem 12 that $A_2$ must lie in $b$, which is impossible.

Thus $e$ is the only optical line through $A_5$ containing an element of $b$ and if $e$ also intersected $a$ it would have to coincide with $d$, since $d$ is the only optical line through $A_3$ which intersects $a$.

Thus if $A_5$ did not coincide with $A_4$ then $A_5$ could not lie in the inertia plane defined by $a$ and $b$.

Thus the inertia plane defined by $c$ and $e$ would be distinct from the inertia plane defined by $a$ and $b$.

Now let $g$ be the optical line through $A_1$ parallel to $f$; then $g$ is a before-parallel of $f$, since $A_1$ is *before* $A_2$.

Then $g$ could not coincide with $b$ for in that case we should have two optical lines $a$ and $f$ both through $A_2$ and both parallel to $b$, which is impossible.

Now $A_3$ lies in the optical line $b$ which intersects $g$ in $A_1$ and so if $A_3$ should lie in the inertia plane defined by $f$ and $g$, then $b$ would have to intersect $f$.

But the only optical line through $A_1$ intersecting $f$ is $c$ and so if $A_3$ should lie in the inertia plane defined by $f$ and $g$, then $b$ would have to coincide with $c$, which is impossible.

Thus $A_3$ would not lie in this inertia plane which therefore would be distinct from the inertia planes defined by $a$ and $b$ and by $c$ and $e$, which both contain $A_3$.

But the inertia planes defined by $a$ and $b$, by $c$ and $e$, and by $f$ and $g$ all contain the two elements $A_1$ and $A_2$, while the two first-mentioned inertia planes also contain $A_3$, which would not be contained by the inertia plane defined by $f$ and $g$.

This is contrary to Post. XIII and so the assumption that $A_5$ is distinct from $A_4$ must be abandoned.

Thus $A_5$ coincides with $A_4$ and therefore the optical line $d$ coincides with the after-parallel of $c$ through $A_3$.

This proves the theorem.

### THEOREM 31

*If $a$, $b$, $c$, $d$, etc. be a set of parallel optical lines which all intersect one optical line $l$ in elements $A$, $B$, $C$, $D$, etc., then through any element of one of the set of optical lines $a, b, c, d$, etc. other than the elements $A, B, C, D$, etc. there is one optical line which intersects each one of the set $a$, $b$, $c$, $d$, etc. and is parallel to $l$.*

Since the elements $A$, $B$, $C$, $D$, etc. are elements of one optical line $l$, therefore of any two of these elements one is *after* the other by Theorem 9.

Thus of any two of the parallel optical lines $a$, $b$, $c$, $d$, etc. one is an after-parallel of the other.

If then one of these optical lines be selected (say $b$) and any element in it (say $X$) distinct from $B$ there will be

> one optical line through $X$ intersecting $a$,
> one optical line through $X$ intersecting $c$,
> one optical line through $X$ intersecting $d$, etc.

But by Theorem 30 all these are parallel to $l$ and since they all go through $X$ they must all be identical.

Also for each element of $b$ there is one such optical line and since any pair of such optical lines are parallel to $l$ they are also parallel to one another.

This theorem shows that an inertia plane contains two sets of parallel optical lines which may be called the *generators* of the inertia plane.

Any generator of one set intersects every generator of the other set but does not intersect any one of its own set.

Also we see that through any element of an inertia plane there are two optical lines lying in the inertia plane and of these two one belongs to one set and the other to the other set.

## THEOREM 32

*Through any element of an inertia plane there are only two distinct optical lines which lie in the inertia plane.*

We have already seen that there are two optical lines which pass through any element of an inertia plane and lie in the inertia plane.

We have now to prove that there cannot be more than two.

Let $A_1$ be any particular element of an inertia plane and let $a$ and $b$ be the two generators of the inertia plane passing through $A_1$.

Suppose, if possible, that a third optical line $c$ passes through $A_1$ and lies in the inertia plane.

Let $A_2$ be an element of $c$ after $A_1$, then $A_2$ must lie in the inertia plane and so there would be two generators of the inertia plane passing through $A_2$ and parallel respectively to $a$ and $b$.

The optical line parallel to $a$ would meet $b$ in some element, $A_3$ say, and the optical line parallel to $b$ would meet $a$ in $A_4$ say.

But if $A_1$, $A_2$ and $A_3$ were all distinct we should have three elements lying in pairs in three distinct optical lines, which is impossible by Theorem 14.

Similarly if $A_1$, $A_2$ and $A_4$ were all distinct.

Thus any optical line through $A_1$ and lying in the inertia plane must coincide either with $a$ or $b$.

## THEOREM 33

*If an inertia plane contain an optical line $a$ and an element $A_1$ which does not lie in the optical line, then $A_1$ is either* before *or* after *an element of $a$.*

There are two optical lines in the inertia plane which pass through $A_1$.

Of these two, one which we shall call $b$ intersects $a$, while the other does not intersect it.

If $b$ intersects $a$ in an element $A_2$, then $A_2$ must be distinct from $A_1$ since $A_1$ does not lie in $a$.

But both $A_1$ and $A_2$ lie in the optical line $b$ and so the one is *after* the other.

Thus $A_1$ is either *before* or *after* $A_2$: an element of the optical line $a$.

### THEOREM 34

*If two elements be such that one is after the other, but does not lie in an optical line with it, then there are an infinite number of inertia planes containing the two elements.*

Let $A_1$ and $A_2$ be the two elements and let $A_2$ be *after* $A_1$.

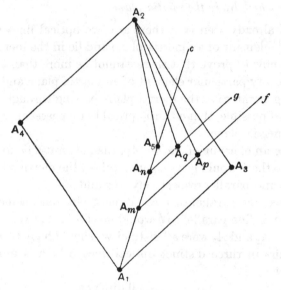

Fig. 7.

Then by Theorem 5 there is at least one other distinct element which is a member both of $\alpha_1$ and of $\beta_2$. Call such an element $A_3$.

Then $A_2$ is in $\alpha_3$ and so both $A_1A_3$ and $A_3A_2$ are optical lines.

But $A_1$ is not in the optical line $A_3A_2$ but is *before* $A_3$ an element of it and so we may take a before-parallel to $A_3A_2$ through $A_1$.

Then through $A_2$ there is one single optical line intersecting this before-parallel in some element, say $A_4$.

Then $A_4A_2$ will be an after-parallel of $A_1A_3$ by Theorem 30.

Now $A_1A_3$ and $A_1A_4$ are two distinct optical lines through $A_1$ and by Theorem 20 there are at least three distinct optical lines containing $A_1$ so that there must be at least one other. Let $c$ be such an optical line.

Then $A_2$ is not in $c$ but is *after* $A_1$ an element of $c$ and so by Post. IX (*b*) there is *one single element* (say $A_5$) common to the optical line $c$ and the sub-set $\beta_2$.

Then $A_1 A_5$ and $A_5 A_2$ are distinct optical lines and since $A_2$ is *after* $A_5$ we may take an after-parallel to $A_1 A_5$ through $A_2$, which together with $A_1 A_5$ will determine an inertia plane containing the given elements.

Let $A_m$ and $A_n$ be any two distinct elements of the optical line $A_1 A_5$ which are *after* $A_1$ and *before* $A_5$.

Then, $A_m$ and $A_n$ being elements which are *after* $A_1$ and not in the optical line $A_1 A_3$, we may take after-parallels to $A_1 A_3$ through $A_m$ and $A_n$. Call these $f$ and $g$ respectively.

Then $A_2$ cannot be an element of $f$ for then we should have the three elements $A_m$, $A_5$ and $A_2$ lying in pairs in three distinct optical lines, which is impossible by Theorem 14.

But $A_5$ is *after* $A_m$ and $A_2$ is *after* $A_5$ and so by Post. III $A_2$ is *after* $A_m$ an element of $f$.

Thus by Post. IX (*b*) there is *one single element* (say $A_p$) common to the optical line $f$ and the sub-set $\beta_2$.

Similarly $A_2$ cannot be an element of $g$ but is *after* $A_n$ an element of $g$ and so there is *one single element*, say $A_q$, common to the optical line $g$ and the sub-set $\beta_2$.

Now $A_m$ and $A_n$ being both elements of the optical line $A_1 A_5$, the one must be *after* the other, and since $f$ and $g$ are both after-parallels of $A_1 A_3$ it follows by Theorem 24 that the one is an after-parallel of the other.

Thus $f$ and $g$ can have no element in common and so $A_p$ and $A_q$ must be distinct.

Further, $A_p$ and $A_q$ cannot both lie in the same optical line through $A_2$, for since $f$ and $g$ are both after-parallels of $A_1 A_3$ therefore by Theorem 31 this hypothetical optical line would also intersect $A_1 A_3$ and would therefore have to be identical with $A_3 A_2$. Thus the optical line $A_1 A_5$ or $c$ would have to be parallel to $A_3 A_2$ and so be identical with $A_1 A_4$, contrary to hypothesis.

Thus the optical lines $A_p A_2$ and $A_q A_2$ must be distinct.

Further, either of them, say $A_p A_2$, must be distinct from $A_3 A_2$ for then $A_3 A_2$ would contain $A_p$ an element of $f$, and since $f$ is an after-parallel of $A_1 A_3$ therefore again $A_1 A_5$ would have to be identical with $A_1 A_4$, contrary to hypothesis.

Again, either of the optical lines $A_p A_2$ or $A_q A_2$ must be distinct from

$A_5 A_2$, for, if we take $A_p A_2$, we should have the three elements $A_m$, $A_p$ and $A_5$ lying in pairs in three distinct optical lines, which is impossible.

Similarly corresponding to each element of the optical line $A_1 A_5$ which is *after* $A_1$ and *before* $A_5$ we may take an after-parallel to $A_1 A_3$ which will have one element in common with the sub-set $\beta_2$ which determines a distinct optical line through $A_2$.

Since there are an infinite number of elements in the optical line $A_1 A_5$ which are *after* $A_1$ and *before* $A_5$, it follows that there are an infinite number of optical lines through $A_2$ which are all distinct.

Since $A_1$ and $A_2$ are not in one optical line therefore $A_1$ cannot lie in any of these optical lines through $A_2$.

But $A_1$ is *before* $A_2$ and so by Post. X (*a*) a before-parallel to each of these optical lines may be taken through $A_1$ and the pair of parallel optical lines will determine an inertia plane containing $A_1$ and $A_2$.

Also since the number of optical lines through $A_2$ is infinite, and since by Theorem 32 only two optical lines pass through any element of an inertia plane and lie in the inertia plane, it follows that there are an infinite number of inertia planes containing the two elements $A_1$ and $A_2$.

## Theorem 35

*If two distinct elements be such that the one is neither* before *nor* after *the other, then there are an infinite number of inertia planes containing the two elements.*

Let $A_1$ and $A_2$ be the two elements.

Then by Post. VI (*a*) and Post. XI (*a*) there are at least two other distinct elements which are members both of $\alpha_1$ and $\alpha_2$.

Let $A_3$ and $A_5$ be two such elements.

Then $A_1 A_3$, $A_1 A_5$, $A_2 A_3$, $A_2 A_5$ are distinct optical lines.

Let $A_m$ and $A_n$ be any two distinct elements of the optical line $A_1 A_5$ which are *after* $A_1$ and *before* $A_5$.

Then $A_m$ and $A_n$ being elements which are *after* $A_1$ and not in the optical line $A_1 A_3$, we may take after-parallels to $A_1 A_3$ through $A_m$ and $A_n$. Call these $f$ and $g$ respectively.

Then $A_2$ cannot be an element of $f$, for then we should have the three elements $A_m$, $A_5$ and $A_2$ lying in pairs in three distinct optical lines, which is impossible by Theorem 14.

But since $f$ is an after-parallel of $A_1 A_3$ it follows by Theorem 21 (*a*) that $A_1A_3$ is a before-parallel of $f$ and so $A_3$ is *before* some element of $f$ or there is an element of $f$ which is *after* $A_3$.

But $A_3$ is *after* $A_2$ and so by Post. III there is an element of $f$ which is *after* $A_2$, or $A_2$ is *before* an element of $f$.

Thus by Post. IX $(a)$ there is *one single element*, say $A_p$, which is an element both of the optical line $f$ and the sub-set $\alpha_2$.

Similarly $A_2$ cannot be an element of $g$ but is *before* an element of $g$ and so there is *one single element*, say $A_q$, common to the optical line $g$ and the sub-set $\alpha_2$.

Now $A_m$ and $A_n$ being both elements of the optical line $A_1A_5$, the one must be *after* the other, and since $f$ and $g$ are both after-parallels of $A_1A_3$ it follows by Theorem 24 that the one is an after-parallel of the other.

Thus $f$ and $g$ can have no element in common and so $A_p$ and $A_q$ must be distinct.

Further, $A_p$ and $A_q$ cannot both lie in the same optical line through $A_2$, for since $f$ and $g$ are both after-parallels of $A_1A_3$ it follows by Theorem 31 that this hypothetical optical line would also intersect $A_1A_3$ and would therefore have to be identical with $A_2A_3$.

Thus $A_2A_3$ would by Theorem 30 require to be either a before- or after-parallel of $A_1A_5$.

But $A_3$ is *after* $A_1$ and $A_2$ is *before* $A_5$ and so one element of $A_2A_3$ is *after* an element of $A_1A_5$ while another element of $A_2A_3$ is *before* an element of $A_1A_5$.

Thus $A_2A_3$ cannot be either a before or after-parallel of $A_1A_5$ and so $A_p$ and $A_q$ cannot both lie in the same optical line through $A_2$.

Thus the optical lines $A_2A_p$ and $A_2A_q$ must be distinct.

Further, either of them must be distinct from $A_2A_3$, for otherwise $A_2A_3$ would, again, require to be an after-parallel of $A_1A_5$, which we showed to be impossible.

Again, either of the optical lines $A_2A_p$, $A_2A_q$ must be distinct from $A_2A_5$, for if we take for instance the case of $A_2A_p$, we should then have the three elements $A_m$, $A_p$ and $A_5$ lying in pairs in three distinct optical lines, which is impossible.

Similarly corresponding to each element of the optical line $A_1A_5$ which is *after* $A_1$ and *before* $A_5$ we may take an after-parallel to $A_1A_3$ which will have one element in common with the sub-set $\alpha_2$ which determines a distinct optical line through $A_2$.

Since there are an infinite number of elements in the optical line $A_1A_5$ which are *after* $A_1$ and *before* $A_5$, it follows that there are an infinite number of optical lines through $A_2$ which are all distinct.

Since $A_1$ is neither *before* nor *after* $A_2$ and is distinct from it, therefore $A_1$ cannot lie in any of the optical lines through $A_2$.

Now there is only one element common to the optical line $A_1A_5$ and the sub-set $\alpha_2$, namely the element $A_5$, and $A_m$ cannot be *after* $A_2$ since otherwise, by Theorem 1, $A_m$ would require to lie in $\alpha_2$. But $A_p$ is *after* $A_2$ and, since it lies in an optical line with $A_m$, $A_p$ must be *after* $A_m$. But $A_m$ is *after* $A_1$ and so $A_p$ is *after* $A_1$. Similarly $A_q$ is *after* $A_1$.

Thus $A_1$ is not an element of any of the optical lines through $A_2$ but is *before* elements of those which we have obtained, and so by Post. X (a) there is *one single optical line* containing $A_1$ and such that each element of it is *before* an element of any particular one of the optical lines through $A_2$ which we have obtained.

Each of these pairs of parallel optical lines determines an inertia plane containing $A_1$ and $A_2$ and, since the number of optical lines through $A_2$ is infinite, and since by Theorem 32 there are only two optical lines which pass through any element of an inertia plane and lie in the inertia plane, it follows that there are an infinite number of inertia planes containing the two elements $A_1$ and $A_2$.

### REMARKS

The last two theorems showed that an infinite number of inertia planes contain any pair of elements which *do not* lie in an optical line.

Further, Theorem 29 showed that an infinite number of inertia planes contain a given optical line and so contain any two elements which *do* lie in an optical line.

It is easy to show that if two or more distinct inertia planes contain an optical line there is no other element which they have in common which does not lie in the optical line.

Thus if we consider two inertia planes $P$ and $Q$ which both contain an optical line $a$, and suppose, if possible, that they have also an element $A$ in common which does not lie in the optical line, then another optical line $b$ through $A$ must exist which is parallel to $a$.

The optical line $b$ must lie in the inertia plane $P$ and also in the inertia plane $Q$, and $b$ must be either a before- or after-parallel of $a$, since $A$ is either *before* or *after* an element of $a$ (Theorem 33).

Thus $a$ and $b$ determine an inertia plane which would be identical both with $P$ and $Q$, which could therefore not be distinct, contrary to hypothesis.

Thus if two or more inertia planes have an optical line in common they can have no other element outside the optical line in common.

We have also seen by Post. XIII that any inertia plane which contains two elements which are common to two distinct inertia planes, contains all elements common to them.

These remarks prepare the way for the following definitions and for Post. XIV.

*Definitions.* If two inertia planes contain two elements in common, then the aggregate of all elements common to the two inertia planes will be called a *general line.*

If two inertia planes contain two elements in common, of which one is *after* the other, but does not lie in the same optical line with it, then the aggregate of all elements common to the two inertia planes will be called an *inertia line.**

If two inertia planes contain two elements in common, of which one is neither *before* nor *after* the other, then the aggregate of all elements common to the two inertia planes will be called a *separation line.*†

POSTULATE XIV. (a) **If $a$ be any inertia line and $A_1$ be any element of the set, then there is one single element common to the inertia line $a$ and the sub-set $\alpha_1$.**

(b) **If $a$ be any inertia line and $A_1$ be any element of the set, then there is one single element common to the inertia line $a$ and the sub-set $\beta_1$.**

### THEOREM 36

*An inertia line in any inertia plane has one single element in common with each optical line in the inertia plane.*

Let $a$ be the inertia line and let $A_1$ be an element in any optical line $b$ in the inertia plane which we shall call $P$.

Then by Post. XIV (a) there is *one single element,* say $A_2$, common to the inertia line $a$ and the sub-set $\alpha_1$.

Also by Post. XIV (b) there is *one single element,* say $A_3$, common to the inertia line $a$ and the sub-set $\beta_1$.

Now if $A_1$ lay in $a$, both $A_2$ and $A_3$ must coincide with $A_1$ since, if $A_2$ were distinct from $A_1$ we should have the two elements $A_1$ and $A_2$

---

* The name "inertia line" has been adopted because an inertia line represents the time path of an unaccelerated particle.

† The name "separation line" has been adopted because a single particle cannot occupy more than one element of a separation line, so that if particles $P$ and $Q$ occupy distinct elements of a separation line they must be *separate* particles.

in $a$ which both lay in $\alpha_1$, contrary to Post. XIV $(a)$ which asserts that there is only *one single element* common to the inertia line $a$ and the sub-set $\alpha_1$.

Thus if $A_1$ lie in $a$, then $A_2$ must coincide with $A_1$.

Similarly if $A_1$ lie in $a$, then $A_3$ must coincide with $A_1$.

Suppose now that $A_1$ does not lie in $a$, then both $A_2$ and $A_3$ must be distinct from $A_1$.

Then we must have $A_1$ *after* $A_3$ and $A_2$ *after* $A_1$ and therefore $A_2$ *after* $A_3$; so that $A_2$ and $A_3$ must be distinct.

Also $A_2$ could not lie in $\alpha_3$, for then we should have the two distinct elements $A_2$ and $A_3$ both common to the inertia line $a$ and the sub-set $\alpha_3$ contrary to Post. XIV $(a)$. Thus $A_2$ and $A_3$ cannot lie in the same optical line.

But since $A_2$ is in $\alpha_1$ and $A_3$ in $\beta_1$ it follows that $A_1$ and $A_2$ lie in an optical line through $A_1$, and also $A_3$ and $A_1$ lie in an optical line through $A_1$, and these optical lines are distinct and both lie in $P$.

Now by Theorem 32 there are only two distinct optical lines in the inertia plane which pass through $A_1$, and so one of them must be $A_1 A_2$ and the other $A_3 A_1$, and since $b$ must be identical with one of these optical lines, it follows that $a$ and $b$ must have one single element in common.

## THEOREM 37

*Of any two distinct elements of an inertia line one is* after *the other.*

Let $A_1$ and $A_2$ be any two distinct elements of the inertia line $a$, and let $b$ be one of the two optical lines in an inertia plane containing $a$ which pass through $A_1$.

Now of the two optical lines in this inertia plane which pass through $A_2$, the one is parallel to $b$ and the other intersects it in some element, say $A_3$.

Now $A_1$ and $A_2$ being distinct cannot both lie in $\alpha_3$ by Post. XIV $(a)$ and they cannot both lie in $\beta_3$ by Post. XIV $(b)$.

Thus one of the two elements $A_1$ and $A_2$ must lie in $\alpha_3$ and the other in $\beta_3$, and so one of them must be *after* $A_3$ and the other *before* $A_3$.

Thus by Post. III one of the two elements $A_1$ and $A_2$ must be *after* the other.

From the definition of a separation line it contains a pair of elements one of which is neither *before* nor *after* the other.

Thus it follows from the above theorem that *no inertia line can be a separation line and no separation line can be an inertia line.*

## Theorem 38

*If $A_1$ be any element in an inertia line $a$, there is at least one other element in the inertia line which is after $A_1$ and also at least one other element in it which is before $A_1$.*

Let $b$ be one of the two optical lines through $A_1$ in any inertia plane which contains $a$ and let $A_2$ be any element in $b$ which is *after $A_1$.*

Then by Post. XIV ($a$) there is one single element, say $A_3$, common to the inertia line $a$ and the sub-set $\alpha_2$.

Then $A_3$ cannot be identical with $A_2$ since then we should have two elements common to the inertia line $a$ and the optical line $b$, contrary to Theorem 36.

Thus $A_3$ is *after $A_2$* and $A_2$ is *after $A_1$* and therefore $A_3$ is *after $A_1$* and is an element of the inertia line $a$.

Similarly if we take any element $A_4$ in the optical line $b$ and *before $A_1$* there will by Post. XIV ($b$) be one single element, say $A_5$, common to the inertia line $a$ and the sub-set $\beta_4$.

Then $A_1$ will be *after $A_4$* and $A_4$ *after $A_5$* and therefore $A_1$ *after $A_5$.*

Thus $A_5$ is *before $A_1$* and is an element of the inertia line $a$.

## Theorem 39

*If $A_1$ and $A_2$ be any two distinct elements of an inertia line $a$, there is at least one other distinct element of $a$ which is after one of the two elements and before the other.*

By Theorem 37 one of the two elements $A_1$ and $A_2$ is *after* the other. We shall suppose that $A_2$ is *after $A_1$.*

Let $b$ and $c$ be the two optical lines through $A_1$ in any inertia plane containing $a$.

Then the optical line through $A_2$ parallel to $c$ will be an after-parallel and will intersect $b$ in some element $A_3$ which must be *after $A_1$.*

Now let $A_4$ be any element in $b$ which is *after $A_1$* and *before $A_3$* and consider the optical line through $A_4$ parallel to $c$.

This will be an after-parallel of $c$ but a before-parallel of $A_3 A_2$ and must intersect the inertia line $a$ in some element, say $A_5$.

Then $A_5$ cannot be *before* any element of $c$ and therefore is not *before $A_1$.*

Also $A_5$ cannot be *after* any element of $A_3 A_2$ and therefore is not *after $A_2$.*

Thus by Theorem 37 $A_5$ must be *after $A_1$* and *before $A_2$* and lies in the inertia line $a$.

*It follows from the above results that there are an infinite number of elements in any inertia line.*

POSTULATE XV.  **If two general lines, one of which is a separation line and the other is not, lie in the same inertia plane, then they have an element in common.**

Since there are an infinite number of optical lines in an inertia plane, and since only two of them pass through any given element, and since by Post. XV each of them has an element in common with any separation line lying in the inertia plane, it follows that there are an infinite number of elements in any separation line.

Further, since as we have remarked in connexion with Theorem 37 no inertia line can be a separation line, it follows that *no element of a separation line is either* before *or* after *another element of it.*

## THEOREM 40

*If $A_1$ and $A_2$ be two distinct elements one of which is neither* before *nor* after *the other, and if a and b be the two optical lines through $A_1$ in an inertia plane containing the two elements, then $A_2$ is* before *an element of one of these optical lines and* after *an element of the other.*

By Theorem 33 $A_2$ must be either *before* or *after* an element of $a$ and also must be either *before* or *after* an element of $b$; but $A_2$ cannot lie either in $a$ or $b$ since it is distinct from $A_1$ and is neither *before* nor *after* it.

Suppose first that $A_2$ is *before* an element of $a$.

Then one of the two optical lines through $A_2$ in the inertia plane will intersect $a$ in some element, say $A_3$, while the other optical line through $A_2$ in the inertia plane will intersect $b$ in some element, say $A_4$.

Then $A_2$ must be *before* $A_3$ since $A_2$ cannot either lie in $a$ or be *after* any element of it.

But $A_3$ cannot either coincide with $A_1$ or be *before* $A_1$, for then we should have $A_2$ before $A_1$, contrary to hypothesis.

Thus $A_3$ must be *after* $A_1$.

But $A_1$ is an element of $b$ and so the optical line $A_2A_3$ (which since it intersects $a$ must be parallel to $b$) must be an after-parallel of $b$.

Thus $A_2$ must be *after* an element of $b$, and since $A_2$ must be either *before* or *after* $A_4$, it follows that $A_2$ is *after* $A_4$.

In a similar manner we may prove that if $A_2$ be *before* an element of $b$ it must be *after* an element of $a$.

Also in an analogous manner we may show that if $A_2$ be *after* an element of $b$ it must be *before* an element of $a$, and if $A_2$ be *after* an element of $a$ it must be *before* an element of $b$.

Thus $A_2$ must be *before* an element of one of the optical lines $a$ and $b$ and *after* an element of the other.

*Definition.* An element in an inertia plane will be said to be *between* a pair of parallel optical lines in the inertia plane if it be *after* an element of the one optical line and *before* an element of the other and does not lie in either optical line.

## THEOREM 41

*If $A_1$ and $A_2$ be any two distinct elements of a separation line, there is at least one other element of the separation line which lies between a pair of parallel optical lines through $A_1$ and $A_2$ respectively in an inertia plane containing the separation line.*

Let $a_1$ and $b_1$ be the two optical lines passing through $A_1$ in any inertia plane containing the separation line.

Then, since $A_2$ is neither *before* nor *after* $A_1$, it follows that $A_2$ is *before* an element of one of the two optical lines $a_1$ and $b_1$ and is *after* an element of the other. (Theorem 40.)

Suppose that $A_2$ is *before* an element of $a_1$.

Then it is *after* an element of $b_1$.

Let $a_2$ and $b_2$ be the two optical lines through $A_2$ parallel respectively to $a_1$ and $b_1$.

Then $a_2$ and $b_2$ lie in the inertia plane and since $A_2$ is *before* an element of $a_1$ therefore $a_2$ is a before-parallel of $a_1$.

Similarly since $A_2$ is *after* an element of $b_1$ it follows that $b_2$ is an after-parallel of $b_1$.

Further, $b_2$ must intersect $a_1$ in some element, say $A_3$, which must be *after* $A_2$ since $a_1$ is an after-parallel of $a_2$.

Let $A_4$ be any element of $b_2$ which is *after* $A_2$ and *before* $A_3$ and consider the optical line through $A_4$ parallel to $a_1$.

We shall call this optical line $a'$.

Then since $A_4$ is *before* $A_3$ it follows that $a'$ is a before-parallel of $a_1$ and since $A_4$ is *after* $A_2$ therefore $a'$ is an after-parallel of $a_2$.

Also $a'$ lies in the inertia plane.

Thus by Post. XV $a'$ must have an element in common with the separation line $A_1 A_2$.

Call this element $A_5$.

Then since $a'$ is a before-parallel of $a_1$ therefore $A_5$ is *before* an element of $a_1$ and since $a'$ is an after-parallel of $a_2$ therefore $A_5$ is *after* an element of $a_2$.

Thus $A_5$ is between the parallel optical lines $a_1$ and $a_2$.

### THEOREM 42

*If $A_1$, $A_2$ and $A_3$ be three elements in a separation line and if $A_3$ lies between a pair of parallel optical lines through $A_1$ and $A_2$ in an inertia plane containing the separation line, then $A_3$ also lies between a second pair of parallel optical lines through $A_1$ and $A_2$ in the inertia plane.*

Let $a_1$ and $a_2$ be a pair of parallel optical lines through $A_1$ and $A_2$ respectively in the inertia plane and suppose that $A_3$ lies between them.

We may without loss of generality suppose that $A_3$ is *after* an element of $a_2$ and *before* an element of $a_1$.

Let $b_1$ be the second optical line which passes through $A_1$ in the inertia plane and let $b_2$ be the second optical line which passes through $A_2$ in the inertia plane.

Then, since $a_1$ and $a_2$ are parallel, $b_1$ and $b_2$ are also parallel.

But since $A_3$ and $A_1$ lie in a separation line, $A_3$ is neither *before* nor *after* $A_1$, and since $A_3$ is *before* an element of $a_1$ therefore by Theorem 40 $A_3$ is *after* an element of $b_1$.

Similarly $A_3$ is neither *before* nor *after* $A_2$ and, since $A_3$ is *after* an element of $a_2$, therefore, by Theorem 40, $A_3$ is *before* an element of $b_2$.

Thus $A_3$ is between the parallel optical lines $b_1$ and $b_2$ passing through $A_1$ and $A_2$ respectively in the inertia plane.

Since there are only two optical lines in an inertia plane which pass through a given element of it, it follows directly from the above theorem that *if $A_1$, $A_2$ and $A_3$ be three elements in a separation line and if $A_3$ lies between a pair of parallel optical lines through $A_1$ and $A_2$ in an inertia plane containing the separation line, then $A_2$ does not lie between a pair of parallel optical lines through $A_1$ and $A_3$ in the inertia plane.*

Similarly $A_1$ does not lie between a pair of parallel optical lines through $A_2$ and $A_3$ in the inertia plane.

### THEOREM 43

*If $A_1$ and $A_2$ be any two elements of a separation line, there is at least one other element of the separation line such that $A_2$ lies between a pair of parallel optical lines through $A_1$ and that element in an inertia plane containing the separation line.*

Using the notation employed in Theorem 41 let us take any element, say $A_6$, in the optical line $b_2$ and *before* $A_2$ and consider the optical line through $A_6$ parallel to $a_2$.

Call this optical line $a''$.

Then since $A_6$ is *before* $A_2$ therefore $a''$ is a before-parallel of $a_2$, and since $a_2$ is a before-parallel of $a_1$ therefore $a''$ is also a before-parallel of $a_1$.

Further, $a''$ lies in the inertia plane and so by Post. XV it has an element in common with the separation line.

Call this element $A_7$.

Then $A_2$ is *before* $A_3$ an element of $a_1$ and is *after* $A_6$ an element of $a''$.

Thus $A_2$ is between the parallel optical lines $a_1$ and $a''$ passing through $A_1$ and the element $A_7$ respectively and lying in the inertia plane.

## THEOREM 44

*Of any three distinct elements of a separation line in a given inertia plane there is one which lies between a pair of parallel optical lines through the other two and in the inertia plane.*

Let $A_1$, $A_2$ and $A_3$ be any three distinct elements in the separation line.

Then, since there are two optical lines in an inertia plane passing through any element of it, let us select one of those passing through one of these elements, say $A_1$, and the parallel optical lines through $A_2$ and $A_3$.

Call these optical lines $a_1$, $a_2$ and $a_3$ respectively.

Then $a_1$, $a_2$ and $a_3$ all intersect any generator of the inertia plane belonging to the opposite set.

Let $b$ be such a generator and suppose that $a_1$, $a_2$ and $a_3$ intersect $b$ in the elements $A_1'$, $A_2'$ and $A_3'$ respectively.

Then $A_1'$, $A_2'$ and $A_3'$ being all elements of the optical line $b$, and being all distinct, it follows that of any two of them one must be *after* the other.

Thus remembering that Post. III must be satisfied it follows that either

$$A_2' \text{ is } after \text{ } A_1' \text{ and } A_3' \text{ } after \text{ } A_2' \quad (1),$$
or $\quad A_2' \text{ is } after \text{ } A_3' \text{ and } A_1' \text{ } after \text{ } A_2' \quad (2),$
or $\quad A_3' \text{ is } after \text{ } A_1' \text{ and } A_2' \text{ } after \text{ } A_3' \quad (3),$
or $\quad A_3' \text{ is } after \text{ } A_2' \text{ and } A_1' \text{ } after \text{ } A_3' \quad (4),$
or $\quad A_1' \text{ is } after \text{ } A_2' \text{ and } A_3' \text{ } after \text{ } A_1' \quad (5),$
or $\quad A_1' \text{ is } after \text{ } A_3' \text{ and } A_2' \text{ } after \text{ } A_1' \quad (6).$

In case (1) $a_2$ is an after-parallel of $a_1$ and a before-parallel of $a_3$ and so each element of $a_2$ is between the parallel optical lines $a_1$ and $a_3$.

Thus $A_2$ is between a pair of parallel optical lines through $A_1$ and $A_3$ in the inertia plane.

Similarly in case (2) $a_2$ is an after-parallel of $a_3$ and a before-parallel of $a_1$ and therefore again $A_2$ is between a pair of parallel optical lines through $A_1$ and $A_3$ in the inertia plane.

In a similar manner in cases (3) and (4) $A_3$ is between a pair of parallel optical lines through $A_1$ and $A_2$ in the inertia plane; while in cases (5) and (6) $A_1$ is between a pair of parallel optical lines through $A_2$ and $A_3$ in the inertia plane.

Thus in all cases one of the three elements is between a pair of parallel optical lines through the other two and in the inertia plane.

### Theorem 45

*If $A$ be an element of an optical line $a$ and if $B$ be an element which is neither before nor after any element of $a$, then no element of the separation line $AB$, with the exception of $A$, is either before or after any element of $a$.*

Let $C$ be any element of the separation line $AB$ other than $A$, and let $c$ be an optical line through $C$ parallel to $a$.

Suppose, if possible, that $C$ is either *before* or *after* some element of $a$.

Then $c$ would be either a before- or after-parallel of $a$ and accordingly $c$ and $a$ would be generators of an inertia plane which would contain the two elements $A$ and $C$ of the separation line $AB$ and would therefore contain every element of $AB$.

Thus the element $B$ would lie in an inertia plane containing the optical line $a$, and therefore, by Theorem 33, $B$ would be either *before* or *after* an element of $a$, contrary to hypothesis.

Thus the assumption that any element of the separation line $AB$, other than $A$, is either *before* or *after* any element of $a$ leads to a contradiction and therefore is not true and so no element of $AB$ with the exception of $A$ is either *before* or *after* any element of $a$.

### Sets of three elements which determine inertia planes

Let $A_1$, $A_2$ and $A_3$ be three distinct elements which do not all lie in one general line, then $A_1$ and $A_2$ must lie in one general line, $A_2$ and $A_3$ in a second and $A_3$ and $A_1$ in a third.

These three general lines need not however lie in one inertia plane, although they do in certain cases.

In these latter cases the three elements determine the inertia plane containing them, since if they should lie in two distinct inertia planes they would lie in one general line, contrary to hypothesis.

It is important to have criteria by which we can say that a set of three elements does lie in one inertia plane.

CASE I. Three elements $A_1$, $A_2$, $A_3$ lie in one inertia plane if $A_1$ and $A_2$ lie in an optical line while $A_3$ is an element not in the optical line but *before* some element of it, or *after* some element of it.

This is clearly true, since, if $A_1$ and $A_2$ lie in the optical line $a$, while $A_3$ does not lie in $a$ but is *before* some element of it, then there is a before-parallel optical line, say $b$ containing $A_3$, and so $a$ and $b$ are a pair of parallel generators of an inertia plane, containing $A_1$, $A_2$ and $A_3$ and which is determined by them.

Similarly if $A_3$ be *after* some element of $a$ there is a definite after-parallel optical line $b$ containing $A_3$, and the two optical lines $a$ and $b$ are a pair of parallel generators of an inertia plane containing $A_1$, $A_2$ and $A_3$ and which is determined by them.

CASE II. Three elements $A_1$, $A_2$, $A_3$ lie in one inertia plane if $A_1$ and $A_2$ lie in an inertia line and $A_3$ be *any* element outside the inertia line.

This can also be readily seen to hold since if $a$ denote the inertia line containing $A_1$ and $A_2$ then by Post. XIV $(a)$ there is one single element, say $A_4$, common to the inertia line $a$ and the sub-set $\alpha_3$, and by Post. XIV $(b)$ there is one single element, say $A_5$, common to the inertia line $a$ and the sub-set $\beta_3$.

Thus $A_3$ and $A_4$ lie in one optical line while $A_3$ and $A_5$ lie in another optical line.

These two optical lines may be taken as generators of opposite sets of an inertia plane containing $A_3$, $A_4$ and $A_5$.

But since this inertia plane contains the two elements $A_4$ and $A_5$ of the inertia line $a$, it must contain every element of $a$ and therefore contains $A_1$ and $A_2$.

Thus the three elements $A_1$, $A_2$ and $A_3$ lie in one inertia plane which is determined by them.

CASE III. Three elements $A_1$, $A_2$, $A_3$ lie in one inertia plane if $A_1$ and $A_2$ lie in a separation line and if $A_3$ be an element not in the separation line but *before* at least two elements of it or *after* at least two elements of it.

In order to show this let $a$ be the separation line containing $A_1$ and $A_2$ and suppose $A_3$ is *before* the elements $A_4$ and $A_5$ of $a$ which are supposed distinct.

Then $A_3$ and $A_4$ must lie either in an optical line or an inertia line since $A_4$ is *after* $A_3$, and similarly $A_3$ and $A_5$ must lie either in an optical line or an inertia line and the two general lines $A_3A_4$ and $A_3A_5$ are distinct.

If $A_3A_4$ and $A_3A_5$ be both optical lines, then they may be taken as generators of opposite sets of an inertia plane containing $A_3$, $A_4$ and $A_5$.

But this inertia plane, since it contains the two distinct elements $A_4$ and $A_5$ of the separation line $a$, must contain every element of it and so must contain $A_1$ and $A_2$.

Thus $A_1$, $A_2$ and $A_3$ lie in one inertia plane which is determined by them.

We shall suppose next that at least one of the general lines $A_3A_4$ and $A_3A_5$ is an inertia line.

Let us say that $A_3A_4$ is an inertia line.

Then by Case II the three elements $A_3$, $A_4$ and $A_5$ lie in one inertia plane which is determined by them.

But since this inertia plane contains the two elements $A_4$ and $A_5$ of the separation line $a$, therefore it contains every element of $a$ and so must contain $A_1$ and $A_2$.

Thus $A_1$, $A_2$ and $A_3$ lie in one inertia plane which is determined by them.

The case when $A_3$ is *after* two distinct elements of $a$ is quite analogous.

If $A_1$ and $A_2$ lie in an optical line $a$ while $A_3$ is an element which is neither *before* nor *after* any element of $a$, then the three elements do not lie in one inertia plane, for by Theorem 45 no element of the general line $A_1A_3$ with the exception of $A_1$ is either *before* or *after* any element of $a$.

But if $A_1$, $A_2$ and $A_3$ should lie in an inertia plane there would be two optical lines through $A_2$ in the inertia plane and both of these would have an element in common with the separation line $A_1A_3$.

Thus there would be at least two elements of $A_1A_3$ which would be *before* or *after* $A_2$, contrary to Theorem 45.

Thus $A_1$, $A_2$ and $A_3$ do not lie in one inertia plane.

If $A_1$ and $A_2$ lie in a separation line $a$, while $A_3$ is *before* one *single* element of $a$ or *after* one *single* element of $a$, then the three elements $A_1$, $A_2$, $A_3$ cannot lie in one inertia plane.

This is easily seen, for if we suppose that they do all lie in one inertia plane, there are two optical lines through $A_3$ in the inertia plane which have each an element in common with $a$.

If these elements be called $A_4$ and $A_5$ then, since $a$ is a separation line, $A_4$ is neither *before* nor *after* $A_5$ and so $A_4$ and $A_5$ must be either both *before* or both *after* $A_3$, contrary to the hypothesis that there is only *one single* element of $a$ which $A_3$ is *after* or *before*.

If $A_1$ and $A_2$ lie in a separation line $a$, while $A_3$ does not lie in $a$ and is neither *before* nor *after* any element of $a$, it is also evident from the above considerations that the three elements $A_1$, $A_2$, $A_3$ cannot lie in one inertia plane.

We have not however as yet proved the possibility of this last case, but shall do so hereafter (Theorem 99). Till then no use will be made of it, and it is merely mentioned here for the sake of completeness.

*Definition.* If an inertia plane have its two sets of generators respectively parallel to the two sets of generators of another distinct inertia plane, then the two inertia planes will be said to be *parallel* to one another.

It is clear that if $P$ be an inertia plane and $A$ be any element outside it, then there is one single inertia plane containing $A$, and parallel to $P$; for there is one single optical line through $A$ parallel to the one set of generators of $P$ and one single optical line through $A$ parallel to the other set of generators of $P$.

These are generators of opposite sets of an inertia plane containing $A$ and determine that inertia plane, which is therefore unique.

It is further clear that two parallel inertia planes can have no element in common, for if the element $A$ lies outside the inertia plane $P$ and if $a$ be an optical line passing through $A$ and parallel to a generator of $P$, then $a$ can have no element in common with $P$ since otherwise it would require to lie entirely in $P$, contrary to the hypothesis that $A$ is outside $P$.

Similarly any optical line $b$ which intersects $a$ and is parallel to a generator of $P$ of the opposite set can have no element in common with $P$.

But if $Q$ be the inertia plane passing through $A$ and parallel to $P$, every element of $Q$ must lie in an optical line such as $b$ and so $P$ and $Q$ can have no element in common.

It is also clear from the definition that *two distinct inertia planes which are parallel to the same inertia plane are parallel to one another*; since distinct optical lines which are parallel to the same optical line are parallel to one another.

## THEOREM 46

*If an inertia plane P have one element in common with each of a pair of parallel inertia planes Q and R, then, if P have a second element in common with Q, it has also a second element in common with R.*

If $P$ and $Q$ have two elements in common they must have a general line in common which we may call $a$.

Let $B_1$ be the element which by hypothesis $P$ and $R$ have in common.

Then if $a$ be an inertia or separation line it follows by Theorem 36 and Post. XV that both the optical lines through $B_1$ in the inertia plane $P$ have an element in common with $a$, while if $a$ be an optical line one of the optical lines through $B_1$ in $P$ has an element in common with $a$.

Thus in all cases at least one of the optical lines through $B_1$ in the inertia plane $P$ has an element in common with $a$.

Let $A_1$ be such an element.

Suppose first that $a$ is an optical line.

Then $a$ is one of the generators of $Q$ and since the inertia plane $R$ is parallel to $Q$ and since $B_1$ lies in $R$ there will be one of the generators of $R$ passing through $B_1$ and parallel to $a$.

Since $A_1$ and $B_1$ lie in an optical line and are distinct, the one must be *after* the other and so this parallel to $a$ through $B_1$ must be either a before- or after-parallel of $a$.

Let us denote it by $b$.

Then $a$ and $b$ determine an inertia plane which contains three distinct elements of $P$ which are not all in one general line and so this inertia plane must be identical with $P$.

Thus since it contains the optical line $b$ it follows that $P$ has a second element in common with $R$.

Suppose next that $a$ is an inertia or separation line and let $c$ be one of the generators of $Q$ which pass through $A_1$.

Then since $R$ is parallel to $Q$ and since $B_1$ lies in $R$ there will be one of the generators of $R$ passing through $B_1$ and parallel to $c$.

Since $A_1$ and $B_1$ lie in an optical line and are distinct, the one must be *after* the other and so this parallel to $c$ through $B_1$ must be a before- or after-parallel.

Let $C$ be any element of $c$ distinct from $A_1$ and let an optical line through $C$ intersect the optical line through $B_1$ parallel to $c$ in the element $D$.

Then by Theorem 30 the optical line $CD$ must be a before- or after-parallel of the optical line $A_1 B_1$.

Let the second optical line through $C$ in the inertia plane $Q$ meet $a$ in the element $A_2$.

The element $A_2$ must exist since $a$ is an inertia or separation line.

Since the optical line $CA_2$ must be a generator of $Q$ of the opposite set to $c$, there must be an optical line through $D$ in the inertia plane $R$ which is parallel to $CA_2$ and is a generator of $R$ of the opposite set to $DB_1$.

Since $C$ and $D$ lie in an optical line and are distinct, the one must be *after* the other and so this parallel to $CA_2$ through $D$ must be a before- or after-parallel.

Let an optical line through $A_2$ intersect the optical line through $D$ parallel to $CA_2$ in the element $B_2$.

Then by Theorem 30 the optical line $A_2B_2$ must be a before- or after-parallel of $CD$ and $CD$ is a before- or after-parallel of $A_1B_1$, and so if $A_1B_1$ and $A_2B_2$ be distinct they must be parallel to one another.

Now the optical lines $CA_1$ and $CA_2$ are distinct from the inertia or separation line $a$ and are also distinct from one another.

Also the element $C$ cannot lie in $a$ since then $CA_1$ would have to be an inertia or separation line.

Thus the elements $A_1$ and $A_2$ are distinct and since they lie in an inertia or separation line they cannot lie in one optical line.

Thus $A_2B_2$ is distinct from $A_1B_1$ and is therefore parallel to it.

Also since the general line $a$ and the optical line $A_1B_1$ lie in the inertia plane $P$ and since the element $A_2$ does not lie in $A_1B_1$ it follows by Theorem 33 that $A_2$ is either *before* or *after* some element of $A_1B_1$.

Thus $A_2B_2$ must be either a before- or after-parallel of $A_1B_1$ and so the optical lines $A_1B_1$ and $A_2B_2$ lie in an inertia plane containing the general line $a$ and the element $B_1$.

This inertia plane must therefore be identical with $P$ and it contains the element $B_2$ in common with $R$ where $B_2$ is distinct from $B_1$.

Thus the theorem holds in all cases.

## REMARKS

It follows directly from this theorem that if two distinct inertia planes $P$ and $Q$ have a general line in common and, if further, $P$ has one element in common with an inertia plane $R$ which is parallel to $Q$, then $P$ and $R$ have a general line in common.

Further, since $Q$ and $R$ can have no element in common, it follows that these two general lines have no element in common.

Again if $Q$ and $R$ be two parallel inertia planes and if $a$ be any general line in $Q$, then there is at least one inertia plane containing $a$ and another general line in $R$.

This may be shown in the following way:

Let $A_1$ be any element of $a$ and let $f$ be any inertia line in $R$.

Then by Post. XIV $(a)$ there is one single element common to the inertia line $f$ and the sub-set $\alpha_1$. Let $B$ be this element and let $A_2$ be any element of $f$ which is *after B*.

Then $A_2$ is *after $A_1$* but does not lie in $\alpha_1$ and so $A_1$ and $A_2$ lie in an inertia line.

Thus $A_2$ and $a$ lie in an inertia plane, say $P$, which by Theorem 46 must contain a second element in common with $R$.

Thus $P$ contains $a$ and another general line in $R$.

It is easy to see that there are really an infinite number of inertia planes which have this property of $P$.

We have seen that if two distinct optical lines intersect a pair of optical lines one of which is an after-parallel of the other, then of the two first-mentioned optical lines one is an after-parallel of the other (Theorem 30).

We have also seen that it is impossible for an optical line to intersect a pair of neutral-parallel optical lines.

Thus we may state the following definition:

*Definition.* If two distinct optical lines intersect a pair of optical lines one of which is an after-parallel of the other, then the four optical lines will be said to form an *optical parallelogram.*

It is evident that an optical parallelogram must lie in an inertia plane.

The elements of intersection will be spoken of as the *corners* of the optical parallelogram.

A pair of corners which lie in one optical line will be spoken of as *adjacent.*

A pair of corners which do not lie in one optical line will be spoken of as *opposite.*

A general line passing through a pair of opposite corners of an optical parallelogram will be spoken of as a *diagonal line* of the optical parallelogram.

We make a distinction between two optical parallelograms having a *diagonal line* in common and having a *diagonal* in common.

When we speak of two optical parallelograms having a *diagonal line* in common we shall mean that a pair of opposite corners of each of the optical parallelograms lie in the same general line.

When, on the other hand, we speak of two optical parallelograms having a *diagonal* in common, we mean that they have a pair of opposite corners in common.

It is obvious that an optical parallelogram has two diagonal lines and it is easy to see that *one of these must be an inertia line, and the other a separation line.*

For if we call the four optical lines $a$, $b$, $c$ and $d$, and if $a$ be an after-parallel of $b$ while $c$ is an after-parallel of $d$, then the intersection element of $a$ and $c$ must be *after* the intersection element of $d$ and $b$ so that these two intersection elements lie in an inertia line.

Further, if we denote the intersection element of $a$ and $c$ by $A_1$, that of $a$ and $d$ by $A_2$, that of $c$ and $b$ by $A_3$ and that of $d$ and $b$ by $A_4$ it follows by Theorem 13 (*b*) that if $A_3$ were either *before* or *after* $A_2$ then $A_3$ would have to lie in the optical line $A_2A_1$, or $a$ contrary to hypothesis.

Thus $A_3$ is neither *before* nor *after* $A_2$ and so $A_2$ and $A_3$ lie in a separation line.

*Definition.* If a general line $a$ have *one single element* in common with a general line $b$, then $a$ will be said to *intersect b*.

Since a general line does not intersect itself and since we may have two optical parallelograms in the same inertia plane having a diagonal line in common, it is permissible to speak of two optical parallelograms in the same inertia plane whose diagonal lines of one kind or the other do not intersect.

This prepares the way for Post. XVI.

POSTULATE XVI. **If two optical parallelograms lie in the same inertia plane, then if their diagonal lines of one kind do not intersect, their diagonal lines of the other kind do not intersect.**

### THEOREM 47

*If a be any general line in an inertia plane P and A be any element of the inertia plane which is not in the general line, then there is one single general line through A in the inertia plane which does not intersect a.*

Let $Q$ be any other inertia plane distinct from $P$ and containing the general line $a$, and let $R$ be an inertia plane passing through $A$ and parallel to $Q$.

Then by Theorem 46 $P$ and $R$ will have a general line in common which can have no element in common with $a$, and so there is at least one general line through $A$ in the inertia plane $P$ which does not intersect $a$.

We must next show that there is only one such general line.

Consider first the case where $a$ is an optical line.

Then of the two optical lines through $A$ in the inertia plane $P$ we know that one is parallel to $a$ while the other intersects it.

Further by Theorem 36 any inertia line through $A$ in the inertia plane $P$ must intersect $a$.

Also by Post. XV any separation line through $A$ in the inertia plane $P$ must intersect $a$.

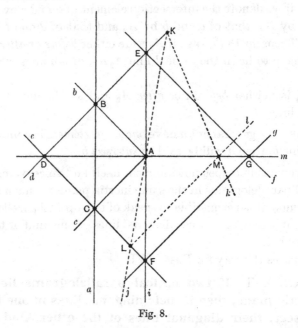

Fig. 8.

Thus if $a$ be an optical line there is one single general line through $A$ in the inertia plane $P$ which does not intersect $a$.

Consider next the cases where $a$ is an inertia or a separation line.

If $a$ be an inertia line, then by Theorem 36 both the optical lines through $A$ in the inertia plane $P$ intersect $a$, while by Post. XV every separation line in $P$ intersects $a$.

Thus when $a$ is an inertia line any general line through $A$ in the inertia plane $P$ which does not intersect $a$ can only be an inertia line.

Also from Post. XV it follows that when $a$ is a separation line any general line through $A$ in the inertia plane $P$ which does not intersect $a$ can only be a separation line.

With these provisos the demonstration of the unique character of the non-intersecting general line is similar in the two cases.

Suppose, if possible, that there are two general lines through $A$ in the inertia plane, say $i$ and $j$, which do not intersect $a$.

Then $i$ and $j$ must intersect in $A$.

Let $b$ and $c$ be the two optical lines through $A$ in the inertia plane and let them intersect $a$ in $B$ and $C$ respectively.

Let $d$ be the second optical line through $B$ in the inertia plane and let $e$ be the second optical line through $C$ in the inertia plane and let $d$ and $e$ intersect in $D$.

Then the optical lines $b$, $c$, $d$ and $e$ form an optical parallelogram.

Let $m$ be the diagonal line through $A$ and $D$.

Let the optical line $d$ intersect $i$ in $E$ and let the optical line $e$ intersect $i$ in $F$.

Let $f$ be the second optical line through $E$ in the inertia plane and let $g$ be the second optical line through $F$ in the inertia plane and let $f$ and $g$ intersect in $G$.

Then the optical lines $f$, $g$, $d$ and $e$ form an optical parallelogram and the diagonal line $i$ is of the same kind as the diagonal line $a$ of the optical parallelogram formed by $b$, $c$, $d$ and $e$.

Thus since the diagonal lines $a$ and $i$ do not intersect it follows by Post. XVI that the diagonal lines of the other kind to the two optical parallelograms do not intersect.

But the two optical parallelograms have the corner $D$ in common and so they must have the diagonal line through $D$ in common.

Thus $G$ must lie in $m$.

Now suppose that the optical line $d$ intersects $j$ in $K$ and that the optical line $e$ intersects $j$ in $L$.

Let $k$ be the second optical line through $K$ in the inertia plane and let $l$ be the second optical line through $L$ in the inertia plane and let $k$ and $l$ intersect in $M$.

Then the optical lines $k$, $l$, $d$ and $e$ form an optical parallelogram and since $j$ is supposed not to intersect $a$ it follows as before that $M$ must lie in $m$.

But now we have the optical parallelograms formed by $f$, $g$, $d$ and $e$, and by $k$, $l$, $d$ and $e$ having the diagonal line $m$ in common, and so, by Post. XVI, their other diagonal lines do not intersect, which is contrary to the hypothesis that $i$ and $j$ intersected in $A$.

Thus the hypothesis that there are two general lines through $A$ in the inertia plane which do not intersect $a$ leads to a contradiction and therefore is not true.

Thus there is in all cases *one single general line* through $A$ in the inertia plane which does not intersect $a$.

## THEOREM 48

*If two inertia planes $P$ and $Q$ have a general line $a$ in common, and if $A$ be any element which does not lie either in $P$ or $Q$, then the inertia planes through $A$ parallel to $P$ and $Q$ respectively have a general line in common.*

Let $R$ and $S$ be the inertia planes through $A$ parallel to $P$ and $Q$ respectively.

Two possibilities are open: either

    (1) $Q$ has one element at least in common with $R$,

or    (2) $Q$ has no element in common with $R$.

Consider first the case where $Q$ has one element at least in common with $R$.

Here, since $Q$ has two elements in common with $P$ and since $P$ and $R$ are parallel, it follows by Theorem 46 that $Q$ has a second element in common with $R$.

Further, since $Q$ and $S$ are parallel and $R$ has two elements in common with $Q$ and has the element $A$ in common with $S$, it follows that $R$ has a second element in common with $S$ and therefore $R$ and $S$ have a general line, say $c$, in common.

Next consider the case where $Q$ has no element in common with $R$.

This case has no analogue in ordinary three-dimensional geometry, but must be considered in our system which is not confined to three dimensions.

We have seen that there is at least one inertia plane containing $a$ and another general line, say $b$, in $R$ since $P$ and $R$ are parallel.

Let $T$ be such an inertia plane, let $B$ be any element in $b$ and let $U$ be the inertia plane through $B$ parallel to $Q$.

Then, since $Q$ and $U$ are parallel and since $T$ contains the general line $a$ and also the element $B$ of $U$, it follows that $T$ contains a general line, say $b'$, in $U$.

But the general lines $b$ and $b'$ both contain the element $B$ and neither of them can intersect $a$.

Thus, since $b$ and $b'$ both lie in one inertia plane $T$, it follows by Theorem 47 that they must be identical, and so $b$ must be common to $U$ and $R$.

Now the inertia planes $S$ and $U$ are both parallel to $Q$ and therefore must be either parallel to one another or else identical.

If they are not identical, the inertia plane $R$ has the general line $b$ in common with $U$ and has the element $A$ in common with $S$.

Thus in either case $R$ and $S$ have a general line in common.

If we consider case (2) of the last theorem it is clear that, if the general line $a$ be an optical line, then since the general line $b$ lies in the same inertia plane $T$ and has no element in common with $a$, it follows by Theorem 47 that $b$ must also be an optical line and be parallel to $a$.

If $c$ be the general line common to $R$ and $S$, then provided $c$ and $b$ are distinct, it follows in a similar manner that $c$ is an optical line parallel to $b$ and therefore also parallel to $a$.

A similar result follows in case (1) and so we always have $c$ parallel to $a$ provided $a$ be an optical line.

Now we have as yet given no definition of the parallelism of any type of general lines except optical lines, but are now in a position to do so.

*Definition.* If $a$ be a general line and $A$ be any element which does not lie in it and if two inertia planes $R$ and $S$ through $A$ are parallel respectively to two others $P$ and $Q$ containing $a$, then the general line which $R$ and $S$ have in common is said to be *parallel* to $a$.

## THEOREM 49

*If $a$ be a general line and $A$ be any element which does not lie in it, then there is one single general line containing $A$ and parallel to $a$.*

Two cases have to be considered:

(1) The element $A$ lies in an inertia plane containing $a$.

(2) The element $A$ does not lie in an inertia plane containing $a$.

Consider first case (1) and let $T$ be the inertia plane containing $A$ and $a$.

Let $P_1, P_2, P_3, P_4$ be any other inertia planes containing $a$, and let $Q_1, Q_2, Q_3, Q_4$ be inertia planes through $A$ parallel to $P_1, P_2, P_3, P_4$ respectively.

Then, since the inertia plane $T$ has the general line $a$ in common with $P_1$ and has the element $A$ in common with $Q_1$, it follows that it has a general line, say $b$, in common with $Q_1$ and $b$ does not intersect $a$.

But, by Theorem 47, there is only one general line through $A$ in the inertia plane $T$ which does not intersect $a$ and so $b$ must be this general line.

Similarly $Q_2, Q_3, Q_4$ must all contain the general line $b$ in common

R                                                                    6

with $T$ and so any pair of the inertia planes $Q_1$, $Q_2$, $Q_3$, $Q_4$ have the same general line $b$ in common.

Thus $b$ is independent of the particular pair of inertia planes $P_1$, $P_2$, $P_3$, $P_4$ which we may select and so there is only one general line through $A$ parallel to $a$.

Suppose next that $A$ does not lie in an inertia plane containing $a$ and suppose that $P_1$, $P_2$, $P_3$, $P_4$ are any inertia planes which are distinct from one another and all contain $a$.

Let $Q_1$, $Q_2$, $Q_3$, $Q_4$ be inertia planes through $A$ and parallel to $P_1$, $P_2$, $P_3$, $P_4$ respectively.

Let $P_n$ be an inertia plane containing $a$ and a general line $b$ in $Q_1$.

Then $b$ is parallel to $a$ and lies in the same inertia plane $P_n$ with it.

If then we take inertia planes $Q_2'$, $Q_3'$, $Q_4'$ through any element of $b$ and parallel to $P_2$, $P_3$, $P_4$ respectively, these will all contain $b$ and will also be respectively parallel to $Q_2$, $Q_3$, $Q_4$ which contain the element $A$.

But the general line $b$ and the element $A$ lie in the inertia plane $Q_1$ and so, by case (1), $Q_2$, $Q_3$, $Q_4$ all have the same general line, say $c$, in common with $Q_1$.

Thus any pair of the inertia planes $Q_1$, $Q_2$, $Q_3$, $Q_4$ have the same general line $c$ in common.

It follows that $c$ is independent of the particular pair of the inertia planes $P_1$, $P_2$, $P_3$, $P_4$ which we may select and so there is only one general line through $A$ parallel to $a$.

Thus the theorem holds in general.

### THEOREM 50

*If two distinct general lines are each parallel to a third, then they are parallel to one another.*

Let $a$ and $b$ be two distinct general lines which are each parallel to the general line $c$.

Let $R_1$ and $R_2$ be two inertia planes each containing $c$ but not containing $a$ or $b$.

Let $P_1$ and $P_2$ be two inertia planes parallel respectively to $R_1$ and $R_2$ and through any element of $a$.

Then $P_1$ and $P_2$ each contain $a$.

Similarly let $Q_1$ and $Q_2$ be two inertia planes parallel respectively to $R_1$ and $R_2$ and containing $b$.

Then $Q_1$ is either parallel to $P_1$ or identical with it, while $Q_2$ is either parallel to $P_2$ or identical with it.

In either case we must have $a$ parallel to $b$.

REMARKS

If $a$ and $b$ be any pair of parallel general lines, it is easy to see that they must be general lines of the same kind, for we know already that two parallel general lines in one inertia plane must be of the same kind, and by two applications of this result it follows that if $a$ and $b$ do not lie in one inertia plane they must also be of the same kind.

THEOREM 51

*If two parallel general lines $a$ and $b$ lie in one inertia plane $R$ and if two other distinct inertia planes $P$ and $Q$ containing $a$ and $b$ respectively have an element $A$ in common, then $P$ and $Q$ have a general line in common which is parallel to $a$ and $b$.*

Let any element in $b$ be selected and let $S$ be the inertia plane through this element and parallel to $P$.

Then the general line $b$ must lie in $S$ and so, since $Q$ contains the general line $b$ and the element $A$, it follows that $P$ and $Q$ contain a general line in common which is parallel to $b$ and therefore also parallel to $a$.

THEOREM 52

*If a pair of non-parallel general lines $a$ and $b$ lie in one inertia plane $P$ and if through an element $A$ not lying in the inertia plane there are two other general lines $c$ and $d$ respectively parallel to $a$ and $b$, then $c$ and $d$ lie in an inertia plane parallel to $P$.*

Let $R$ be any inertia plane distinct from $P$ which contains $a$ but not $A$, and let $S$ be any inertia plane distinct from $P$ which contains $b$ but not $A$.

Let $P'$ be the inertia plane through $A$ parallel to $P$, while $R'$ and $S'$ are the inertia planes through $A$ parallel to $R$ and $S$ respectively.

Then $P'$ and $R'$ have a general line in common which is parallel to $a$ and since it passes through $A$ must be identical with $c$; while $P'$ and $S'$ have a general line in common which is parallel to $b$ and since it passes through $A$ must be identical with $d$.

Thus $c$ and $d$ lie in the inertia plane $P'$ which is parallel to $P$.

THEOREM 53

*If three distinct inertia planes $P$, $Q$ and $R$ and three parallel general lines $a$, $b$ and $c$ be such that $a$ lies in $P$ and $R$, $b$ in $Q$ and $P$ and $c$ in $R$ and $Q$, then if $Q'$ be an inertia plane parallel to $Q$ through any element of $P$ which does not lie in $b$ the inertia planes $R$ and $Q'$ have a general line in common which is parallel to $c$.*

Since the inertia plane $P$ contains two elements in common with $Q$

and one element in common with the parallel inertia plane $Q'$, it follows by Theorem 46 that $P$ and $Q'$ have two elements in common and therefore have a general line in common which is parallel to $b$. Call this general line $d$.

If this general line should happen to coincide with $a$, the result follows directly.

We shall therefore consider the case where it does not coincide with $a$.

Let $A$ be any element in $a$.

Then, in case $a$ be an optical line, the other optical line through $A$ in the inertia plane $P$ will intersect $b$, while, if $a$ be an inertia or separation line, both the optical lines through $A$ in the inertia plane $P$ will intersect $b$.

Thus in all cases there is at least one optical line through $A$ in the inertia plane $P$ which intersects $b$.

Let such an optical line intersect $b$ in $B$ and let an optical line through $B$ in the inertia plane $Q$ intersect $c$ in $C$.

Then $BA$ and $BC$ may be taken as generators of opposite sets of an inertia plane, say $S$, which contains $A$, $B$ and $C$.

Now the general line $a$ is parallel to $b$ and therefore also parallel to $d$, and, since $BA$ passes through $A$, is distinct from $a$, and lies in the inertia plane $P$, it follows that $BA$ intersects $d$ in some element, say $D$, which accordingly lies in the inertia plane $Q'$.

But since $D$ lies in $BA$ it lies in the inertia plane $S$ and thus $S$ contains two elements ($B$ and $C$) in common with $Q$ and an element $D$ in common with the parallel inertia plane $Q'$.

It follows by Theorem 46 that $S$ contains a second element in common with $Q'$ and so $S$ and $Q'$ contain a general line in common which must be parallel to $CB$.

If we denote this general line in $S$ and $Q'$ by $g$, then any general line through $C$ in the inertia plane $S$, with the exception of $CB$, must intersect $g$.

But the element $A$ does not lie in $b$ and so does not lie in the inertia plane $Q$ and therefore does not lie in $CB$.

Thus since the general line $CA$ is distinct from $CB$, and since $CA$ must lie in $S$, it follows that $CA$ must intersect $g$ in some element, say $F$.

But $C$ and $A$ both lie in the inertia plane $R$ which accordingly must contain the general line $CA$ and therefore the element $F$.

Thus since the inertia plane $R$ contains the general line $c$ in common with $Q$ and contains the element $F$ in the parallel inertia plane $Q'$, it follows that $R$ must have a general line in common with $Q'$ and this general line must be parallel to $c$.

## THEOREM 54

*If $P_1$ and $P_2$ be a pair of parallel inertia planes while an inertia plane $Q_1$ has parallel general lines $a$ and $b$ in common with $P_1$ and $P_2$ respectively and if $Q_2$ be an inertia plane parallel to $Q_1$ through some element (say $C$) of $P_2$ which does not lie in $b$, then the inertia planes $P_1$ and $Q_2$ will have a general line in common which is parallel to $a$ and $b$.*

Since $Q_2$ is parallel to $Q_1$ and since $P_2$ has the general line $b$ in common with $Q_1$ and has the element $C$ in common with $Q_2$, it follows, by Theorem 46, that $P_2$ and $Q_2$ have a general line (say $c$) in common which is parallel to $b$ and therefore also to $a$.

Let $A$ be any element of $a$ and let $g$ be any inertia line in $P_2$ which does not coincide with either $b$ or $c$, while $G$ is the one single element common to $g$ and the $\alpha$ sub-set of $A$. Then $AG$ is an optical line.

Let $E$ be any element of $g$ which is *after* $G$ but does not lie either in $b$ or $c$.

Then $AE$ will be an inertia line so that $E$ and the general line $a$ lie in an inertia plane which we shall call $R$.

Then, by Theorem 51, $P_2$ and $R$ have a general line (say $e$) in common which is parallel to $a$, $b$ and $c$.

But now, by Theorem 53, since the three distinct inertia planes $P_2$, $Q_1$ and $R$ and the three parallel general lines $e$, $b$ and $a$ are such that $e$ lies in $P_2$ and $R$, $b$ in $Q_1$ and $P_2$ and $a$ in $R$ and $Q_1$, and since further $Q_2$ is an inertia plane parallel to $Q_1$ through the element $C$ of $P_2$ which does not lie in $b$, it follows that $R$ and $Q_2$ have a general line (say $f$) in common which is parallel to $a$ and therefore also to $e$ and $c$.

Making use of the same theorem a second time, we have the three distinct inertia planes $R$, $P_2$ and $Q_2$ and the three parallel general lines $f$, $e$ and $c$ such that $f$ lies in $R$ and $Q_2$, $e$ in $P_2$ and $R$ and $c$ in $Q_2$ and $P_2$, and so, since $P_1$ is an inertia plane parallel to $P_2$ through an element of $R$ which does not lie in $e$, it follows that the inertia planes $Q_2$ and $P_1$ have a general line (say $d$) in common which is parallel to $c$ and therefore also parallel to $a$ and $b$.

Thus the theorem is proved.

## THEOREM 55

(a) *If $a$ and $b$ be two parallel separation lines in the same inertia plane and if one element of $b$ be* before *an element of $a$, then each element of $b$ is* before *an element of $a$.*

Let $A$ be the element of $b$ which by hypothesis is *before* an element of $a$.

Let the two optical lines through $A$ in the inertia plane be called $c$ and $d$.

Let $B$ be any other element of $b$.

Then by Theorem 40 $B$ must be *before* an element of one of the optical lines $c$ and $d$ and *after* an element of the other.

It will be sufficient to consider the case when $B$ is *before* an element of $c$ and *after* an element of $d$, since the proof in the other case is similar.

Let $e$ and $f$ be the two optical lines through $B$ in the inertia plane and let $e$ be the one which is parallel to $c$. Then $f$ intersects $c$ in some element $C$.

Also $c$ intersects $a$ in some element $D$ (Post. XV) and $D$ must be *after* $A$; for since $A$ is *before* an element of $a$, we should otherwise have one element of $a$ *after* another, contrary to the hypothesis that $a$ is a separation line.

Now, since $B$ is *before* an element of $c$ and cannot also be *after* an element of $c$, and since $C$ lies in the optical line $f$ through $B$, it follows that $C$ is *after* $B$.

Now $C$ cannot be *before* $A$ for then $A$ would be *after* $B$, contrary to the hypothesis that $A$ and $B$ lie in a separation line.

If $C$ be either *before* $D$ or coincident with $D$, then $B$ is *before* $D$ an element of $a$.

Suppose next that $C$ is *after* $D$ and let $E$ be the element in which $f$ intersects $a$.

Let $h$ be the second optical line through $D$ in the inertia plane and let $g$ be the second optical line through $E$ in the inertia plane and let $g$ and $h$ intersect in $F$.

Then the optical lines $c$, $f$, $h$ and $g$ form an optical parallelogram whose diagonal line through $D$ and $E$ is $a$.

Let $j$ be the other diagonal line through $C$ and $F$, then $j$ is an inertia line.

Let the optical lines $d$ and $e$ intersect in $G$.

Then the optical lines $c$, $f$, $d$ and $e$ form an optical parallelogram whose diagonal line through $A$ and $B$ is $b$.

Thus in the two optical parallelograms, since the diagonal lines $a$ and $b$ do not intersect, it follows that the diagonal lines of the other kind do not intersect (Post. XVI).

But the two optical parallelograms have the corner $C$ in common and so they must have a diagonal line in common and so $G$ must lie in $j$.

Also $D$ is *after* $A$ and so $h$ must be an after-parallel of $d$.

But, since $F$ and $G$ are elements of $j$ which is an inertia line, it follows that the one is *after* the other; and since no element of $d$ can be *after* an element of $h$, it follows that $F$ must be *after* $G$.

Thus since $F$ is an element of $g$ and $G$ is an element of $e$, it follows that $g$ is an after-parallel of $e$.

But since $E$ and $B$ lie in the optical line $f$, one of them must be *after* the other, and since $B$ lies in $e$ it cannot be *after* $E$ which is an element of $g$.

Thus $E$ is *after* $B$ and so $B$ is *before* an element of $a$.

Thus in all cases $B$ is *before* an element of $a$.

(b) *If $a$ and $b$ be two parallel separation lines in the same inertia plane and if one element of $b$ be* after *an element of $a$, then each element of $b$ is* after *an element of $a$.*

## Theorem 56

(a) *If $a$ and $b$ be a pair of parallel separation lines in the same inertia plane and if an optical line $c$ intersects $a$ in $A_1$ and $b$ in $B_1$ while a parallel optical line $d$ intersects $a$ in $A_2$ and $b$ in $B_2$, then if $B_1$ is before $A_1$ we have also $B_2$ before $A_2$.*

By Theorem 55, since $B_1$ is *before* $A_1$, therefore $B_2$ is *before* an element of $a$.

But since $A_2$ and $B_2$ are distinct elements in the optical line $d$, therefore one of them is *after* the other.

Further, $B_2$ could not be *after* $A_2$ for then since $B_2$ is *before* an element of $a$ we should have $A_2$ *before* this element of $a$, contrary to the hypothesis that $a$ is a separation line.

Thus $B_2$ must be *before* $A_2$.

(b) *If $a$ and $b$ be a pair of parallel separation lines in the same inertia plane and if an optical line $c$ intersects $a$ in $A_1$ and $b$ in $B_1$ while a parallel optical line $d$ intersects $a$ in $A_2$ and $b$ in $B_2$, then if $B_1$ is after $A_1$ we have also $B_2$ after $A_2$.*

## Theorem 57

(a) *If $a$ and $b$ be a pair of parallel inertia lines in the same inertia plane and if an optical line $c$ intersect $a$ in $A_1$ and $b$ in $B_1$, while a parallel optical line $d$ intersects $a$ in $A_2$ and $b$ in $B_2$; then if $B_1$ is before $A_1$ we have also $B_2$ before $A_2$.*

Since $B_1$ and $B_2$ are elements of an inertia line $b$, one of them must be *after* the other.

We shall first consider the case when $B_2$ is *after* $B_1$.

Let $e$ be the second optical line through $B_2$ in the inertia plane and let $f$ be the second optical line through $B_1$ in the inertia plane.

Then since, by hypothesis, $d$ is parallel to $c$ it follows that $e$ must intersect $c$ in some element $C$, while $d$ must intersect $f$ in some element $F$.

But, since $B_2$ is *after* $B_1$, it follows that $e$ must be an after-parallel of $f$ and $d$ must be an after-parallel of $c$.

Thus, since $B_1$ and $C$ lie in one optical line, it follows that $C$ is *after* $B_1$ and similarly, since $B_2$ and $C$ lie in one optical line, it follows that $B_2$ is *after* $C$.

Let the optical line $e$ intersect $a$ in $D$.

If then $C$ is *before* $A_1$ we shall have $A_1$ in the $\alpha$ sub-set of $C$ and by Post. XIV (*b*) there is *one single element* common to the inertia line $a$ and the $\beta$ sub-set of $C$, and since there are only two optical lines through $C$ in the inertia plane, it follows that this element must be identical with $D$.

Thus $D$ is *before* $C$ and $C$ is *before* $B_2$ and consequently $D$ is *before* $B_2$ and since $D$ and $B_2$ lie in one optical line it follows that $D$ lies in the $\beta$ sub-set of $B_2$.

If $C$ were identical with $A_1$, it would also be identical with $D$ and again $D$ would lie in the $\beta$ sub-set of $B_2$.

But by Post. XIV (*a*) there is one single element common to the inertia line $a$ and the $\alpha$ sub-set of $B_2$ and since there are only two optical lines through $B_2$ in the inertia plane this element must lie in $d$ and must therefore be identical with $A_2$.

Thus since $A_2$ lies in the $\alpha$ sub-set of $B_2$ and is not identical with $B_2$, therefore $B_2$ must be *before* $A_2$.

Thus in case $C$ is either *before* $A_1$ or identical with $A_1$ we have $B_2$ *before* $A_2$.

Next suppose $C$ is *after* $A_1$.

Then the optical lines $e$, $d$, $c$ and $f$ form an optical parallelogram whose diagonal line through $B_1$ and $B_2$ is $b$.

Let $j$ be the other diagonal line through $C$ and $F$.

Then since $b$ is an inertia line, $j$ must be a separation line.

Again let $g$ be the second optical line through $D$ in the inertia plane and let $h$ be the second optical line through $A_1$ in the inertia plane and let $g$ and $h$ intersect in $E$.

Then the optical lines $e$, $g$, $c$ and $h$ form an optical parallelogram whose diagonal line through $A_1$ and $D$ is $a$.

Thus the two optical parallelograms formed by $e$, $d$, $c$ and $f$ and by $e$, $g$, $c$ and $h$ have diagonal lines of one kind, $b$ and $a$, which do not intersect and so by Post. XVI their diagonal lines of the other kind do not intersect.

But the two optical parallelograms have the corner $C$ in common and so they have the diagonal line through $C$ in common.

Thus $E$ lies in $j$ and since $j$ is a separation line $E$ is neither *before* nor *after* $F$.

But since $A_1$ is *after* $B_1$ it follows that $h$ is an after-parallel of $f$ and so $E$ must be *after* an element of $f$.

But since $E$ is neither *before* nor *after* $F$, it follows by Theorem 40 that since $E$ is *after* an element of $f$ it must be *before* an element of $d$.

Thus $g$ is a before-parallel of $d$ and since $D$ and $B_2$ lie in the optical line $e$ which intersects $g$ in $D$ and $d$ in $B_2$, it follows that $D$ is *before* $B_2$.

Thus $D$ lies in the $\beta$ sub-set of $B_2$ and in the optical line $e$.

But by Post. XIV $(a)$ there is *one single element* common to the inertia line $a$ and the $\alpha$ sub-set of $B_2$ and since there are only two optical lines through $B_2$ in the inertia plane it follows that this element must lie in $d$ and is therefore identical with $A_2$.

Thus since $A_2$ is in the $\alpha$ sub-set of $B_2$ and is not identical with $B_2$, therefore $B_2$ is *before* $A_2$.

This proves the theorem provided $B_2$ is *after* $B_1$.

Suppose now that $B_1$ is *after* $B_2$.

Then $c$ must be an after-parallel of $d$ and, since $A_1$ and $A_2$ lie in $c$ and $d$ respectively and, since they both lie in the inertia line $a$, it follows that $A_1$ must be *after* $A_2$.

Suppose now, if possible, that $A_2$ is *before* $B_2$, then reversing the rôles of the inertia lines $a$ and $b$ it would follow from what we have already proved that, $c$ and $d$ being parallel, $A_1$ would have to be *before* $B_1$, contrary to hypothesis.

Thus, since $B_2$ must be either *after* or *before* $A_2$ and cannot be *after*, it follows that $B_2$ is *before* $A_2$.

$(b)$ *If $a$ and $b$ be a pair of parallel inertia lines in the same inertia plane and if an optical line $c$ intersect $a$ in $A_1$ and $b$ in $B_1$ while a parallel optical line $d$ intersects $a$ in $A_2$ and $b$ in $B_2$; then if $B_1$ is after $A_1$ we have also $B_2$ after $A_2$.*

Since a pair of parallel inertia lines always lie in an inertia plane, the words "in the same inertia plane" may be omitted in the enunciation of this theorem.

## THEOREM 58

*If two elements A and B lie in one optical line and if two other elements C and D lie in a parallel optical line in the same inertia plane, then if A be after B and C after D the general lines AD and BC intersect.*

Let $a$ be the optical line containing $A$ and $B$, and let $b$ be the parallel optical line containing $C$ and $D$.

Then the general lines $AD$ and $BC$ cannot be parallel optical lines, for since $B$ is *before* $A$ an optical line through $B$ which intersected $b$ would be a before-parallel of an optical line though $A$ which intersected $b$ and so the element in which the former optical line intersected $b$ would be *before* the element in which the latter optical line intersected $b$.

Further, Theorems 56 and 57 show that $AD$ and $BC$ cannot be either parallel separation lines or parallel inertia lines.

Again $AD$ and $BC$ cannot both be optical lines for we know that two optical lines which intersect a pair of parallel optical lines are themselves parallel.

Thus we are left with the following possibilities as to the general lines $AD$ and $BC$:

(1) One is an optical line and the other an inertia line.

(2) One is an optical line and the other a separation line.

(3) One is a separation line and the other an inertia line.

(4) Both are inertia lines.

(5) Both are separation lines.

In case (1) Theorem 36 shows that the general lines intersect.

In cases (2) and (3) it follows from Post. XV that the general lines intersect.

In cases (4) and (5), since we have shown that the two general lines cannot be parallel, it follows by Theorem 47 that they must intersect.

Thus in all cases the general lines $AD$ and $BC$ intersect.

*Definitions.* If four optical lines form an optical parallelogram, they will be spoken of as the *side lines* of the optical parallelogram.

A pair of side lines which do not intersect will be called *opposite*.

The element of intersection of the diagonal lines will be spoken of as the *centre* of the optical parallelogram.

## THEOREM 59

*If any two distinct elements A and O be taken in an inertia or separation line i in a given inertia plane, then there is one single optical parallelogram in the inertia plane having O as the centre and A as one of its corners.*

Let $a$ and $b$ be the two optical lines through $A$ in the inertia plane while $c$ and $d$ are those through $O$; the optical line $c$ being parallel to $a$ and the optical line $d$ parallel to $b$.

Let $j$ be the second diagonal line of the optical parallelogram formed by $a$, $b$, $c$ and $d$.

Then by Theorem 47 there is one single general line through $O$ and parallel to $j$.

Call this general line $k$ and let $a$ intersect $k$ in $D$ while $b$ intersects $k$ in $C$.

The elements of intersection must exist since $k$, being parallel to $j$, must be an inertia or separation line according as $i$ is a separation or inertia line; while $a$ and $b$ are both optical lines.

Let $e$ be the second optical line through $C$ in the inertia plane, while $f$ is the second optical line through $D$ in the inertia plane and let $e$ and $f$ intersect in $B$.

Then $a$, $b$, $e$ and $f$ form an optical parallelogram in the same inertia plane with that formed by $a$, $b$, $c$ and $d$ and their diagonal lines of one kind $k$ and $j$ do not intersect and so by Post. XVI their diagonal lines of the other kind do not intersect.

But the corner $A$ is common to both optical parallelograms and so the diagonal line $i$ which passes through that corner must be a diagonal line of both optical parallelograms.

Thus $B$ must lie in $i$ and so $O$ is the centre of the optical parallelogram formed by $a$, $b$, $e$ and $f$, while $A$ is one of its corners.

Again, if there were a second optical parallelogram in the inertia plane having $O$ as centre and $A$ one of its corners, then such an optical parallelogram would have $i$ as one of its diagonal lines and so the other diagonal lines of the two optical parallelograms would not intersect.

Further, since the two optical parallelograms have the element $O$ common to these other diagonal lines, the latter must be identical.

But there are only two optical lines, $a$ and $b$, through $A$ in the inertia plane and these intersect $k$ in $D$ and $C$ respectively, which must accordingly be a pair of opposite corners of the second optical parallelogram.

But then the second optical parallelogram would have $e$ and $f$ as its remaining side lines and so could not be distinct from the first optical parallelogram.

Thus there is no second optical parallelogram in the inertia plane having $O$ as centre and $A$ as one of its corners.

## THEOREM 60

*If two optical parallelograms have two opposite corners in common, then they have a common centre.*

Two cases are possible:

(1) The common opposite corners may lie in an inertia line.

(2) The common opposite corners may lie in a separation line.

We shall consider first the case where they lie in an inertia line.

Let $A$ and $B$ be the two common opposite corners of the optical parallelograms: $B$ being *after* $A$.

Let $C$ and $D$ be the other pair of opposite corners of the one optical parallelogram which we shall suppose to lie in an inertia plane $P$, while

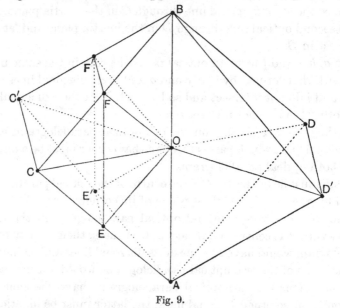

Fig. 9.

$C'$ and $D'$ are the other pair of opposite corners of the other optical parallelogram which we shall suppose to lie in an inertia plane $P'$.

Then $P$ and $P'$ must be distinct if the optical parallelograms are distinct.

Let $O$ be the centre of the optical parallelogram whose corners are $A, B, C, D$, and let $OE$ and $OF$ be optical lines through $O$ parallel to $CB$ and $AC$ respectively and intersecting $AC$ and $CB$ in $E$ and $F$ respectively.

Then $E$, $C$, $F$ and $O$ form the corners of an optical parallelogram in the inertia plane $P$, and this optical parallelogram and the one whose corners are $A$, $C$, $B$ and $D$ have the common diagonal line $CD$ and so their diagonal lines of the other kind do not intersect.

Thus $AB$ and $EF$ are parallel and $EF$ is an inertia line.

Now let $OE'$ and $OF'$ be optical lines through $O$ parallel to $C'B$ and $AC'$ respectively and intersecting $AC'$ and $C'B$ in $E'$ and $F'$ respectively.

Then $AC$ and $AC'$ may be taken as generators of opposite sets of an inertia plane $Q_1$, while $OF$ and $OF'$ will be generators of opposite sets of a parallel inertia plane $Q_2$.

Similarly $BC$ and $BC'$ may be taken as generators of opposite sets of an inertia plane $R_1$, while $OE$ and $OE'$ will be generators of opposite sets of a parallel inertia plane $R_2$.

But $Q_1$ and $R_1$ have the general line $CC'$ in common, while $Q_1$ and $R_2$ have the general line $EE'$ in common and so since $R_1$ and $R_2$ are parallel it follows that $CC'$ and $EE'$ are parallel.

Again since $R_1$ and $Q_2$ have the general line $FF'$ in common and since $Q_1$ and $Q_2$ are parallel, it follows that $FF'$ and $CC'$ are parallel.

Thus $FF'$ is parallel to $EE'$.

But since $EF$ is an inertia line there exists an inertia plane containing $E$, $F$ and $F'$. Let $S$ be this inertia plane.

Then there exists in $S$ a general line through $E$ which is parallel to $FF'$ and, since there can be only one parallel to $FF'$ through $E$, this must be identical with the general line $EE'$.

Thus $E'$ must lie in the inertia plane $S$.

But since $AB$ and $EF$ are parallel and lie in $P$ while $P'$ and $S$ are two other distinct inertia planes containing $AB$ and $EF$ respectively and since $P'$ and $S$ have an element $F'$ in common, it follows by Theorem 51 that the general line $E'F'$ which is common to $P'$ and $S$ is parallel to $AB$.

But now $E'$, $C'$, $F'$ and $O$ form the corners of an optical parallelogram in the inertia plane $P'$, and this optical parallelogram and the one whose corners are $A$, $C'$, $B$ and $D'$ have one pair of diagonal lines, namely $E'F'$ and $AB$, which do not intersect and so their diagonal lines of the other kind do not intersect.

But these latter diagonal lines are $C'O$ and $C'D'$ respectively and so since they have the element $C'$ in common it follows that they are identical.

Thus the element $O$ must lie in $C'D'$ and since it also lies in $AB$ it follows that $O$ is the centre of the optical parallelogram whose corners are $A$, $C'$, $B$, $D'$.

Thus the optical parallelograms having $A$ and $B$ as opposite corners have a common centre $O$.

We have next to consider the case where the common opposite corners lie in a separation line.

Let $A$ and $B$ be the two common opposite corners of the optical parallelograms: $B$ being neither *before* nor *after* $A$.

Let $C$ and $D$ be the other pair of opposite corners of the one optical parallelogram, which we shall suppose to lie in an inertia plane $P$, while $C'$ and $D'$ are the other pair of opposite corners of the other

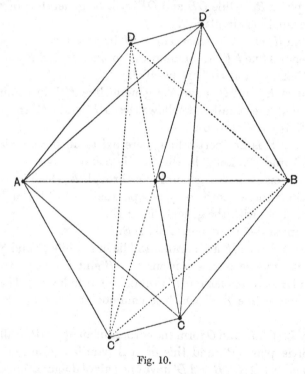

Fig. 10.

optical parallelogram, which we shall suppose to lie in an inertia plane $P'$.

Then $P$ and $P'$ must be distinct if the optical parallelograms are distinct.

We shall further suppose $D$ to be *after* $C$ and $D'$ after $C'$.

Now the following pairs of intersecting optical lines may be taken as generators of opposite sets of certain inertia planes which we shall denote by the following symbols opposite each pair.

| Optical lines | | | | | Inertia plane |
|---|---|---|---|---|---|
| $CA$ and $C'A$ | . | . | . | . | $Q_1$ |
| $BD$ and $BD'$ | . | . | . | . | $Q_2$ |
| $CB$ and $C'B$ | . | . | . | . | $R_1$ |
| $AD$ and $AD'$ | . | . | . | . | $R_2$ |
| $AC'$ and $AD$ | . | . | . | . | $S_1$ |
| $BD'$ and $BC$ | . | . | . | . | $S_2$ |
| $BC'$ and $BD$ | . | . | . | . | $T_1$ |
| $AD'$ and $AC$ | . | . | . | . | $T_2$ |

Of these inertia planes we evidently have those pairs parallel which are represented by the same letters.

Thus the general line $C'D$, since it lies in $S_1$ and $T_1$, must be parallel to the general line $CD'$, since the latter lies in $S_2$ and $T_2$.

Similarly the general line $DD'$, since it lies in $Q_2$ and $R_2$, must be parallel to the general line $C'C$, since the latter lies in $Q_1$ and $R_1$.

But $CD$ is an inertia line and so there is an inertia plane containing $C$, $D$ and $D'$, and if we call this inertia plane $U$ then $U$ contains the general lines $CD'$ and $DD'$ and so $U$ must also contain the general lines through $D$ parallel to $CD'$ and through $C$ parallel to $DD'$.

That is: the inertia plane $U$ must contain $C'D$ and $C'C$.

Thus $U$ must contain $C'$ and therefore contains $C'D'$.

Thus the centres of the two optical parallelograms must lie in the inertia plane $U$ and in the separation line $AB$.

The inertia plane $U$ cannot however have more than one element in common with $AB$, for otherwise it would contain both $A$ and $B$, and since $U$ contains $D$ we should have $U$ identical with $P$; but $U$ contains $D'$ which does not lie in $P$ and so this is impossible.

Thus the element in which $CD$ intersects $AB$ must be identical with the element in which $C'D'$ intersects $AB$, or in other words the two optical parallelograms have a common centre.

Thus the theorem is proved.

## THEOREM 61

*If two optical parallelograms have two adjacent corners in common, then optical lines through the centres of the optical parallelograms and intersecting their common side line intersect it in the same element.*

Let $A$ and $B$ be the two common adjacent corners of two optical parallelograms which we shall suppose to lie in separate inertia planes $P$ and $P'$.

We shall suppose $C$ and $D$ to be the other corners of the optical parallelogram in $P$ and shall suppose $C$ to be opposite to $B$ and $D$ opposite to $A$.

We may further, without limitation of generality, take the diagonal line $CB$ as the inertia diagonal line.

We shall suppose $C'$ and $D'$ to be the remaining corners of the optical parallelogram in $P'$ and we shall take $C'$ opposite to $B$ and $D'$ opposite to $A$.

Let $O$ be the centre of the optical parallelogram in $P$ and let the one optical line through $O$ in the inertia plane $P$ intersect $AB$ in $M$, while the other optical line in $P$ through $O$ intersects $AC$ in $E$.

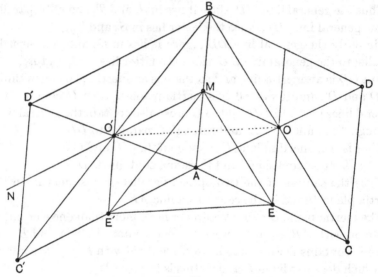

Fig. 11.

Then $A, E, O$ and $M$ form the corners of an optical parallelogram also in the inertia plane $P$.

The optical parallelograms whose corners are $A, E, O, M$ and $A, C, D, B$ have the diagonal line $AD$ in common and so, by Post. XVI, their diagonal lines of the other kind do not intersect.

Thus $EM$ and $CB$ are parallel.

Now let $MN$ be the optical line through $M$ parallel to $AC'$ and let $MN$ intersect the diagonal line $C'B$ in $O'$.

Let $O'E'$ be the optical line through $O'$ parallel to $MA$ and intersecting $AC'$ in $E'$.

Then $O'E'$ is parallel to $OE$ and unless it be a neutral-parallel we have $O'E'$ and $OE$ in one inertia plane.

Now, since $MN$ is an optical line through $M$ which neither intersects $OE$ nor is parallel to it, it follows by Post. XII that there is *one single element* in $MN$ which is neither *before* nor *after* any element of $OE$.

If $O_0$ be this element, we shall suppose first that $O'$ is distinct from $O_0$ and thereby ensure that $O'E'$ and $OE$ lie in one inertia plane.

Call this inertia plane $Q$.

Now, since $MO$ and $MO'$ are respectively parallel to $AE$ and $AE'$ and all four are optical lines, it follows that $M$, $O$ and $O'$ lie in one inertia plane, say $R_1$, while $A$, $E$ and $E'$ lie in a parallel inertia plane, say $R_2$.

But $Q$ has the elements $O$ and $O'$ in common with $R_1$ and has the elements $E$ and $E'$ in common with $R_2$ and so the general lines $OO'$ and $EE'$ are parallel.

We have however further seen that $OB$ and $EM$ are parallel and are both inertia lines.

Thus $O$, $O'$ and $B$ lie in one inertia plane, say $S_1$, while $E$, $E'$ and $M$ lie in a parallel inertia plane, say $S_2$.

But the inertia plane $P'$ has the elements $O'$ and $B$ in common with $S_1$ and has the elements $E'$ and $M$ in common with $S_2$.

Thus $BO'$ and $ME'$ are parallel.

But $BO'$ is the same general line as $BC''$, which is a diagonal line of the optical parallelogram whose corners are $A$, $C''$, $D'$, $B$, while $ME'$ is a diagonal line of the optical parallelogram whose corners are $A$, $E'$, $O'$, $M$ and these diagonal lines do not intersect.

It follows by Post. XVI that their other diagonal lines $AD'$ and $AO'$ do not intersect and so since they have the element $A$ in common they must be identical.

Thus $O'$ must lie in $AD'$ and since it also lies in $BC''$, it follows that $O'$ is the centre of the optical parallelogram whose corners are $A$, $C''$, $D'$, $B$.

Thus the optical lines through the centres $O$ and $O'$ and intersecting $AB$, intersect it in the same element $M$.

Now this same method of proof holds for the case of any optical parallelogram in the inertia plane $P'$ which has $A$ and $B$ as adjacent corners, provided that the diagonal line through $B$ does not intersect $MN$ in $O_0$, and so all such optical parallelograms have their centres in the optical line $MN$.

Again, if we select a second optical parallelogram in the inertia plane $P$ having $A$ and $B$ as adjacent corners but not having $O$ as centre, we may use a similar method of proof and show that all optical parallelo-

R

grams in the inertia plane $P'$ having $A$ and $B$ as adjacent corners have, with *one possible exception*, got their centres in one optical line.

This *one possible exception* is, however, different from the *one possible exception* which we found before and so it follows that no exception exists.

Similar considerations show that all optical parallelograms in the inertia plane $P$, having $A$ and $B$ as adjacent corners, have their centres in one optical line $MO$.

Thus the theorem holds for optical parallelograms in the inertia planes $P$ and $P'$ and will therefore also hold for optical parallelograms in any other inertia planes which contain $A$ and $B$.

*Definition.* If $A$ and $B$ be two distinct elements lying in an inertia line or in a separation line, then the centre of an optical parallelogram of which $A$ and $B$ are a pair of opposite corners will be spoken of as the *mean* of the elements $A$ and $B$.

Theorem 60 shows that if two elements $A$ and $B$ lie in an inertia or separation line their mean is independent of the particular optical parallelogram used to define it.

Since a diagonal line of an optical parallelogram is either an inertia or a separation line, the above definition fails for the case of two distinct elements lying in an optical line.

In this case we adopt the following definition.

*Definition.* If $A$ and $B$ be two distinct elements lying in an optical line, then an optical line through the centre of an optical parallelogram of which $A$ and $B$ are a pair of adjacent corners and intersecting the optical line $AB$, intersects it in an element which will be spoken of as the *mean* of the elements $A$ and $B$.

Theorem 61 shows that if two elements $A$ and $B$ lie in an optical line, their mean is independent of the particular optical parallelogram used to define it.

### REMARKS

If $A$, $B$, $C$ and $D$ be the corners of an optical parallelogram such that $B$ is *after* $A$ and $C$ *after* $B$, then $C$ will be *after* $A$ and so $AC$ will be the inertia diagonal line and $BD$ will be the separation diagonal line.

If $AC$ and $BD$ intersect in $O$, then $O$ is neither *before* nor *after* $B$, since $O$ and $B$ are elements of a separation line.

Now $O$ cannot be *after* $C$, for this would entail $O$ being *after* $B$, and also $O$ cannot be *before* $A$, for this would entail $O$ being *before* $B$.

Thus since $A$, $O$ and $C$ are distinct elements of the one inertia line we must have $O$ *after* $A$ and *before* $C$.

If now an optical line be taken through $O$ parallel to $AD$ and $BC$ and intersecting $AB$ in $E$, then $OE$ must be an after-parallel of $AD$ and a before-parallel of $BC$.

Thus, since $A$, $E$ and $B$ are distinct elements of the optical line $AB$, it follows that $E$ is *after* $A$ and *before* $B$.

We see from these results that the mean of two elements lying either in an inertia or optical line must be *after* the one and *before* the other.

### THEOREM 62

*If $A$, $B$ and $B'$ be three distinct elements in a general line $a$, then the mean of $A$ and $B'$ must be distinct from the mean of $A$ and $B$.*

Let us first take the case where $a$ is an optical line and let $P$ be any inertia plane containing $a$.

Let $a_1$ be any optical line lying in $P$ and parallel to $a$, and let optical lines through $A$, $B$ and $B'$ intersect $a_1$ in the elements $A_1$, $B_1$ and $B_1'$ respectively.

Then $A$, $A_1$, $B_1$, $B$ form the corners of an optical parallelogram, while $A$, $A_1$, $B_1'$, $B'$ form the corners of another optical parallelogram having the two adjacent corners $A$ and $A_1$ in common with the first.

Let $C$ and $C'$ be the centres of these two optical parallelograms respectively.

Then, as we have seen, $C$ and $C'$ must lie in an optical line parallel to $a$.

An optical line through $C$ parallel to $A_1 A$ will intersect $a$ in some element $M$, which is the mean of $A$ and $B$; while an optical line through $C'$ parallel to $A_1 A$ will intersect $a$ in some element $M'$, which is the mean of $A$ and $B'$.

Now $C'$ cannot be identical with $C$, for then the general line $A_1 C$ would be identical with the general line $A_1 C'$, and so $B'$ would have to be identical with $B$: contrary to hypothesis.

Thus $CM$ and $C'M'$ must be distinct and parallel optical lines, and therefore $M'$ must be distinct from $M$, as was to be proved.

Next let us consider the case where $a$ is either an inertia line or a separation line and let $P$ be any inertia plane containing $a$.

Then there is one single optical parallelogram in $P$ having $A$ and $B$ as a pair of opposite corners and a centre, say $C$, whose position in $a$ is independent of $P$ by Theorem 60.

But there is also one single optical parallelogram in $P$ having $A$ and $B'$ as a pair of opposite corners and a centre, say $C'$, whose position in $a$ is also independent of $P$.

Then $C'$ could not be identical with $C$ for, were this the case, we should have two distinct optical parallelograms in $P$ having $C$ as a common centre and $A$ as a common corner, which would be contrary to what we proved in Theorem 59.

Thus, whatever type of general line $a$ may be, the mean of $A$ and $B$ must be distinct from the mean of $A$ and $B'$.

It follows at once from this theorem that if $A$ and $C$ be any two elements in any type of general line, there is not more than one element $B$ such that $C$ is the mean of $A$ and $B$.

### Theorem 63

*If two or more optical parallelograms have a pair of opposite side lines in common, their centres lie in a parallel optical line in the same inertia plane.*

We have already seen in the course of proving Theorem 61 that this result must hold if the two optical parallelograms have a third side in common.

In case this is not so, let $A_1$, $B_1$, $C_1$, $D_1$ be four distinct elements in an optical line $a$ and let $b$ be a parallel optical line in an inertia plane containing $a$.

Let the second optical lines through $A_1$, $B_1$, $C_1$, $D_1$ respectively in the inertia plane intersect $b$ in $A_2$, $B_2$, $C_2$, $D_2$ respectively and let $A_1, B_1, A_2, B_2$ be the corners of one of the optical parallelograms under consideration and $C_1, D_1, C_2, D_2$ the corners of another.

Then $A_1, D_1, A_2, D_2$ is a third optical parallelogram.

Call these optical parallelograms (1), (2) and (3) and let their centres be $O$, $O'$, $O''$ respectively.

Then by the first case $O$ and $O''$ lie in an optical line parallel to $a$ and $b$ since (1) and (3) have the pair of adjacent corners $A_1$ and $A_2$ in common.

Similarly $O'$ and $O''$ lie in an optical line parallel to $a$ and $b$ since (2) and (3) have the pair of adjacent corners $D_1$ and $D_2$ in common.

But there is only one optical line through $O''$ parallel to $a$ and $b$ and so $O$, $O'$ and $O''$ lie in one optical line parallel to $a$ and $b$.

Thus all optical parallelograms having $a$ and $b$ as a pair of opposite side lines must have their centres in the optical line $OO'$.

## Theorem 64

*If two optical parallelograms have a pair of opposite side lines in com-
mon and if one diagonal line of the one optical parallelogram passes
through the centre of the other, then the two optical parallelograms have a
common centre.*

Since the centre of an optical parallelogram is the element of intersection of its diagonal lines, and since, by hypothesis, one diagonal line
of the one optical parallelogram passes through the centre of the other,
it follows that both centres must lie in that diagonal line.

Now we know that in any optical parallelogram the one diagonal line
is an inertia line, while the other is a separation line.

Thus the centres of the two optical parallelograms must lie in an
inertia line or a separation line.

But we have already seen by Theorem 63 that they lie in an optical
line, and since any two distinct elements determine a general line, it
follows that the centres cannot be distinct.

Thus the two optical parallelograms have a common centre.

## Theorem 65

*If two optical parallelograms P and Q in the same inertia plane have a
common centre, then the elements in which a pair of opposite side lines of
P intersect the diagonal lines of Q form the corners of an optical parallelogram with the same centre.*

Let $O$ be the common centre of the two optical parallelograms $P$ and
$Q$ and let $i$ and $j$ be the two diagonal lines of $Q$ while $a$ and $b$ are a pair
of opposite side lines of $P$.

Let $a$ intersect $i$ in $E$ and $j$ in $F$, while $b$ intersects $i$ in $G$ and $j$ in $H$.

Denote the second optical line through $E$ in the inertia plane by $c$,
and suppose it intersects $b$ in $H'$.

Denote the second optical line through $G$ in the inertia plane by $d$,
and suppose it intersects $a$ in $F'$.

Then the optical lines $a$, $c$, $b$ and $d$ form an optical parallelogram one
of whose diagonal lines, namely $i$, passes through $O$ the centre of the
optical parallelogram $P$ of which $a$ and $b$ are opposite side lines, and so
by Theorem 64 these two optical parallelograms have a common
centre $O$.

Thus if $j'$ be the second diagonal line of the optical parallelogram
formed by $a$, $c$, $b$ and $d$, it has the element $O$ in common with $j$.

The two optical parallelograms $Q$ and that formed by $a$, $c$, $b$ and $d$ have however the diagonal line $i$ in common and thus their diagonal lines of one kind do not intersect, and so by Post. XVI their diagonal lines of the other kind do not intersect.

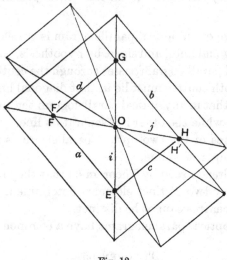

Fig. 12.

But these diagonal lines are $j$ and $j'$ which as we have seen have the element $O$ in common and therefore must be identical.

Thus $F'$ must be identical with $F$ and $H'$ must be identical with $H$ and so the elements $E$, $F$, $G$ and $H$ must form the corners of an optical parallelogram having the same centre as the two original optical parallelograms, as was to be proved.

### Remarks and Definitions

If $a$ and $b$ be any two distinct inertia lines and $A_0$ be any element in $a$ which is not an element of intersection with $b$, then from Post. XIV $(a)$ it follows that there is one single element common to the inertia line $b$ and the $\alpha$ sub-set of $A_0$.

Call this element $B_0$.

Then $B_0$ is distinct from $A_0$ and cannot be an element of intersection of the two inertia lines, for if it were $A_0$ and $B_0$ would lie both in an inertia line and an optical line, which is impossible.

Further, there cannot be an element of intersection of the two inertia lines lying *after* $A_0$ and *before* $B_0$ for, by Theorem 12, any element which is *after* $A_0$ and *before* $B_0$ must lie in the optical line $A_0 B_0$ and so, being distinct from $A_0$, it could not also lie in the inertia line $a$.

Thus any element of intersection of the two inertia lines, if such an element exists, must lie either *before* $A_0$ or *after* $B_0$.

Again from Post. XIV ($a$) it follows that there is one single element, say $A_1$, common to the inertia line $a$ and the $\alpha$ sub-set of $B_0$, and again $A_1$ cannot be an element of intersection of the inertia lines.

Further, any such element, if it exists, must lie either *before* $A_0$ or *after* $A_1$.

Proceeding again in the same way there is one single element, say $B_1$, common to the inertia line $b$ and the $\alpha$ sub-set of $A_1$ and one single element $A_2$ common to the inertia line $a$ and the $\alpha$ sub-set of $B_1$, and so on.

Thus we get an infinite series of elements $A_0$, $A_1$, $A_2$, $A_3$, ... in the inertia line $a$ and another infinite series of elements $B_0$, $B_1$, $B_2$, $B_3$, ... in the inertia line $b$.

An element of intersection of the two inertia lines if such an element exists must lie either *before* $A_0$ or *after* $A_n$, where $n$ is any finite integer whatever.

This process will be spoken of as *taking steps along the inertia line $a$ with respect to the inertia line $b$.*

The passing from $A_0$ to $A_1$ is the first step, the passing from $A_1$ to $A_2$ the second, and so on.

If $X$ be an element which is *after* $A_0$ in the inertia line $a$ and *before* $A_n$ but not *before* $A_{n-1}$, then the element $X$ will be said to be *surpassed* from $A_0$ in $n$ steps taken with respect to $b$.

If $C$ be an element of intersection of the two inertia lines and if $C$ be *after* $A_0$, it is evident from what we have said that $C$ *cannot be surpassed from $A_0$ in any finite number of steps.*

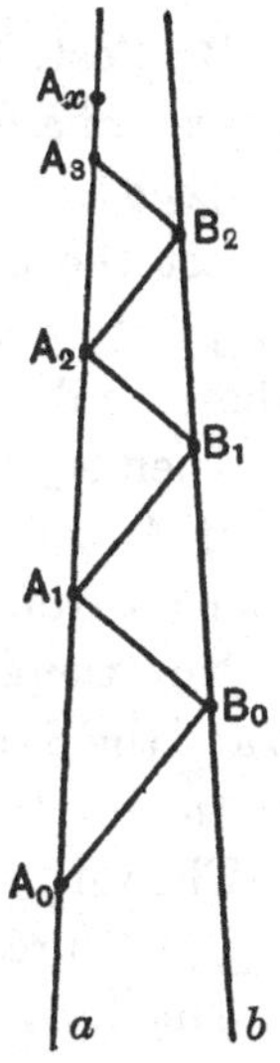

Fig. 13.

These remarks and definitions prepare the way for Post. XVII.

**POSTULATE XVII. If $A_0$ and $A_x$ be two elements of an inertia line $a$ such that $A_x$ is after $A_0$, and if $b$ be a second inertia line which does not intersect $a$ either in $A_0$, $A_x$ or any element both after $A_0$ and before $A_x$, then $A_x$ may be surpassed in a finite number of steps taken from $A_0$ along $a$ with respect to $b$.**

This postulate will be found to take the place of the well-known *axiom of Archimedes*, to which it will be seen to bear a certain resemblance.

It, however, unlike the axiom of Archimedes, contains no reference to congruence.

It follows directly from Post. XVII that if the two inertia lines $a$ and $b$ *do not intersect at all* then $A_x$ may *always* be surpassed in a finite number of steps.

There is also what is equivalent to a ($b$) form of this postulate which, however, is not independent.

It may be stated and proved as follows:

*If $A_0$ and $A_x$ be two elements of an inertia line $a$ such that $A_x$ is before $A_0$, and if $b$ be a second inertia line which does not intersect $a$ either in $A_0$, $A_x$, or any element both before $A_0$ and after $A_x$, then $A_0$ may be reached in a finite number of steps taken along $a$ from an element before $A_x$ in $a$ and with respect to $b$.*

By Post. XVII since $A_0$ is after $A_x$ it follows that $A_0$ may be surpassed in a finite number of steps, say $n$, taken from $A_x$ along $a$ with respect to $b$.

Let the elements marking these steps in $a$ be denoted by $A_{x+1}$, $A_{x+2}$, $A_{x+3}$, ... $A_{x+n}$ and let the elements in $b$ lying in the $\beta$ sub-sets of these be denoted respectively by $B_x$, $B_{x+1}$, $B_{x+2}$, ... $B_{x+n-1}$.

Then $A_0$ may either coincide with $A_{x+n-1}$ or be *after* it.

If $A_0$ coincides with $A_{x+n-1}$, then it is reached in $n-1$ steps taken along $a$ from $A_x$.

Now there is *one single element*, say $B_{x-1}$, common to the inertia line $b$ and the $\beta$ sub-set of $A_x$ and also *one single element*, say $A_{x-1}$, common to the inertia line $a$ and the $\beta$ sub-set of $B_{x-1}$.

Then $A_{x-1}$ is *before* $A_x$ and $A_0$ is reached in $n$ steps taken along $a$ from $A_{x-1}$ with respect to $b$.

This proves the result if $A_0$ coincides with $A_{x+n-1}$.

Suppose next that $A_0$ does not coincide with $A_{x+n-1}$.

Then $A_0$ is *after* $A_{x+n-1}$ and *before* $A_{x+n}$.

Let $B_{-1}$ be the *one single element* common to the inertia line $b$ and the $\beta$ sub-set of $A_0$ and let $A_{-1}$ be the *one single element* common to the inertia line $a$ and the $\beta$ sub-set of $B_{-1}$.

Let $B_{-2}$ be the *one single element* common to the inertia line $b$ and the $\beta$ sub-set of $A_{-1}$ and let $A_{-2}$ be the *one single element* common to the inertia line $a$ and the $\beta$ sub-set of $B_{-2}$, and so on, till we get to an element $A_{-n}$.

Now $B_{-1}$ cannot coincide with $B_{x+n-1}$ for then $A_0$ and $A_{x+n}$ would be two distinct elements of the inertia line $a$ both lying in the $\alpha$ sub-set of $B_{-1}$, contrary to Post. XIV $(a)$.

Further, $A_0$ is *before* $A_{x+n}$ and $B_{-1}$ is *before* $A_0$ and so $B_{-1}$ is *before* $A_{x+n}$.

It follows that $B_{-1}$ cannot be *after* $B_{x+n-1}$, since otherwise, by Theorem 12, $B_{-1}$ would require to lie in the optical line $A_{x+n}B_{x+n-1}$. But $B_{-1}$ is distinct from $B_{x+n-1}$ and both lie in the inertia line $b$ and therefore cannot both lie in one optical line.

It follows that $B_{-1}$ must be *before* $B_{x+n-1}$.

Similarly $B_{x+n-2}$ must be *before* $B_{-1}$.

Reversing the rôles of $a$ and $b$ we get in an analogous way:

$A_{-1}$ is *before* $A_{x+n-1}$ and $A_{x+n-2}$ is *before* $A_{-1}$.

Repeating this reasoning we get:

$A_{-2}$ is *before* $A_{x+n-2}$ and $A_{x+n-3}$ is *before* $A_{-2}$,

........................................................

........................................................

and so we see that $A_{-n}$ is *before* $A_x$ and $A_x$ is *before* $A_{-n+1}$.

Thus $A_{-n}$ is an element in $a$ which is *before* $A_x$, and $A_0$ may be reached in a finite number $n$ of steps taken from $A_{-n}$ with respect to $b$ along $a$.

Thus the result holds in general.

## THEOREM 66

(a) *If $A_0$ and $A_x$ be two elements in an inertia line $a$ which lies in the same inertia plane with another inertia line $b$ which does not intersect $a$ in $A_0$, $A_x$, or any element after the one and before the other, and if an optical line through $A_0$ intersects $b$ in $B_0$ so that $B_0$ is after $A_0$, then a parallel optical line through $A_x$ will intersect $b$ in an element which is after $A_x$.*

We shall first suppose that $A_x$ is *after* $A_0$.

Let the optical line through $A_x$ parallel to $A_0B_0$ intersect $b$ in $B_x$.

Then by Post. XVII $A_x$ may be surpassed in a finite number of steps say $n$, taken from $A_0$ along $a$ with respect to $b$.

Let the elements (including $A_0$) marking these steps in $a$ be $A_0$, $A_1$, $A_2$, ... $A_n$ and let the elements in $b$ lying in the $\alpha$ sub-sets of these be $B_0$, $B_1$, $B_2$, ... $B_n$ respectively.

Then $A_x$ may either coincide with $A_{n-1}$ or be *after* it.

Now the optical line $B_0A_1$ intersects the two optical lines $A_0B_0$ and

$A_1 B_1$ and so these latter two optical lines belong to one set and are therefore parallel.

Similarly $A_1 B_1$ intersects the two optical lines $B_0 A_1$ and $B_1 A_2$ and so these two are also parallel but belong to the other set.

Proceeding thus we see that the optical lines $A_0 B_0$, $A_1 B_1$, $A_2 B_2$, ... $A_n B_n$ belong to one set and are all parallel, while $B_0 A_1$, $B_1 A_2$, $B_2 A_3$, ... $B_{n-1} A_n$ belong to the other set and are all parallel.

But $\qquad\qquad A_1$ lies in the $\alpha$ sub-set of $B_0$,

$$
\begin{array}{ccc}
B_1 & ,, & ,, & A_1, \\
A_2 & ,, & ,, & B_1, \\
B_{n-1} & ,, & ,, & A_{n-1}, \\
A_n & ,, & ,, & B_{n-1}, \\
B_n & ,, & ,, & A_n.
\end{array}
$$

Thus if $A_x$ coincides with $A_{n-1}$, then $B_x$ must coincide with $B_{n-1}$ and therefore $B_x$ must lie in the $\alpha$ sub-set of $A_x$, and since $B_x$ and $A_x$ are distinct it follows that $B_x$ is *after* $A_x$ and the optical lines $A_0 B_0$ and $A_x B_x$ are parallel.

This proves the theorem in this case.

If $A_x$ does not coincide with $A_{n-1}$, then it must be *after* $A_{n-1}$ and *before* $A_n$.

Also since $A_x B_x$ is parallel to $A_0 B_0$ it must be parallel to $A_{n-1} B_{n-1}$ and to $A_n B_n$.

But since $A_x$ is *after* $A_{n-1}$ and *before* $A_n$ it follows that $A_x B_x$ is an after-parallel of $A_{n-1} B_{n-1}$ and a before-parallel of $A_n B_n$.

Further, $A_x B_x$ must intersect the optical line $B_{n-1} A_n$ in some element, say $C$, since $B_{n-1} A_n$ is an optical line of the opposite set to $A_x B_x$ and so $C$ must be *after* $B_{n-1}$ and *before* $A_n$.

Thus $B_{n-1}$ must lie in the $\beta$ sub-set of $C$, while $A_n$ lies in the $\alpha$ sub-set of $C$.

But by Post. XIV (*a*) there is one single element common to the inertia line $b$ and the $\alpha$ sub-set of $C$ and this must lie in the other optical line through $C$ in the inertia plane; that is to say in the optical line $A_x B_x$ and must therefore be identical with $B_x$.

Similarly by Post. XIV (*b*) there is *one single element* common to the inertia line $a$ and the $\beta$ sub-set of $C$ and this must be identical with $A_x$.

Thus $C$ is *after* $A_x$ and *before* $B_x$ and therefore $B_x$ is *after* $A_x$.

Thus the theorem is proved for all cases in which $A_x$ is *after* $A_0$.

A similar method shows that the theorem is true when $A_x$ is *before* $A_0$ except that the corresponding (*b*) form takes the place of Post. XVII.

Thus the theorem holds in general.

(b) *If $A_0$ and $A_x$ be two elements in an inertia line a which lies in the same inertia plane with another inertia line b which does not intersect a in $A_0$, $A_x$, or any element before the one and after the other, and if an optical line through $A_0$ intersects b in $B_0$ so that $B_0$ is before $A_0$, then a parallel optical line through $A_x$ will intersect b in an element which is before $A_x$.*

### Theorem 67

(a) *If $A_0$ and $A_x$ be two elements in a separation line a which lies in the same inertia plane with another separation line b which does not intersect a in $A_0$, $A_x$ or any element lying between a pair of parallel optical lines through $A_0$ and $A_x$ in the inertia plane, and if an optical line through $A_0$ intersects b in $B_0$ so that $B_0$ is after $A_0$, then a parallel optical line through $A_x$ will intersect b in an element which is after $A_x$.*

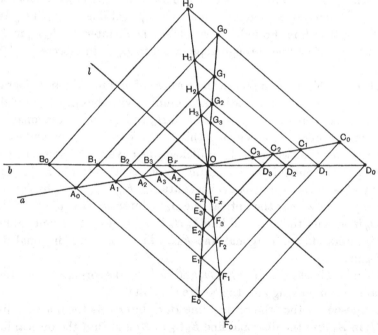

Fig. 14.

In case the separation lines $a$ and $b$ do not intersect at all, then since they lie in one inertia plane they are parallel and the result follows directly from Theorem 56 (b).

We shall therefore consider the case in which an element of intersection of $a$ and $b$ does exist and we shall denote this element by $O$.

We shall suppose first that $A_x$ is between a pair of parallel optical lines through $A_0$ and $O$ in the inertia plane.

Now let $l$ be the optical line through $O$ parallel to $A_0 B_0$.

It will be sufficient to consider the case where $l$ is an after-parallel of $A_0 B_0$, since the case of a before-parallel is quite analogous.

If $A_x B_x$ be the optical line through $A_x$ parallel to $l$ and meeting $b$ in the element $B_x$, then $A_x B_x$ will be an after-parallel of $A_0 B_0$ and a before-parallel of $l$.

Now, by Theorem 59, there exists a definite optical parallelogram in the inertia plane having $O$ as centre and $B_0$ as one of its corners, so that $b$ is one of its diagonal lines.

Let $D_0$ be the corner opposite $B_0$ and let the optical line $A_0 B_0$ intersect the other diagonal line in $F_0$ while the second optical line through $B_0$ in the inertia plane intersects the same diagonal line in $H_0$.

Then $B_0$, $F_0$, $D_0$ and $H_0$ are the corners of the optical parallelogram.

Let the separation line $a$ intersect the optical line $D_0 H_0$ in $C_0$ and let the optical line through $C_0$ parallel to $D_0 F_0$ intersect $B_0 F_0$ in $E_0$, while the optical line through $A_0$ parallel to $D_0 F_0$ intersects $D_0 H_0$ in $G_0$.

Then $A_0$, $E_0$, $C_0$ and $G_0$ are the corners of an optical parallelogram having a pair of opposite side lines in common with the optical parallelogram whose corners are $B_0$, $F_0$, $D_0$ and $H_0$ and having its diagonal line $a$ passing through $O$ the centre of this optical parallelogram, and so, by Theorem 64, the two optical parallelograms have a common centre $O$.

Denote the optical parallelogram whose corners are $B_0$, $F_0$, $D_0$ and $H_0$ by $P_0$ and the one whose corners are $A_0$, $E_0$, $C_0$ and $G_0$ by $Q_0$.

Suppose now that the optical line $A_0 G_0$ intersects the diagonal line $B_0 D_0$ in $B_1$ and the diagonal line $F_0 H_0$ in $H_1$ and that the optical line $E_0 C_0$ intersects the diagonal line $F_0 H_0$ in $F_1$ and the diagonal line $B_0 D_0$ in $D_1$.

Then by Theorem 65, $B_1$, $F_1$, $D_1$ and $H_1$ form the corners of an optical parallelogram having also the centre $O$. Call it $P_1$.

Suppose now that the optical line $B_1 F_1$ intersects the diagonal line $A_0 C_0$ in $A_1$ and the diagonal line $E_0 G_0$ in $E_1$ and that the optical line $D_1 H_1$ intersects the diagonal line $A_0 C_0$ in $C_1$ and the diagonal line $E_0 G_0$ in $G_1$.

Then, by Theorem 65, $A_1$, $E_1$, $C_1$, $G_1$ form the corners of an optical parallelogram $Q_1$ which bears the same relation to the optical parallelogram $P_1$ whose corners are $B_1$, $F_1$, $D_1$ and $H_1$ as the optical parallelogram $Q_0$ to the optical parallelogram $P_0$.

This construction may be repeated indefinitely, and we obtain a series of parallel optical lines $B_0 F_0$, $B_1 F_1$, $B_2 F_2$, $B_3 F_3$, etc., intersecting the separation line $b$ in the elements $B_0$, $B_1$, $B_2$, $B_3$, etc., and the other diagonal line of the optical parallelogram $P_0$ in the elements $F_0$, $F_1$, $F_2$, $F_3$, etc.

Further, these same optical lines intersect the separation line $a$ in the elements $A_0$, $A_1$, $A_2$, $A_3$, etc., and the other diagonal line of the optical parallelogram $Q_0$ in the elements $E_0$, $E_1$, $E_2$, $E_3$, etc.

Again we have another set of parallel optical lines $A_0 B_1$, $A_1 B_2$, $A_2 B_3$, $A_3 B_4$, etc., and a further set $E_0 F_1$, $E_1 F_2$, $E_2 F_3$, $E_3 F_4$, etc.

Now by hypothesis $l$ is an after-parallel of $A_0 B_0$ and, since $O F_0$ is an inertia line, it follows that $F_0$ is *before O*.

Similarly $E_0$ is *before O*.

But since $b$ is a separation line and $B_0$ is *after* $A_0$ we must also have $B_1$ *after* $A_0$.

It follows that $B_1 F_1$ is an after-parallel of $B_0 F_0$ and, since $E_0 F_1$ is an optical line, we must have $F_1$ *after* $E_0$, so that $F_1$ lies in the $\alpha$ sub-set of $E_0$.

Also since $F_0 F_1$ is an inertia line we must have $F_1$ *after* $F_0$ so that $F_0$ is not an element of the optical line $E_0 F_1$ but is *before* an element of it.

Thus, since $F_0 E_0$ is an optical line, we must have $E_0$ *after* $F_0$ and so $E_0$ must lie in the $\alpha$ sub-set of $F_0$.

Also, from what we showed on p. 103, the element $O$ of intersection of the two inertia lines $F_0 H_0$ and $E_0 G_0$ cannot lie *before* $F_1$ and, since we already know that $F_0$ is *before O*, it follows that $F_1$ is also *before O*.

Thus the optical line $l$ must be an after-parallel of $B_1 F_1$: that is to say $l$ is an after-parallel of $A_1 B_1$ and so $E_1$ is also *before O*.

But we saw that $B_1$ must be *after* $A_0$ and so since $a$ is a separation line we must have $B_1$ *after* $A_1$.

By repetition of this reasoning we can show that:

$$B_2 F_2 \text{ is an after-parallel of } B_1 F_1$$
$$B_3 F_3 \qquad \text{,,} \qquad \text{,,} \qquad B_2 F_2$$
.............................................
.............................................

while the optical line $l$ is an after-parallel of all these.

Also we can show that

$$E_1 \text{ lies in the } \alpha \text{ sub-set of } F_1 \text{ while } F_2 \text{ lies in the } \alpha \text{ sub-set of } E_1$$
$$E_2 \qquad \text{,,} \qquad \text{,,} \qquad F_2 \text{ ,, } F_3 \qquad \text{,,} \qquad \text{,,} \qquad E_2$$
.............................................
.............................................

Thus $F_1$, $F_2$, $F_3$, ... mark steps taken along the inertia line $F_0O$ with respect to the inertia line $E_0O$.

Now let the optical line $A_xB_x$ intersect $F_0O$ in $F_x$ and $E_0O$ in $E_x$.

Then, by hypothesis, $A_xB_x$ is a before-parallel of $l$ and it follows that both $F_x$ and $E_x$ are *before* $O$.

Thus, by Post. XVII, $F_x$ may be surpassed in a finite number $n$ of steps taken from $F_0$ along $F_0O$ with respect to $E_0O$.

Now we have

$$B_1 \text{ after } A_1,$$
$$B_2 \quad ,, \quad A_2,$$
$$\dots\dots\dots$$
$$\dots\dots\dots$$

If then $F_x$ should happen to coincide with $F_{n-1}$ we should have $A_x$ coinciding with $A_{n-1}$ and $B_x$ coinciding with $B_{n-1}$ and accordingly we should have $B_x$ *after* $A_x$.

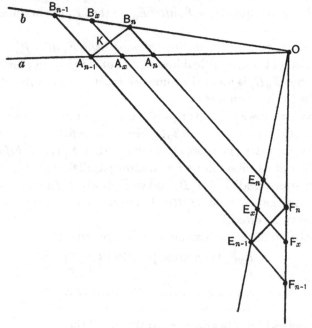

Fig. 15.

Suppose next that $F_x$ does not coincide with $F_{n-1}$, but is *after* $F_{n-1}$ and *before* $F_n$.

Then $B_xF_x$ will be an after-parallel of $B_{n-1}F_{n-1}$ and a before-parallel of $B_nF_n$.

Let $B_x F_x$ intersect $A_{n-1} B_n$ in the element $K$.

Then, since $A_{n-1} K$ is an optical line, we must have $K$ *after* $A_{n-1}$ and also $K$ *before* $B_n$.

Since $A_{n-1}$ and $A_x$ lie in the separation line $a$, we must have $K$ *after* $A_x$: while, since $B_n$ and $B_x$ lie in the separation line $b$, we must have $B_x$ *after* $K$.

It follows by Post. III that $B_x$ must be *after* $A_x$ as was to be proved.

Now we started out by considering the case where $A_x$ is between a pair of parallel optical lines through $A_0$ and $O$ in the inertia plane; if instead we had taken the case where $A_0$ is between a pair of parallel optical lines through $A_x$ and $O$ in the inertia plane, then the supposition that $A_x$ was *after* $B_x$ would, in a similar manner, lead to the conclusion that $A_0$ was *after* $B_0$, contrary to the hypothesis that $B_0$ is *after* $A_0$.

Also, since $A_x$ and $B_x$ could not coincide without the separation lines being identical, it follows that we must also in this case have $B_x$ *after* $A_x$.

Thus the theorem holds in general.

(b) *If $A_0$ and $A_x$ be two elements in a separation line $a$ which lies in the same inertia plane with another separation line $b$ which does not intersect $a$ in $A_0$, $A_x$ or any element lying between a pair of parallel optical lines through $A_0$ and $A_x$ in the inertia plane, and if an optical line through $A_0$ intersects $b$ in $B_0$ so that $B_0$ is before $A_0$, then a parallel optical line through $A_x$ will intersect $b$ in an element which is before $A_x$.*

### Theorem 68

*If two elements $A$ and $B$ lie in one optical line and if two other elements $C$ and $D$ lie in a parallel optical line in the same inertia plane, then if $A$ be after $B$ and $C$ after $D$ the element of intersection of the general lines $AD$ and $BC$ (which was proved in Theorem 58 to exist) lies between the two given optical lines.*

Let $a$ be the optical line containing $A$ and $B$, and let $b$ be the parallel optical line containing $C$ and $D$.

Since one of the optical lines must be an after-parallel of the other and since it is immaterial which of them, we shall suppose that $a$ is an after-parallel of $b$.

Now the general lines $AD$ and $BC$ cannot *both* be optical lines since two optical lines which intersect a pair of parallel optical lines are themselves parallel and have no element of intersection.

One of them however may be an optical line.

Suppose first that $BC$ is an optical line and that $E$ is the element of intersection of $AD$ and $BC$.

Then, since $a$ is an after-parallel of $b$ and $CB$ is an optical line, therefore $B$ is *after* $C$.

But $C$ is *after* $D$ and therefore $B$ is *after* $D$, and since $A$ is *after* $B$ it follows that $A$ is *after* $D$.

Thus, since $AD$ cannot be an optical line and has one element which is *after* another, it must be an inertia line.

Now, since $C$ is *after* $D$ and lies in an optical line containing $D$, it follows that $D$ is in the $\beta$ sub-set of $C$; and since $E$ lies in the second optical line through $C$ in the inertia plane, it follows by Post. XIV ($a$) that $E$ must be in the $\alpha$ sub-set of $C$.

Thus, since $E$ cannot be identical with $C$, it follows that $E$ is *after* $C$.

Similarly, since $A$ is *after* $B$ and $A$ and $B$ lie in an optical line, it follows that $A$ is in the $\alpha$ sub-set of $B$; and since $E$ lies in the second optical line through $B$ in the inertia plane, it follows by Post. XIV ($b$) that $E$ must be in the $\beta$ sub-set of $B$.

Thus since $E$ cannot be identical with $B$, it follows that $E$ is *before* $B$.

This proves that $E$ lies between $a$ and $b$.

Suppose secondly that $AD$ is an optical line and again let $E$ be the element of intersection of $AD$ and $BC$.

Let the optical line through $C$ parallel to $DA$ intersect $a$ in $F$.

Then $C$ being *after* $D$ it follows that $F$ must be *after* $A$ and since $A$ is *after* $B$ therefore $F$ must be *after* $B$.

Now $F$ must be *after* $C$ and therefore $C$ lies in the $\beta$ sub-set of $F$, as does also $B$.

But if $C$ were either *before* or *after* $B$ then, by Theorem 13 ($b$), $C$ would have to lie in $a$ which is impossible.

Thus $C$ is neither *before* nor *after* $B$, so that $CB$ must be a separation line.

Now $D$ cannot be *after* $E$, for since $C$ is *after* $D$ we should then have $C$ *after* $E$ which is impossible since $C$ and $E$ lie in a separation line.

But since $D$ and $E$ are distinct elements of an optical line, the one must be *after* the other and thus $E$ must be *after* $D$.

Again $E$ cannot be *after* $A$, for since $A$ is *after* $B$ we should then have $E$ *after* $B$ which is impossible since $E$ and $B$ are elements of a separation line.

But $E$ must be either *before* or *after* $A$ since $E$ and $A$ are distinct elements of an optical line, and since $E$ cannot be *after*, it must be *before* $A$.

Thus again in this case $E$ lies between $a$ and $b$.

Next take the case where one of the two general lines $AD$ and $BC$ is an inertia line and the other a separation line.

If $BC$ be a separation line and $E$ be the element of intersection with $AD$, then $E$ is neither *before* nor *after* $C$ and also neither *before* nor *after* $B$.

But $E$ cannot be *before* $D$, for since $D$ is *before* $C$ we should then have $C$ *after* $E$, which is impossible.

Thus since $D$ and $E$ are distinct elements of an inertia line, we must have $E$ *after* $D$.

Again $E$ cannot be *after* $A$, for since $A$ is *after* $B$ we should then have $E$ *after* $B$, which is impossible.

Thus since $A$ and $E$ are distinct elements of an inertia line, we must have $E$ *before* $A$.

Thus again in this case $E$ lies between $a$ and $b$.

If $BC$ is an inertia line we must have $B$ *after* $C$, since $a$ is an after-parallel of $b$.

Since then $C$ is *after* $D$ we must have $B$ *after* $D$, and since $A$ is *after* $B$ we must have $A$ *after* $D$.

But $AD$ could not be an optical line, for, since $B$ is *after* $D$ and *before* $A$, it would then follow by Theorem 12 that $B$ must itself be an element of $AD$; which is impossible. Thus $AD$ must be an inertia line.

Accordingly we shall next take the case where both the general lines $AD$ and $BC$ are inertia lines and $E$ is their element of intersection.

By Theorem 66, if $A$ were *before* $E$ then $C$ being *after* $D$ would imply that $B$ was *after* $A$, contrary to hypothesis; while if $D$ were *after* $E$ then $A$ being *after* $B$ would imply that $D$ was *after* $C$, contrary again to hypothesis.

Thus since $E$ cannot be identical with either $A$ or $D$, it follows that $E$ must be *after* $D$ and *before* $A$ and so $E$ lies between $a$ and $b$.

Finally we have the case where $AD$ and $BC$ are both separation lines and $E$ their element of intersection.

Let $c$ be an optical line through $E$ parallel to $a$ and $b$.

First suppose, if possible, that $c$ is an after-parallel of $a$; then $c$ would also be an after-parallel of $b$ since $a$ is an after-parallel of $b$.

Thus $AD$ and $BC$ would intersect in an element which was not between $a$ and $b$ and did not lie either in $a$ or $b$, and so by Theorem 67, $A$ being *after* $B$ would imply that $D$ was *after* $C$, contrary to hypothesis.

The same would hold if we supposed $c$ to be a before-parallel of $b$.

Thus $c$ cannot be an after-parallel of $a$ and cannot be identical with $a$ and therefore must be a before-parallel of $a$.

Also $c$ cannot be a before-parallel of $b$ and cannot be identical with $b$, and thus $c$ must be an after-parallel of $b$.

Thus the element $E$ must be *after* an element of $b$ and *before* an element of $a$ and so $E$ lies between $a$ and $b$.

This exhausts all the possibilities and so we see that the theorem holds in general.

THEOREM 69

*If two elements $A$ and $B$ lie in one optical line and if two other elements $C$ and $D$ lie in a parallel optical line in the same inertia plane, then if $A$ be after $B$ and if the general lines $AD$ and $BC$ intersect in an element $E$ lying between the parallel optical lines, we must also have $C$ after $D$.*

Let $a$ be the optical line containing $A$ and $B$, and let $b$ be the parallel optical line containing $C$ and $D$.

Then one of the optical lines $a$ and $b$ is an after-parallel of the other, but as the demonstration is quite analogous in the two cases we shall only consider that in which $a$ is an after-parallel of $b$.

We must therefore have $E$ *after* an element of $b$ and *before* an element of $a$.

Now $AD$ and $BC$ cannot both be optical lines since two optical lines which both intersect a pair of parallel optical lines are themselves parallel and so the element $E$ could not exist.

We may however have one of them an optical line and shall first consider the case in which $AD$ is such.

In this case $E$ is *before* $A$ and therefore $E$ lies in the $\beta$ sub-set of $A$, as does also $B$.

But $E$ cannot be either *after* or *before* $B$, for otherwise, by Theorem 13 ($b$), $E$ would require to lie in the optical line $a$ and so $E$ could not lie between $a$ and $b$.

It follows that $BE$ must be a separation line.

Thus $C$ can be neither *before* nor *after* $E$.

But $D$ is *before* $E$ and so if $C$ were *before* $D$ we should have $C$ *before* $E$, which is impossible.

Further, $C$ cannot coincide with $D$ and therefore $C$ must be *after* $D$.

We shall next consider the case where $BC$ is an optical line.

Then we have $B$ *after* $E$, and since $A$ is *after* $B$ it follows that $A$ is *after* $E$ and so $AE$ is an inertia line.

Again $E$ is *after* $C$ and so $E$ lies in the $\alpha$ sub-set of $C$ and therefore by Post. XIV $(b)$ $D$ must lie in the $\beta$ sub-set of $C$.

Thus since $C$ and $D$ cannot be identical, we must have $C$ *after* $D$.

We shall next consider the case where one of the general lines $BC$ and $AD$ is an inertia line and the other a separation line.

Now if $BC$ were an inertia line we should have $B$ *after* $E$ and so, since $A$ is *after* $B$, we should have also $A$ *after* $E$.

Thus in this case both general lines would be inertia lines and so we must suppose instead that $BC$ is a separation line and $AD$ an inertia line.

Then since $E$ cannot be *before* any element of $b$, and since it must be either *before* or *after* $D$ it follows that $E$ must be *after* $D$.

But $D$ cannot be *after* $C$, for then we should have $E$ *after* $C$, which is impossible since $C$ and $E$ lie in a separation line.

Thus since $C$ and $D$ cannot be identical, we must have $C$ *after* $D$.

We have next to consider the cases where the general lines $BC$ and $AD$ are both separation lines and where they are both inertia lines.

The constructions and demonstrations are analogous in both cases up to a certain point.

By Theorem 59 there is an optical parallelogram in the inertia plane having $E$ as centre and $B$ as one of its corners.

Let $C'$ be the corner opposite to $B$ and let the optical line through $C'$ in the inertia plane and of the opposite set to $AB$ intersect $AB$ in the element $G$.

Then $GE$ is the other diagonal line of the optical parallelogram.

Let the second optical line through $B$ in the inertia plane intersect $GE$ in $F$.

Then $B$, $F$, $C'$ and $G$ are the corners of the optical parallelogram.

Let $AE$ intersect the optical line $FC'$ in $D'$; let an optical line through $A$ parallel to $BF$ intersect $FC'$ in $H$, and let an optical line through $D'$ parallel to $C'G$ intersect $BG$ in $I$.

Then $A$, $H$, $D'$ and $I$ are the corners of an optical parallelogram having a pair of opposite side lines in common with the optical parallelogram whose corners are $B$, $F$, $C'$ and $G$ and having one of its diagonal lines $AD'$ passing through $E$ the centre of this optical parallelogram.

It follows from Theorem 64 that these two optical parallelograms have a common centre.

Let $AH$ intersect $BC'$ in $A_1$ and $FG$ in $F_1$ and let $ID'$ intersect $BC'$ in $C_1$ and $FG$ in $G_1$.

Then by Theorem 65 the elements $A_1$, $F_1$, $C_1$ and $G_1$ form the corners of another optical parallelogram with the same centre.

Suppose now first that $AE$ and $BE$ are both separation lines, then $EG$ and $EI$ are both inertia lines, and by hypothesis $E$ is *before* an element of $BG$ and so $E$ must be *before* $G$ and also *before* $I$.

Also, since $B$ and $A_1$ lie in a separation line and since $A$ is *after* $B$, it follows that $A$ must also be *after* $A_1$.

Thus $A_1 G_1$ must be a before-parallel of $BG$ and so $G_1$ must be *before* $G$.

Thus $G_1 D'$ must be a before-parallel of $GC'$, and since $C'$ and $D'$ lie in an optical line we must have $C'$ *after* $D'$.

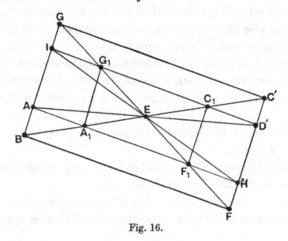

Fig. 16.

Now $E$ being the centre of the optical parallelogram whose corners are $B$, $G$, $C'$ and $F$ and being *before* an element of $BG$ must be *after* an element of $FC'$.

Thus $E$ is between the parallel optical lines $BG$ and $FC'$.

Now the optical line $b$ containing $C$ and $D$ may either coincide with $FC'$, in which case $C$ is *after* $D$, or else $b$ may be a before-parallel of $FC'$ or an after-parallel of $FC'$.

In any case, however, if $e$ be an optical line through $E$ parallel to $a$ and $b$, then $FC'$ and $b$ are each before-parallels of $e$, so that in no case can $E$ lie between $FC'$ and $b$.

Thus by Theorem 67 since $C'$ is *after* $D'$ we must have $C$ *after* $D$.

Suppose next that $AE$ and $BE$ are both inertia lines, then $EG$ and $EI$ are both separation lines, and by hypothesis $E$ is *before* an element of $BG$, so $E$ is *before* $A$ and also *before* $B$.

Also, since $B$ and $A_1$ lie in an inertia line and since $B$ is in the $\beta$ sub-set of $A$ and distinct from it, therefore $A_1$ must be in the $\alpha$ sub-set of $A$, and since $B$ and $A$ are distinct, $A$ and $A_1$ must also be distinct and therefore $A_1$ is *after* $A$.

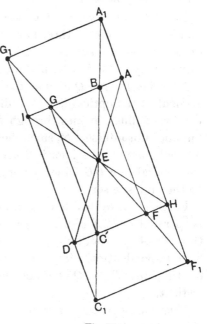

Thus $A_1 G_1$ must be an after-parallel of $AI$, and since $G_1$ and $I$ lie in an optical line we must have $G_1$ *after* $I$.

But since $G_1$ and $G$ lie in a separation line, the one is neither *before* nor *after* the other and so $G$ must also be *after* $I$.

Thus $GC'$ must be an after-parallel of $ID'$, and since $C'$ and $D'$ lie in an optical line we must have $C'$ *after* $D'$.

From this point the demonstration is similar to that of the case where $AE$ and $BE$ are both separation lines, except that the reference is to Theorem 66 instead of Theorem 67.

Fig. 17.

This exhausts all the possibilities, and so the theorem holds in general.


### Theorem 70

*If $A$, $B$ and $C$ be three elements in a separation line and if $B$ be between a pair of parallel optical lines through $A$ and $C$ in an inertia plane containing the separation line, then $B$ is also between a pair of parallel optical lines through $A$ and $C$ in any other inertia plane containing the separation line.*

Let $a$ be an optical line through $A$, and $c$ a parallel optical line through $C$; both lying in the given inertia plane, say $P$, and such that $B$ lies between $a$ and $c$.

We may suppose that $B$ is *before* an element of $a$ and *after* an element of $c$ without any essential loss of generality.

Let an optical line through $B$ in the inertia plane, and of the opposite set to $a$ and $c$, intersect $a$ in $D$ and $c$ in $E$.

Then $D$ must be *after* $B$, and $E$ must be *before* $B$.

Further, since $A$, $B$ and $C$ lie in a separation line, we must have $D$ *after* $A$ and $E$ *before* $C$.

Now let $Q$ be any other inertia plane containing the separation line, and let $a'$, $b'$ and $c'$ be three parallel optical lines through $A$, $B$ and $C$ respectively in the inertia plane $Q$.

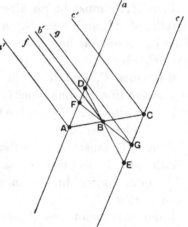

Now the element $D$ is *after* $B$, an element of the optical line $b'$, while the optical line $a$ passes through $D$ but does not intersect $b'$, since then it would have to be identical with the optical line $DB$ which belongs to the opposite set.

Further, the optical line $a$ cannot be parallel to $b'$, for since $a$ passes through $A$ it would in that case have to be identical with $a'$ and the inertia planes $P$ and $Q$ could not be distinct.

Fig. 18.

Thus each element of $a$ is not *after* an element of $b'$, and so by Post. XII (*b*) there is one single element of $a$, say $F$, which is neither *after* nor *before* any element of $b'$.

Thus by Theorem 22 there is one single optical line containing $F$ and such that no element of it is either *before* or *after* any element of $b'$.

If $f$ be this optical line, then $f$ is a neutral-parallel of $b'$.

But since $a'$ and $b'$ lie in the inertia plane $Q$ and are parallel, the one must be an after-parallel of the other and so $a'$ cannot be identical with $f$.

Thus $F$ must be either *after* or *before* $A$ and cannot be identical with it.

Now the general line $FB$ lies in the inertia plane $P$ and is clearly a separation line since $F$ is neither *before* nor *after* $B$.

Let $FB$ intersect the optical line $c$ in $G$.

Then, by Theorem 45, $G$ is neither *before* nor *after* any element of $b'$, and so if an optical line $g$ be taken through $G$ parallel to $b'$ it will be a neutral-parallel.

Now, by Theorem 69, since $B$ lies between the parallel optical lines $a$ and $c$ passing through $A$ and $C$ respectively and lying in the inertia plane $P$, it follows that if $F$ be *after* $A$ then $C$ is *after* $G$; while if $A$ be *after* $F$ then $G$ is *after* $C$.

If however $F$ be *after* $A$, then $a'$ must be a before-parallel of $f$, and therefore, by Theorem 26 (*a*), $a'$ must be a before-parallel of $b'$.

Then we shall have also $c'$ an after-parallel of $g$, and therefore, by Theorem 26 (*b*), $c'$ must be an after-parallel of $b'$.

Thus $B$ will be *after* an element of $a'$ and *before* an element of $c'$: that is, $B$ will be between the parallel optical lines $a'$ and $c'$ passing through $A$ and $C$ respectively in the inertia plane $Q$.

Similarly if $F$ be *before* $A$, then $a'$ must be an after-parallel of $f$, and therefore, by Theorem 26 (*b*), $a'$ must be an after-parallel of $b'$.

We shall in that case have also $c'$ a before-parallel of $g$, and therefore, by Theorem 26 (*a*), $c'$ must be a before-parallel of $b'$.

Thus again we shall have $B$ between the parallel optical lines $a'$ and $c'$ passing through $A$ and $C$ respectively in the inertia plane $Q$.

Thus the theorem is proved.

### REMARKS

If $A$, $B$ and $C$ be three elements in an optical or inertia line $l$, and if $B$ be between a pair of parallel optical lines through $A$ and $C$ in an inertia plane containing $l$, then it is easy to see that $B$ is also between a pair of parallel optical lines through $A$ and $C$ in any other inertia plane containing $l$.

This follows directly from the consideration that, in this case, of any two of the three elements $A$, $B$, $C$, one is *after* the other.

We accordingly introduce the following definition.

*Definition.* If three distinct elements lie in a general line and if one of them lies between a pair of parallel optical lines through the other two in an inertia plane containing the general line, then the element which is between the parallel optical lines will be said to be *linearly between* the other two elements.

The above definition is so framed as to apply to all three types of general line and for this reason is more complicated than it need be if we were dealing only with optical or inertia lines.

For the case of elements lying in either of these types of general line, one element is linearly between two other elements if it be *after* the one and *before* the other.

In the case of elements lying in a separation line, however, no one is either *before* or *after* another and so we have to fall back on our definition involving parallel optical lines.

The distinction between the three cases is interesting.

Thus if the three elements $A$, $B$ and $C$ lie in a general line $a$, and if $B$

be linearly between $A$ and $C$, then, in case $a$ be an inertia line, we must have either $B$ *after* $A$ and $C$ *after* $B$ or else $B$ *after* $C$ and $A$ *after* $B$, and similarly when $a$ is an optical line.

If $a$ be an inertia line and $B$ be *after* $A$ and $C$ *after* $B$, then $B$ will be *before* elements of both optical lines through $C$ and *after* elements of both optical lines through $A$ in any inertia plane containing $a$.

If $a$ be an optical line and $B$ be *after* $A$ and $C$ *after* $B$, then, *apart from a itself*, there is only one optical line through any element of $a$ in any inertia plane containing $a$, and so we should have $B$ *before* an element of the optical line through $C$ and *after* an element of the parallel optical line through $A$.

If $a$ be a separation line, however, we should have $B$ *before* an element of one of the optical lines through $C$ and *after* an element of the parallel optical line through $A$ and also *after* an element of the second optical line through $C$ and *before* an element of the parallel optical line through $A$.

The distinctions are perhaps exhibited more clearly by the following figures:

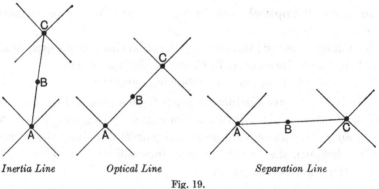

*Inertia Line*          *Optical Line*                    *Separation Line*

Fig. 19.

From Theorem 70 it follows that the property of one element being linearly between two others is independent of the particular inertia plane in which the elements are considered as lying and so may be regarded as a relation of the one element to the other two.

This relation has been defined in terms of the relations *before* and *after*, not only for the cases where the three elements considered are such that of any two of them one is *after* the other; but also for the case of elements in a separation line when this is no longer so.

It is thus possible to state certain general results which hold for all three types of general line involving the conception linearly between.

Peano has given some eleven axioms of the *straight line* which are as follows:

(1) There is at least one point.

(2) If $A$ is any point, there is a point distinct from $A$.

(3) If $A$ is a point, there is no point lying between $A$ and $A$.

(4) If $A$ and $B$ are distinct points, there is at least one point lying between $A$ and $B$.

(5) If the point $C$ lies between $A$ and $B$, it also lies between $B$ and $A$.

(6) The point $A$ does not lie between the points $A$ and $B$.

*Definition.* If $A$ and $B$ are points, the symbol $AB$ represents the class of points such as $C$ with the property that $C$ lies between $A$ and $B$.

*Definition.* If $A$ and $B$ are points, the symbol $A'B$ represents the class of points such as $C$ with the property that $B$ lies between $A$ and $C$. Thus $A'B$ is the prolongation of the line beyond $B$, and $B'A$ its prolongation beyond $A$.

(7) If $A$ and $B$ are distinct points, there exists at least one member of $A'B$.

(8) If $A$ and $D$ are distinct points, and $C$ is a member of $AD$ and $B$ of $AC$, then $B$ is a member of $AD$.

(9) If $A$ and $D$ are distinct points, and $B$ and $C$ are members of $AD$, then either $B$ is a member of $AC$, or $B$ is identical with $C$, or $B$ is a member of $CD$.

(10) If $A$ and $B$ are distinct points, and $C$ and $D$ are members of $A'B$, then either $C$ is identical with $D$, or $C$ is a member of $BD$, or $D$ is a member of $BC$.

(11) If $A, B, C, D$ are points, and $B$ is a member of $AC$ and $C$ of $BD$, then $C$ is a member of $AD$.

*Definition.* The straight line possessing $A$ and $B$, symbolised by str. $(A, B)$, is composed of the three classes $A'B$, $AB$, $B'A$ together with the points $A$ and $B$ themselves.

Of these axioms the writer has succeeded in proving nos. 6 and 9 from the others, so that they are really redundant.*

It is easy to see, with our definition of linearly between, that corresponding results hold for all three types of "general line".

As regards axioms (1) and (2) which we shall express thus:

(1) *There is at least one element,*

and (2) *If $A$ be any element there is an element distinct from $A$,*

* *Messenger of Mathematics*, vol. XIII, pp. 121–123 and 134.

the first follows from our preliminary statement on p. 27, while the second follows directly from Posts. II and I and also from Post. V.

As regards axiom (3) we shall put it in the form:

(3) *If A is an element, there is no element lying linearly between A and A.*

This follows from the definition of linearly between.

(4) *If A and B are distinct elements, there is at least one element lying linearly between A and B.*

From our remarks at the end of Theorem 35 it appears that there are an infinite number of inertia planes containing any two distinct elements and accordingly any two distinct elements lie in a general line.

If A and B lie in an optical line, then Theorem 11 shows that there is at least one element which is *after* the one and *before* the other and is therefore linearly between them.

If A and B lie in an inertia line, the same result follows from Theorem 39; while if they lie in a separation line, it follows from Theorem 41.

(5) *If the element C lies linearly between A and B, it also lies linearly between B and A.*

This follows from the definition of linearly between.

(6) *The element A does not lie linearly between the elements A and B.*

This follows from the definition of what is meant by an element lying between a pair of parallel optical lines in an inertia plane. According to this definition the element must not lie in either optical line.

(7) *If A and B are distinct elements, there is at least one element such that B lies linearly between it and A.*

If A and B lie in an optical line or an inertia line, one of them must be *after* the other.

If it be the element A which is *after* B, then Theorems 7 and 38 show that there is at least one element of the general line which is *before* B, and so B lies linearly between it and A.

Similarly if A be *before* B there is an element of the general line which is *after* B, and so B is linearly between it and A.

If A and B lie in a separation line, the result follows from Theorem 43.

(8) *If A and D are distinct elements and C is linearly between A and D, and B linearly between A and C, then B is linearly between A and D.*

This is readily seen to be true if we take a set of parallel optical lines

$a$, $b$, $c$ and $d$ through $A$, $B$, $C$ and $D$ respectively in any inertia plane containing the four elements.

Let these optical lines intersect an optical line $f$ of the opposite set in $A'$, $B'$, $C'$ and $D'$ respectively.

Remembering that Post. III must be satisfied, it is clear that we must have either:

(i) *$C'$ after $D'$ and $A'$ after $C'$* together with *$B'$ after $C'$* and *$A'$ after $B'$*;

or (ii) *$C'$ before $D'$ and $A'$ before $C'$* together with *$B'$ before $C'$* and *$A'$ before $B'$*.

In case (i) it follows by Post. III that $B'$ is *after $D'$* and consequently since $B'$ is *before $A'$* we have $B$ linearly between $A$ and $D$.

Similarly in case (ii) we have $B'$ *before $D'$* and *after $A'$*, and therefore again $B$ linearly between $A$ and $D$.

(9) *If $A$ and $D$ are distinct elements and $B$ and $C$ are each linearly between $A$ and $D$, then either $B$ is linearly between $A$ and $C$ or $B$ is identical with $C$ or $B$ is linearly between $C$ and $D$.*

This result may be deduced in a similar manner to the last.

We must have either

(i) *$B'$ after $D'$ and $A'$ after $B'$* together with *$C'$ after $D'$* and *$A'$ after $C'$*;

or (ii) *$B'$ before $D'$ and $A'$ before $B'$* together with *$C'$ before $D'$* and *$A'$ before $C'$*.

Then the elements $B'$ and $C'$ must either be identical or else the one is *after* the other.

In case (i) if $B'$ be *after $C'$*, since also $B'$ is *before $A'$*, we have $B$ linearly between $A$ and $C$.

If $B'$ is identical with $C'$, then $B$ is identical with $C$.

If $C'$ be *after $B'$*, then since also $D'$ is *before $B'$* we have $B$ linearly between $C$ and $D$.

Similarly in case (ii) we must either have $B$ linearly between $A$ and $C$ or $B$ identical with $C$ or $B$ linearly between $C$ and $D$.

(10) *If $A$ and $B$ are distinct elements and if $B$ is linearly between $A$ and $C$ and also linearly between $A$ and $D$, then either $C$ is identical with $D$, or $C$ is linearly between $B$ and $D$, or $D$ is linearly between $B$ and $C$.*

This result may also be deduced in a similar way. We must have either:

(i) *$B'$ after $C'$ and $A'$ after $B'$* together with *$B'$ after $D'$*;

or (ii) *$B'$ before $C'$ and $A'$ before $B'$* together with *$B'$ before $D'$*.

Then the elements $C'$ and $D'$ must either be identical or else the one is *after* the other.

In case (i) if $C'$ is *after* $D'$, then since $C'$ is *before* $B'$ we have $C$ linearly between $B$ and $D$.

If $C'$ is identical with $D'$, then $C$ is identical with $D$.

If $D'$ is *after* $C'$, then since $D'$ is *before* $B'$ we have $D$ linearly between $B$ and $C$.

Similarly in case (ii) we must either have $C$ linearly between $B$ and $D$ or $C$ identical with $D$, or $D$ linearly between $B$ and $C$.

(11) *If $A, B, C, D$ are elements and $B$ is linearly between $A$ and $C$, and $C$ is linearly between $B$ and $D$, then $C$ is linearly between $A$ and $D$.*

This result may also be deduced in a similar way. We must have either:

    (i) *$B'$ after $C'$ and $A'$ after $B'$ together with $C'$ after $D'$*;

or  (ii) *$B'$ before $C'$ and $A'$ before $B'$ together with $C'$ before $D'$.*

In case (i) since $B'$ is *after* $C'$ and $A'$ *after* $B'$, it follows by Post. III that $A'$ is *after* $C'$, and so $C$ must be linearly between $A$ and $D$.

Similarly in case (ii) we must also have $C$ linearly between $A$ and $D$.

Thus all these axioms of Peano hold for the general line.

### THEOREM 71

(a) *If $A_0$ and $A_x$ be two elements in a general line $a$ which lies in the same inertia plane with another general line $b$ which intersects $a$ in the element $C$ such that either $A_0$ is linearly between $C$ and $A_x$, or $A_x$ is linearly between $C$ and $A_0$, and if an optical line through $A_0$ intersects $b$ in $B_0$ so that $B_0$ is after $A_0$, then a parallel optical line through $A_x$ will intersect $b$ in an element which is after $A_x$.*

We have already proved special cases of this in Theorems 66 and 67, and have now to prove the general theorem.

The optical line through $A_x$ parallel to $A_0 B_0$ must intersect $b$ since $b$ intersects $A_0 B_0$ in $B_0$.

Let the element of intersection of this optical line through $A_x$ with $b$ be $B_x$.

Then $B_x$ cannot be identical with $A_x$, for then the general lines $a$ and $b$ would have two distinct elements $C$ and $A_x$ in common and would therefore be identical, which is impossible since $a$ and $b$ intersect by hypothesis.

Further, if $A_x$ were *after* $B_x$ the general lines $a$ and $b$ would intersect in some element between the parallel optical lines (Theorem 68).

That is to say in some element linearly between $A_0$ and $A_x$.

But $a$ and $b$ have only one element $C$ in common, so that if $A_x$ were *after* $B_x$ we should require $C$ to be linearly between $A_0$ and $A_x$, contrary to the hypothesis that either $A_0$ is linearly between $C$ and $A_x$ or $A_x$ is linearly between $C$ and $A_0$.

Thus $B_x$ must be *after* $A_x$.

(b) *If $A_0$ and $A_x$ be two elements in a general line $a$ which lies in the same inertia plane with another general line $b$ which intersects $a$ in the element $C$ such that either $A_0$ is linearly between $C$ and $A_x$ or $A_x$ is linearly between $C$ and $A_0$, and if an optical line through $A_0$ intersects $b$ in $B_0$ so that $B_0$ is before $A_0$, then a parallel optical line through $A_x$ will intersect $b$ in an element which is before $A_x$.*

*Definition.* We shall speak of a general line $l$ as being *co-directional* with a general line $m$ when $l$ is either parallel to $m$ or identical with it.

## Theorem 72

*If three parallel general lines $a$, $b$ and $c$ in one inertia plane $P$ intersect a general line $d_1$ in $A_1$, $B_1$ and $C_1$ respectively and intersect a second general line $d_2$ in $A_2$, $B_2$ and $C_2$ respectively, then if $B_1$ is linearly between $A_1$ and $C_1$ we shall also have $B_2$ linearly between $A_2$ and $C_2$.*

If $a$, $b$ and $c$ be optical lines, then we must either have $b$ an after-parallel of $c$ and a before-parallel of $a$, or else have $b$ an after-parallel of $a$ and a before-parallel of $c$.

In either case $B_2$ will be *after* an element of one of the pair of parallel optical lines $a$ and $c$ and *before* an element of the other.

Thus, as $B_2$ cannot lie in either $a$ or $c$, and as these optical lines pass through $A_2$ and $C_2$ respectively, it follows that $B_2$ is linearly between $A_2$ and $C_2$.

Next consider the cases where $a$, $b$ and $c$ are separation or inertia lines: the methods of proof being similar in the two cases.

Let parallel optical lines in $P$ and passing through $A_1$ and $C_1$ intersect $b$ in $B_1'$ and $B_1''$ respectively.

Let optical lines co-directional with these and passing through $A_2$ and $C_2$ intersect $b$ in $B_2'$ and $B_2''$ respectively.

Now as $B_1$ is supposed to be linearly between $A_1$ and $C_1$ it must lie between the parallel optical lines $A_1 B_1'$ and $B_1'' C_1$.

It follows therefore, from Theorem 69, that if $A_1$ is *after $B_1'$* we shall also have $B_1''$ *after $C_1$*.

If, on the other hand, $B_1'$ is *after $A_1$* we shall also have $C_1$ *after $B_1''$*.

In the first of these cases, that is to say when $A_1$ is *after $B_1'$* and $B_1''$ *after $C_1$*, by Theorem 56 or Theorem 57 (according as $a$, $b$ and $c$ are separation or inertia lines), it follows that $A_2$ is *after $B_2'$* and $B_2''$ is *after $C_2$*.

If $A_2B_2'$ is distinct from $B_2''C_2$ it follows, by Theorem 68, that $B_2$ must lie between the parallel optical lines $A_2B_2'$ and $B_2''C_2$; so that $B_2$ is linearly between $A_2$ and $C_2$.

If $A_2B_2'$ and $B_2''C_2$ are not distinct optical lines, then $B_2'$ and $B_2''$ will both coincide with $B$, which will be *after $C_2$* and *before $A_2$*; so that $B_2$ will still be linearly between $A_2$ and $C_2$.

The same result follows in a similar way in the case where $B_1'$ is *after $A_1$* and $C_1$ *after $B_1''$*.

Thus the theorem holds in all cases.

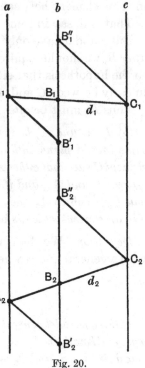

Fig. 20.

## THEOREM 73

(a) *If an element B be linearly between two elements A and C and if another element D be* before *both A and C but not in the general line AC, then DB is an inertia line and B is* after *D.*

Consider first the case where $AC$ is a separation line.

Let a general line through $B$ parallel to $CD$ intersect $AD$ in $E$.

Then since $B$ is linearly between $A$ and $C$ we must have $E$ linearly between $D$ and $A$.

Thus since $D$ is *before $A$* it follows that $E$ is *after $D$* and *before $A$*.

But $EB$ must be an inertia line or an optical line, according as $DC$ is an inertia line or an optical line, and so $B$ must be either *before* or *after $E$*.

But $B$ cannot be *before $E$* for then, since $E$ is *before $A$*, we should have $B$ *before $A$*, contrary to the hypothesis that $A$ and $B$ lie in a separation line.

Thus $B$ must be *after* $E$ and, since $E$ is *after* $D$, it follows that $B$ is *after* $D$.

Thus $DB$ is either an optical or an inertia line.

But if $DB$ were an optical line, then since $E$ is *after* $D$ and *before* $B$ it would follow that $E$ must lie in $DB$, which is impossible since $BE$ is parallel to $CD$.

Thus $DB$ must be an inertia line.

Next consider the case where $AC$ is an optical or inertia line.

We then must have either $C$ *after* $A$ or $A$ *after* $C$ and it is sufficient to consider the case where $C$ is *after* $A$.

Then $B$ must be *after* $A$ and *before* $C$.

But $A$ is *after* $D$ and so $B$ must be *after* $D$.

Thus again $DB$ must be either an optical line or an inertia line.

If $DB$ were an optical line, then since $A$ is *after* $D$ and *before* $B$ the element $A$ would have to lie in $DB$ and so $D$, $A$ and $C$ would all lie in one general line, contrary to hypothesis.

Thus again $DB$ must be an inertia line, and so the theorem is proved.

(b) *If an element B be linearly between two elements A and C and if another element D be after both A and C but not in the general line AC, then DB is an inertia line and B is before D.*

### REMARKS

A somewhat analogous result is the following:

*If an element B be after an element A and before an element C, and if D be another distinct element such that DA and DC are both separation lines, then DB is also a separation line.*

This may be proved as follows:

The element $B$ cannot be *before* $D$; for, since $A$ is *before* $B$, we should have $A$ *before* $D$, contrary to the hypothesis that $DA$ is a separation line.

Similarly $B$ cannot be *after* $D$; for, since $C$ is *after* $B$, we should have $C$ *after* $D$, contrary to the hypothesis that $DC$ is a separation line.

Thus $B$ is neither *before* nor *after* $D$, so that $DB$ must be a separation line, as was to be proved.

The following two theorems are special cases of Theorems 76 and 77, but as the proofs of the general theorems are reduced to depend on these special cases, the latter are treated separately.

## Theorem 74

*If A, B and C be three distinct elements in an inertia plane such that AB and AC are distinct optical lines, and if D be an element linearly between A and B while E is an element linearly between A and C, then there exists an element which lies both linearly between C and D and also linearly between B and E.*

It will be sufficient to consider the case where $A$ is *after* $B$, since the case where $A$ is *before* $B$ may be treated in an analogous manner.

Since $E$ is linearly between $A$ and $C$, therefore $E$ is between the optical line $AB$ and a parallel optical line through $C$.

Thus, $A$ being supposed *after* $B$, this optical line through $C$ will intersect the general line $BE$ in some element, say $G$, such that $G$ is *after* $C$ (Theorem 69).

But since $D$ is linearly between $A$ and $B$ we must, in these circumstances, have $D$ after $B$.

Thus, since $G$ is *after* $C$, it follows by Theorem 68 that the general lines $BG$ and $DC$ intersect in some element, say $F$, which is between the parallel optical lines $DB$ and $CG$.

That is, $F$ is linearly between $C$ and $D$, and is the element of intersection of $CD$ and $BE$.

By taking an optical line through $B$ parallel to $AC$ we may prove in an analogous manner that $F$ is linearly between $B$ and $E$.

## Theorem 75

*If A, B and C be three distinct elements in an inertia plane such that AB and AC are distinct optical lines and if D be an element linearly between A and B while F is an element linearly between C and D, then there exists an element, say E, which lies linearly between A and C and such that F lies linearly between B and E.*

As in the previous theorem, it will be sufficient to consider only the case where $A$ is *after* $B$.

Under these circumstances we should have $D$ after $B$ and so, since $F$ is linearly between $C$ and $D$, we should have $F$ between the optical line $AB$ and a parallel optical line through $C$.

Thus, by Theorem 69, this optical line through $C$ will intersect $BF$ in some element, say $G$, such that $G$ is *after* $C$.

But, since $A$ is *after* $B$ and $G$ is *after* $C$, therefore, by Theorem 68, the general lines $AC$ and $BG$ intersect in some element, say $E$, such that $E$ is between the parallel optical lines $AB$ and $CG$.

Thus $E$ is linearly between $A$ and $C$, where $E$ is the element of intersection of $AC$ and $BF$.

It then follows, as in the last theorem, that $F$ lies linearly between $B$ and $E$.

THEOREM 76

*If $A$, $B$ and $C$ be three elements in an inertia plane which do not all lie in one general line and if $D$ be an element linearly between $A$ and $B$, while $E$ is an element linearly between $A$ and $C$, there exists an element which lies both linearly between $B$ and $E$ and linearly between $C$ and $D$.*

Let $V$ be the inertia plane containing $A$, $B$ and $C$ and let $a$ be any inertia line through $A$ which does not lie in $V$.

Let $b$ and $c$ be inertia lines parallel to $a$ and passing through $B$ and $C$ respectively.

Then $b$ and $c$ lie in one inertia plane, say $P_{bc}$, $c$ and $a$ in a second inertia plane, say $P_{ca}$, and $a$ and $b$ in a third inertia plane, say $P_{ab}$.

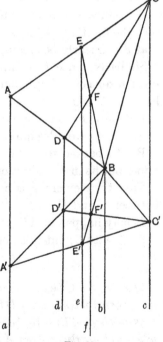

Fig. 21.

Let one of the optical lines through $B$ in the inertia plane $P_{ab}$ intersect $a$ in $A'$ and let one of the optical lines through $A'$ in the inertia plane $P_{ca}$ intersect $c$ in $C'$.

Then $A'B$ and $A'C'$ may be taken as generators of opposite sets of an inertia plane, say $S$, containing $B$, $C'$ and $A'$.

Let $d$ be the inertia line through $D$ parallel to $a$ and let $e$ be the inertia line through $E$ parallel to $a$.

Then $d$ will lie in $P_{ab}$ and, since $D$ is linearly between $A$ and $B$, it follows by Theorem 72 that $d$ must intersect $A'B$ in some element, say $D'$, such that $D'$ is linearly between $A'$ and $B$.

Similarly $e$ will lie in $P_{ca}$ and, since $E$ is linearly between $A$ and $C$, it follows that $e$ will intersect $A'C'$ in some element, say $E'$, such that $E'$ is linearly between $A'$ and $C'$.

But, since $A'B$ and $A'C'$ are two distinct optical lines in the inertia plane $S$, it follows by Theorem 74 that there exists an element, say

R

$F'$, which lies both linearly between $B$ and $E'$ and linearly between $C'$ and $D'$.

Now, since $b$ and $c$ are parallel inertia lines lying in the inertia plane $P_{bc}$, it follows that there is an inertia plane, say $P_{bf}$, containing $b$ and the element $F'$, and an inertia plane, say $P_{cf}$, containing $c$ and the element $F'$ and these inertia planes must, by Theorem 51, have a general line, say $f$, in common.

Further, $f$ must be parallel to $b$ and $c$ and must therefore be an inertia line.

But the inertia plane $P_{bf}$ contains the general line $BF'$ and must therefore contain $E'$ and the inertia line $e$ which passes through $E'$ and is parallel to $b$.

Thus $P_{bf}$ contains the element $E$ and therefore contains the general line $BE$.

Similarly $P_{cf}$ contains the general line $CD$.

But, since $F'$ is linearly between $B$ and $E'$, it follows by Theorem 72 that the inertia line $f$ must intersect $BE$ in some element, say $F$, such that $F$ is linearly between $B$ and $E$.

Similarly, since $F'$ is linearly between $C'$ and $D'$, it follows that $f$ must intersect $CD$ in some element, say $\bar{F}$, such that $\bar{F}$ is linearly between $C$ and $D$.

But both $F$ and $\bar{F}$ must lie in $V$ and so, if they were distinct, the inertia line $f$ would require to lie in $V$.

But $f$ is parallel to $a$, of which only one element lies in $V$ and therefore $f$ does not lie in $V$ and accordingly $\bar{F}$ must be identical with $F$.

Thus the element $F$ is both linearly between $B$ and $E$ and linearly between $C$ and $D$.

It may happen in this and the next theorem that $A'$ coincides with $A$ or $C'$ with $C$, but this does not affect the validity.

### THEOREM 77

*If $A$, $B$ and $C$ be three elements in an inertia plane which do not all lie in one general line and if $D$ be an element linearly between $A$ and $B$ while $F$ is an element linearly between $C$ and $D$, there exists an element, say $E$, which is linearly between $A$ and $C$ and such that $F$ is linearly between $B$ and $E$.*

Let $V$ be the inertia plane containing $A$, $B$ and $C$ and let $a$ be any inertia line through $A$ which does not lie in $V$, while $b$ and $c$ are inertia lines parallel to $a$ through $B$ and $C$ respectively.

Let $P_{bc}$, $P_{ca}$, $P_{ab}$, $A'$, $C'$, $S$, $d$, $D'$, have the same significance as in the last theorem, and let $P_{cd}$ be the inertia plane containing the parallel inertia lines $c$ and $d$.

Let $f$ be an inertia line through $F$ parallel to $c$ and $d$ and which will also lie in $P_{cd}$.

Since $F$ is linearly between $C$ and $D$ it follows, by Theorem 72, that $f$ will intersect $C'D'$ in some element, say $F'$, such that $F'$ is linearly between $C'$ and $D'$.

But, as in the last theorem, $D'$ is linearly between $A'$ and $B$ and so, since $A'B$ and $A'C'$ are two distinct optical lines in the inertia plane $S$, it follows, by Theorem 75, that there exists an element, say $E'$, which lies linearly between $A'$ and $C'$ and such that $F'$ lies linearly between $B$ and $E'$.

If now we denote the inertia plane containing $b$ and $f$ by $P_{bf}$, then $P_{bf}$ contains the element $E'$ in common with the inertia plane $P_{ca}$.

But, since $b$ lies in $P_{bc}$ and $P_{bf}$ while the parallel inertia line $c$ lies in $P_c$ and $P_{ca}$, it follows, by Theorem 51, that $P_{bf}$ and $P_{ca}$ have a general line, say $e$, in common which passes through $E'$ and is parallel to $b$ and $c$ and is therefore an inertia line.

Now since $a$ is also parallel to $e$ and lies in the same inertia plane $P_{ca}$ with it and, since $E'$ is linearly between $A'$ and $C'$, it follows, by Theorem 72, that $e$ must intersect $AC$ in some element, say $E$, such that $E$ is linearly between $A$ and $C$.

Again, since $b, f$ and $e$ all lie in the inertia plane $P_{bf}$ and, since $F'$ is linearly between $B$ and $E'$, it follows, by Theorem 72, that $BF$ must intersect $e$ in some element $\bar{E}$ such that $F$ is linearly between $B$ and $\bar{E}$.

But both $E$ and $\bar{E}$ must lie in $V$ and so, if they were distinct, the inertia line $e$ would require to lie in $V$.

But $e$ is parallel to $a$, of which only one element lies in $V$ and therefore $e$ does not lie in $V$ and accordingly $\bar{E}$ must be identical with $E$.

Thus the element $E$ is linearly between $A$ and $C$ and is such that $F$ is linearly between $B$ and $E$.

## REMARKS

Peano has given the following three axioms of the plane:

(12) If $r$ is a straight line, there exists a point which does not lie on $r$.

(13) If $A, B, C$ are three non-collinear points and $D$ lies on the segment $BC$, and $E$ on the segment $AD$, there exists a point $F$ on both the segment $AC$ and the prolongation $B'E$.

(14) If $A, B, C$ are three non-collinear points and $D$ lies on the

segment $BC$ and $F$ on the segment $AC$, there exists a point $E$ lying on both the segments $AD$ and $BF$.

Now since there is always an element outside any general line it follows that the analogue of Peano's axiom (12) holds for our geometry.

Further, provided the three elements $A$, $B$, $C$ lie in an inertia plane, Theorem 76 corresponds to Peano's axiom (14) while Theorem 77 corresponds to his axiom (13).

Also Theorem 47 corresponds to the axiom of parallels in Euclidean geometry so far as an inertia plane is concerned.

An inertia plane however differs from a Euclidean plane, since there are three types of general line in the former and only one type of straight line in the latter.

Further, although closed figures exist in an inertia plane, there is no closed figure which corresponds to a circle.

How this comes about will be seen hereafter.

### Theorem 78

*If $A$, $B$ and $C$ be three elements in an inertia plane $P$ which do not all lie in one general line and if $D$ be the mean of $A$ and $B$, then a general line through $D$ parallel to $BC$ intersects $AC$ in an element which is the mean of $A$ and $C$.*

The general line $BC$, which we shall denote by $a$, may be either: (i) an optical line, (ii) a separation line, or (iii) an inertia line.

As regards case (i): let $a'$ be an optical line through $A$ parallel to $a$, and let $AK$ be the second optical line which passes through $A$ and lies in the inertia plane $P$, and let it intersect $a$ in the element $K$.

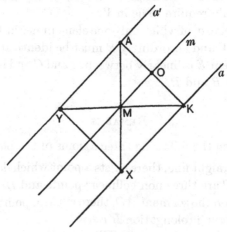

Fig. 22.

Let $X$ be any other element in $a$ and let the optical line through $X$ parallel to $KA$ intersect $a'$ in $Y$. Then $A, K, X, Y$ are the corners of an optical parallelogram whose diagonal lines $AX$ and $YK$ intersect in some element, say $M$, which is the mean both of $A$ and $X$ and of $Y$ and $K$.

An optical line $m$ through $M$ parallel to $a$ will intersect $AK$ in an element (say $O$) which is the mean of $A$ and $K$.

The position of $O$ is independent of the position of $X$ in the optical line $a$, so that if $B$ and $C$ be any two elements in $a$, and $D$ be the mean of $A$ and $B$ the general line through $D$ parallel to $BC$ will be identical with $m$ and will intersect $AC$ in an element which is the mean of $A$ and $C$.

Case (ii) may be proved as follows:

Let any inertia line in $P$ which passes through $A$ intersect $a$ in some element $K$, and let $X$ be any other element in $a$.

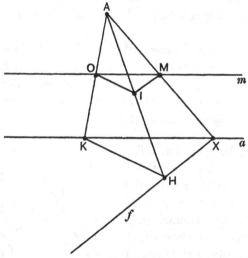

Fig. 23.

The element $A$ will be either *before* or *after* $K$, but the method of proof is analogous in both cases and so we shall consider the one where $A$ is *after* $K$.

Now, since $a$ is a separation line, $K$ will be neither *before* nor *after* $X$.

If then $Q$ be any inertia plane containing $a$ but distinct from $P$, the element $K$ will be *before* an element of one of the optical lines in $Q$ which pass through $X$ and *after* an element of the other.

Let $f$ be the optical line through $X$ in the inertia plane $Q$ such that

$K$ is *after* an element of it and let the optical line through $K$ in $Q$ of the opposite set to $f$ intersect $f$ in $H$. Then $K$ is *after* $H$.

But since $A$ is supposed to be *after* $K$ we shall have $A$ *after* $H$ and $AH$ will be an inertia line.

Thus $A$, $H$ and $X$ will lie in an inertia plane and $A$, $H$ and $K$ will lie in another inertia plane.

Let $O$ be the mean of $A$ and $K$.

Then, by case (i), an optical line through $O$ parallel to $KH$ will intersect $AH$ in some element $I$ such that $I$ is the mean of $A$ and $H$.

Also, by case (i), an optical line through $I$ parallel to $HX$ will intersect $AX$ in some element $M$ which will be the mean of $A$ and $X$.

But now $I$ does not lie in $Q$ and so $O$, $I$ and $M$ lie in an inertia plane, say $Q'$, which will be parallel to $Q$.

Thus, since $P$ has the separation line $a$ in common with $Q$ and has the general line $OM$ (which we shall denote by $m$) in common with the parallel inertia plane $Q'$, it follows that $m$ is a separation line parallel to $a$ and intersecting $AX$ in an element $M$ which is the mean of $A$ and $X$.

For all positions of $X$ the mean of $A$ and $X$ lies on the separation line $m$ which passes through $O$ which is the mean of $A$ and $K$.

Thus if $B$ and $C$ be any two elements in the separation line $a$ and $D$ be the mean of $A$ and $B$, then $D$ will lie in the separation line $m$, which will also pass through the mean of $A$ and $C$ and is parallel to $BC$.

Case (iii) may be proved as follows:

Let any inertia line in $P$ which passes through $A$ and is not parallel to $a$ intersect $a$ in some element $K$ and let $X$ be any other element in $a$.

Let $R$ be any inertia plane distinct from $P$ but containing the general line $AX$.

Now there are two optical lines in $R$ which pass through $X$ and one at least of these optical lines must be distinct from $AX$. Let $l$ be such an optical line.

Now, since $a$ is an inertia line, it follows that $a$ and $l$ lie in an inertia plane which we shall denote by $Q$.

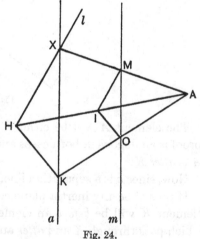

Fig. 24.

Let the optical line through $K$ in the inertia plane $Q$ of the opposite set to $l$ intersect $l$ in the element $H$.

Then, since $AK$ is an inertia line, it follows that the three elements $H$, $A$ and $K$ lie in an inertia plane.

Let $O$ be the mean of $A$ and $K$.

Then, by case (i), an optical line through $O$ parallel to $KH$ will intersect $AH$ in some element $I$ such that $I$ is the mean of $A$ and $H$.

Also, by case (i), an optical line through $I$ parallel to $HX$ will intersect $AX$ in some element $M$ which will be the mean of $A$ and $X$.

But now $I$ does not lie in $Q$ and so $O$, $I$ and $M$ lie in an inertia plane, say $Q'$, which will be parallel to $Q$.

Thus, since $P$ has the inertia line $a$ in common with $Q$ and has the general line $OM$ (which we shall denote by $m$) in common with the parallel inertia plane $Q'$, it follows that $m$ is an inertia line parallel to $a$ and intersecting $AX$ in an element $M$ which is the mean of $A$ and $X$.

For all positions of $X$ the mean of $A$ and $X$ lies on the inertia line $m$ which passes through $O$ which is the mean of $A$ and $K$.

If then $B$ and $C$ be any two elements in the inertia line $a$ and $D$ be the mean of $A$ and $B$, it will lie in the inertia line $m$, which will also pass through the mean of $A$ and $C$ and is parallel to $BC$.

Thus in all cases the theorem holds.

Since there is only one general line through $D$ parallel to $BC$ and this must pass through the mean of $A$ and $C$, it follows directly that, if $E$ be the mean of $A$ and $C$, then the general line $DE$ is parallel to $BC$.

*Definition.* If two parallel general lines in an inertia plane be both intersected by another pair of parallel general lines, then the four general lines will be said to form a *general parallelogram in the inertia plane*.

It will be seen hereafter that it is necessary to extend the meaning of the phrase *general parallelogram* to the case of figures which do not lie in an inertia plane and so the words "*in an inertia plane*" are important.

The general lines which form a general parallelogram in an inertia plane will be called the *side lines* of the general parallelogram.

A pair of parallel side lines will be said to be *opposite*.

The elements of intersection of pairs of side lines which are not parallel will be called the *corners* of the general parallelogram.

A pair of corners which do not lie in the same side line will be said to be *opposite*.

A general line passing through a pair of opposite corners will be called a *diagonal line* of the general parallelogram.

It is clear that a general parallelogram in an inertia plane has two diagonal lines.

Further, it is clear that an optical parallelogram is a particular case of a general parallelogram in an inertia plane.

### THEOREM 79

*If we have a general parallelogram in an inertia plane, then:*

(1) *The two diagonal lines intersect in an element which is the mean of either pair of opposite corners.*

(2) *A general line through the element of intersection of the diagonal lines and parallel to either pair of side lines, intersects either of the other side lines in an element which is the mean of the pair of corners through which that side line passes.*

Let $A$, $B$, $C$, $D$ be the corners of the general parallelogram:

$A$ and $C$ being one pair of opposite corners and $B$ and $D$ the other pair.

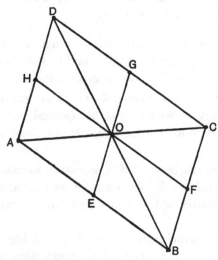

Fig. 25.

Let $E$ be the mean of $A$ and $B$,

,, $F$ ,, $B$ ,, $C$,

,, $G$ ,, $C$ ,, $D$,

,, $H$ ,, $D$ ,, $A$.

If a general line be taken through $E$ parallel to $BC$ and $AD$, then by Theorem 78 it will intersect $AC$ in an element which is the mean of $A$ and $C$ and therefore will intersect $CD$ in an element which is the mean of $C$ and $D$. That is in the element $G$.

Similarly the general line $FH$ will pass through the mean of $A$ and $C$.

Thus the element of intersection of $EG$ and $FH$ (which we shall call $O$) is the mean of $A$ and $C$. Similarly $O$ is the mean of $B$ and $D$. Thus the mean of $A$ and $C$ is identical with the mean of $B$ and $D$; or the diagonal lines intersect in an element which is the mean of either pair of opposite corners. The second part of the theorem also holds.

## Theorem 80

*If $A$, $B$, $C$, $D$ be the corners of a general parallelogram in an inertia plane; $AB$ and $DC$ being one pair of parallel side lines and $BC$ and $AD$ the other pair of parallel side lines, then if $E$ be the mean of $A$ and $B$ while $F$ is the mean of $D$ and $C$, the general lines $AF$ and $EC$ are parallel to one another.*

Since the general line $AF$ is not parallel to $BC$, it must intersect $BC$ in some element, say $G$.

Now by Theorem 79 a general line through the intersection of the diagonal lines and parallel to $BC$ will intersect $AB$ in the mean of $A$ and $B$, and will intersect $DC$ in the mean of $D$ and $C$.

Thus the general line $EF$ is parallel to $BC$.

But since $A$, $B$ and $G$ are three elements in an inertia plane which do not all lie in one general line and since $E$ is the mean of $A$ and $B$ while $EF$ is parallel to $BG$, it follows by Theorem 78 that $F$ is the mean of $A$ and $G$.

Similarly since $FC$ is parallel to $AB$ it follows that $C$ is the mean of $G$ and $B$.

But since $E$ is the mean of $B$ and $A$ while $C$ is the mean of $B$ and $G$, it follows by Theorem 78 that $EC$ is parallel to $AG$: that is, $EC$ is parallel to $AF$, as was to be proved.

## Theorem 81

*If three parallel general lines $a$, $b$ and $c$ in one inertia plane intersect a general line $d_1$ in $A_1$, $B_1$ and $C_1$ respectively and intersect a second general line $d_2$ in $A_2$, $B_2$ and $C_2$ respectively, and if $B_1$ be the mean of $A_1$ and $C_1$, then $B_2$ will be the mean of $A_2$ and $C_2$.*

If $A_2$ should happen to coincide with $A_1$, or if $C_2$ should happen to coincide with $C_1$, the result follows directly from Theorem 78.

If $d_2$ should happen to be parallel to $d_1$, then the result follows from Theorem 79 (2).

In any other case let a general line through $A_1$ parallel to $d_2$ intersect $b$ in $B$ and $c$ in $C$.

Then, by Theorem 78, $B$ is the mean of $A_1$ and $C$ and so, by Theorem 79 (2), $B_2$ will be the mean of $A_2$ and $C_2$.

## REMARKS

If $A_0$ and $A_n$ be two distinct elements in a general line $a$, we can always find $n-1$ elements $A_1$, $A_2$, ... $A_{n-1}$ in $a$ (where $n-1$ is any integer) such that:

$$A_1 \text{ is the mean of } A_0 \text{ and } A_2,$$
$$A_2 \text{ is the mean of } A_1 \text{ and } A_3,$$
$$\dots\dots\dots\dots\dots\dots\dots\dots\dots\dots\dots$$
$$\dots\dots\dots\dots\dots\dots\dots\dots\dots\dots$$
$$A_{n-1} \text{ is the mean of } A_{n-2} \text{ and } A_n.$$

For let $P$ be any inertia plane containing $a$ and let $b$ be any general line in $P$ which passes through $A_0$ and is distinct from $a$.

Let $A_1'$ be any element in $b$ distinct from $A_0$ and let $A_2'$, $A_3'$, ... $A'_{n-1}$, $A_n'$ be other elements in $b$ such that:

$$A_1' \text{ is the mean of } A_0 \text{ and } A_2',$$
$$A_2' \text{ is the mean of } A_1' \text{ and } A_3',$$
$$\dots\dots\dots\dots\dots\dots\dots\dots\dots\dots\dots$$
$$\dots\dots\dots\dots\dots\dots\dots\dots\dots\dots$$
$$A'_{n-1} \text{ is the mean of } A'_{n-2} \text{ and } A_n'.$$

Let general lines through $A_1'$, $A_2'$, ... $A'_{n-1}$ parallel to $A_n'A_n$ intersect $a$ in the elements $A_1$, $A_2$, ... $A_{n-1}$.

Then, by Theorem 81, it follows that:

$$A_1 \text{ is the mean of } A_0 \text{ and } A_2,$$
$$A_2 \text{ is the mean of } A_1 \text{ and } A_3,$$
$$\dots\dots\dots\dots\dots\dots\dots\dots\dots\dots\dots$$
$$\dots\dots\dots\dots\dots\dots\dots\dots\dots\dots$$
$$A_{n-1} \text{ is the mean of } A_{n-2} \text{ and } A_n,$$

and so the $n-1$ elements $A_1$, $A_2$, ... $A_{n-1}$ can be found as stated.

## Theorem 82

(a) *If A be any element in an optical line a and A' be any element in a neutral-parallel optical line a', then, if B be any element in a which is after A, the general line through B parallel to AA' intersects a' in an element which is after A'.*

Since $A$ and $A'$ lie in the neutral-parallel optical lines $a$ and $a'$ respectively, it follows that $A$ is neither *before* nor *after* $A'$ and so there is at least one element which is common to the $\alpha$ sub-sets of $A$ and $A'$.

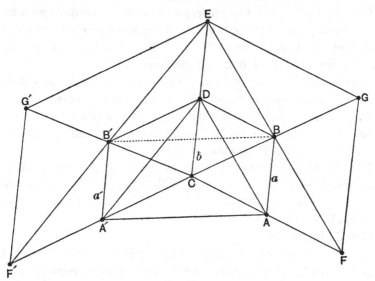

Fig. 26.

Let $C$ be such an element and let $b$ be the optical line through $C$ parallel to $a$ or $a'$.

Then since $C$ is *after* both $A$ and $A'$, it follows that $b$ is an after-parallel of both $a$ and $a'$ and accordingly $b$ and $a$ lie in one inertia plane while $b$ and $a'$ lie in another.

Let the optical line through $B$ parallel to $AC$ intersect $b$ in the element $D$ and let the optical line through $D$ parallel to $CA'$ intersect $a'$ in the element $B'$.

Then $A, C, D, B$ form the corners of an optical parallelogram in an inertia plane which we shall call $P$, while $A', C, D, B'$ form the corners of another optical parallelogram in another inertia plane which we shall call $P'$.

Now, since $B$ is *after* $A$ and $C$ is also *after* $A$, while $AC$ and $AB$ are both optical lines, it follows that the diagonal line $CB$ is a separation line and accordingly the diagonal line $AD$ is an inertia line having $D$ *after* $A$.

Further, $D$ must be *after* $C$ and, since $C$ is *after* $A'$, it follows that the diagonal line $A'D$ is an inertia line having $D$ *after* $A'$, and accordingly the diagonal line $CB'$ is a separation line.

Thus, since $C$ is *after* $A'$, we must also have $B'$ *after* $A'$.

Let the general line through $B$ parallel to $AD$ intersect $b$ in $E$ and $CA$ in $F$, and let the optical lines through $E$ and $F$ respectively parallel to $CF$ and $CE$ intersect one another in $G$.

Then $F$, $C$, $E$, $G$ are the corners of an optical parallelogram in the same inertia plane as the optical parallelogram whose corners are $A$, $C$, $D$, $B$ and the diagonal lines $FE$ and $AD$ do not intersect and so the diagonal lines $CG$ and $CB$ do not intersect.

Thus $B$ must lie in $CG$ and since it also lies in $FE$ it follows that $B$ is the centre of the optical parallelogram whose corners are $F$, $C$, $E$, $G$.

Now let the general line through $E$ parallel to $DA'$ intersect $CA'$ in $F''$ and let the optical lines through $E$ and $F''$ respectively parallel to $CF''$ and $CE$ intersect one another in $G'$.

Then $F'$, $C$, $E$, $G'$ are the corners of an optical parallelogram in the same inertia plane as the optical parallelogram whose corners are $A'$, $C$, $D$, $B'$ and the diagonal lines $F'E$ and $A'D$ do not intersect and so the diagonal lines $CG'$ and $CB'$ do not intersect.

Thus $B'$ lies in $CG'$.

But the optical parallelograms whose corners are $F$, $C$, $E$, $G$ and $F'$, $C$, $E$, $G'$ have the pair of adjacent corners $C$ and $E$ in common and the optical line $BD$ through the centre of the first of these intersects $CE$ in $D$, and so it follows by Theorem 61 that the centre of the second optical parallelogram lies in the optical line through $D$ parallel to $CF'$ and $EG'$.

Thus the centre of the optical parallelogram whose corners are $F'$, $C$, $E$, $G'$ lies in $DB'$.

But this centre also lies in $CG'$ and therefore it must be the element $B'$.

Thus $B'$ must lie in $F'E$.

But we saw that $AD$ and $A'D$ were both inertia lines and so they lie in an inertia plane, say $Q_1$, while $BE$ and $B'E$ which are respectively parallel to these must lie in a parallel inertia plane, say $Q_2$.

Further, $AC$ and $A'C$ are both optical lines and so they lie in an inertia plane, say $R_1$, while $BD$ and $B'D$ which are respectively parallel to these must lie in a parallel inertia plane, say $R_2$.

But the general lines $AA'$ and $BB'$ lie in the parallel inertia planes

$Q_1$ and $Q_2$ respectively and also in the parallel inertia planes $R_1$ and $R_2$ respectively, and since these inertia planes are distinct it follows that $BB'$ is parallel to $AA'$.

Thus the parallel to $AA'$ through $B$ intersects $a'$ in the element $B'$ which is *after $A'$.*

(b) *If $A$ be any element in an optical line $a$ and $A'$ be any element in a neutral-parallel optical line $a'$, then if $B$ be any element in $a$ which is* before *$A$, the general line through $B$ parallel to $AA'$ intersects $a'$ in an element which is* before *$A'$.*

## Theorem 83

*If $A$ and $B$ be two elements lying respectively in the two neutral-parallel optical lines $a$ and $b$, and if $A'$ be a second and distinct element in $a$, there is only one general line through $A'$ and intersecting $b$ which does not intersect the general line $AB$.*

We have seen by Theorem 82 ($a$ and $b$) that the general line through $A'$ parallel to $AB$ must intersect $b$.

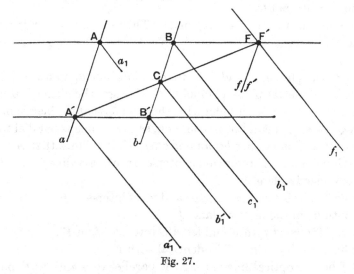

Fig. 27.

Let $B'$ be the element of intersection.

Then the general lines $AB$ and $A'B'$, being parallel, cannot intersect.

Let any other general line through $A'$ and intersecting $b$ intersect it in the element $C$.

Then if $C$ should coincide with $B$ the general lines $A'C$ and $AB$ have the element $B$ in common and therefore intersect.

Suppose next that $C$ does not coincide with $B$.

Let $P_1$ be any inertia plane containing $a$ and let $P_2$ be the parallel inertia plane containing $b$.

Let $a_1$ be any inertia line through $A$ in the inertia plane $P_1$ and let $Q$ be the inertia plane containing $a_1$ and the element $B$.

Then $Q$ must contain a general line, say $b_1$, in common with $P_2$ and the general lines $a_1$ and $b_1$ must be parallel.

Again let $a_1'$ be a general line through $A'$ parallel to $a_1$.

Then $a_1'$ must lie in the inertia plane $P_1$ and must be an inertia line.

Thus the general line $a_1'$ and the element $B'$ must lie in an inertia plane, say $Q'$, and since $a_1'$ is parallel to $a_1$ and $A'B'$ is parallel to $AB$, it follows by Theorem 52 that $Q'$ is parallel to $Q$.

But the inertia plane $Q'$ contains the general line $a_1'$ in $P_1$ and the element $B'$ in $P_2$ and therefore since $P_1$ and $P_2$ are parallel it follows that $Q'$ and $P_2$ contain a general line, say $b_1'$, in common, which will be parallel to $a_1'$.

Again, since $a_1'$ is an inertia line, there is an inertia plane containing $a_1'$ and the element $C$.

If we call this inertia plane $R$, then by Theorem 51 the inertia planes $P_2$ and $R$ have a general line, say $c_1$, in common and $c_1$ is parallel to $a_1'$ and $b_1'$.

Thus since $c_1$ lies in $P_2$ and $R$, $b_1'$ in $Q'$ and $P_2$, and $a_1'$ in $R$ and $Q'$, and since $Q$ is an inertia plane parallel to $Q'$ through the element $B$ of $P_2$ which does not lie in $b_1'$, it follows by Theorem 53 that the inertia planes $R$ and $Q$ have a general line in common, say $f_1$, which is parallel to $a_1'$.

Now since $C$ is neither *before* nor *after* $A'$, it follows that $A'C$ is a separation line and therefore must intersect the inertia line $f_1$ since both lie in one inertia plane $R$.

Similarly $AB$ is a separation line and must intersect the inertia line $f_1$ since both lie in the inertia plane $Q$.

Let $AB$ intersect $f_1$ in $F$ and let $A'C$ intersect $f_1$ in $F'$.

We have to show that $F'$ is identical with $F$.

Let $f$ be the optical line through $F$ parallel to $a$ and let $f'$ be the optical line through $F'$ parallel to $a$.

Then since $B$ is neither *before* nor *after* any element of $a$, it follows by Theorem 45 that no element of the general line $AB$ with the exception of $A$ is either *before* or *after* any element of $a$; and similarly no element of the general line $A'C$ with the exception of $A'$ is either *before* or *after* any element of $a$.

But $F$ cannot be identical with $A$, for this would require $C$ to lie in

$P_1$, which is impossible, and $F'$ cannot be identical with $A'$ since $F'$ and $A'$ lie in parallel inertia planes $Q$ and $Q'$.

Thus $F$ is neither *before* nor *after* any element of $a$ and $F'$ is neither *before* nor *after* any element of $a$.

It follows that $f$ is a neutral-parallel of $a$ and also $f'$ is a neutral-parallel of $a$.

Suppose now, if possible, that $F'$ is distinct from $F$; then since $F$ and $F'$ lie in the inertia line $f_1$, it would follow that the one was *after* the other.

Also if they were distinct, since they both lie in the same inertia line they could not also lie in one optical line and so the optical lines $f$ and $f'$ would be distinct and the one would be an after-parallel of the other.

But we have seen that $f$ and $f'$ are each neutral-parallels of $a$ and so it would follow by Theorem 28 that they were neutrally parallel to one another.

But one optical line cannot be both a neutral-parallel and an after-parallel of another optical line and so the supposition that $F'$ is distinct from $F$ leads to a contradiction and therefore is not true.

Thus $F'$ is identical with $F$ and therefore the general line $A'C$ intersects the general line $AB$.

Thus there is no general line through $A'$ and intersecting $b$ which does not also intersect $AB$, except the parallel general line $A'B'$.

### THEOREM 84

*If $a$ and $b$ be two neutral-parallel optical lines and if one general line intersects $a$ in $A$ and $b$ in $B$, while a second general line intersects $a$ in $A'$ and $b$ in $B'$, then an optical line through any element of $AB$ and parallel to $a$ or $b$ intersects $A'B'$.*

Let $D$ be any element of $AB$ and let $d$ be an optical line through $D$ parallel to $a$ or $b$.

We have to show that $d$ intersects $A'B'$.

If $D$ should coincide with either $A$ or $B$, no proof is required.

If $A'B'$ be parallel to $AB$, then the result follows directly by Theorem 82 ($a$ and $b$).

If $A'B'$ be not parallel to $AB$, then by Theorem 83 the general lines $AB$ and $A'B'$ must intersect in some element, say $C$.

Now, the general lines $AB$ and $A'B'$ being supposed distinct, $C$ must be distinct from at least one of the elements $A$ and $B$ and without limitation of generality we may suppose that $C$ is distinct from $B$.

Let $Q$ be any inertia plane containing the optical line $b$ and let $b_1$ be any inertia line through $B$ in $Q$.

Let $b_1'$ be the parallel inertia line through $B'$ which will also lie in $Q$.

Let $P$ be the inertia plane containing $b_1$ and $C$, while $R$ is the inertia plane containing $b_1'$ and $C$.

Then by Theorem 51 $P$ and $R$ have a general line, say $c_1$, in common, which is parallel to $b_1$ and $b_1'$.

Suppose that $D$ is not identical with $B$ and let $Q'$ be the inertia plane through $D$ and parallel to $Q$.

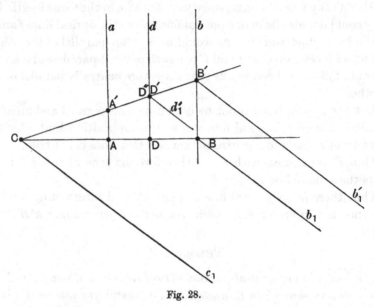

Fig. 28.

Then we have the three distinct inertia planes $P$, $Q$ and $R$ and the three parallel general lines $c_1$, $b_1$ and $b_1'$, such that $c_1$ lies in $P$ and $R$, $b_1$ in $Q$ and $P$, and $b_1'$ in $R$ and $Q$, while $Q'$ is an inertia plane parallel to $Q$ through an element of $P$ which does not lie in $b_1$, and so by Theorem 53 the inertia planes $R$ and $Q'$ have a general line in common which is parallel to $b_1'$.

Call this general line $d_1'$.

Then $d_1'$ is an inertia line.

Now the optical line $d$ must lie in $Q'$ and must therefore intersect $d_1'$ in some element, say $D'$.

Also $A'B'$ being a separation line in the inertia plane $R$ must intersect the inertia line $d_1'$ in some element, say $D''$.

We have to show that $D''$ is identical with $D'$.

Suppose if possible that $D''$ is distinct from $D'$ and let $d''$ be the optical line through $D''$ parallel to $b$.

Then since by Theorem 45 $D$ is neither *before* nor *after* any element of $b$, it follows that $d$ is a neutral-parallel of $b$.

Similarly $d''$ is a nuetral-parallel of $b$ and so if $D'$ and $D''$ were distinct and did not lie in the same optical line, it would follow by Theorem 28 that $d''$ was a neutral-parallel of $d$.

But $D'$ and $D''$ lie in $d_1'$, which is an inertia line, and so if $D'$ and $D''$ were distinct one of them would have to be *after* the other and so $d$ and $d''$ could not be neutral-parallels.

Thus the supposition that $D''$ is distinct from $D'$ leads to a contradiction and so $D''$ must be identical with $D'$.

Thus the optical line $d$ intersects $A'B'$ in $D'$, which proves the theorem.

## THEOREM 85

*If $a$ and $b$ be two neutral-parallel optical lines and $E$ be any element in a separation line $AB$ which intersects $a$ in $A$ and $b$ in $B$, and if $A'B'$ be any other separation line intersecting $a$ in $A'$ and $b$ in $B'$ but not parallel to $AB$, then $E$ either lies in $A'B'$ or in a separation line parallel to $A'B'$ which intersects both $a$ and $b$.*

If $E$ does not lie in $A'B'$, then by Theorem 84 an optical line through $E$ parallel to $a$ or $b$ intersects $A'B'$ in some element, say $E'$, which is either *before* or *after* $E$.

Thus by Theorem 82 the general line through $E$ parallel to $A'B'$ intersects $a$ and similarly it intersects $b$.

Thus $E$ must lie in a separation line parallel to $A'B'$ and intersecting both $a$ and $b$ when it does not lie in $A'B'$ itself.

## REMARKS

If $a$ and $b$ be two neutral-parallel optical lines and if $c$ and $d$ be any two non-parallel separation lines intersecting both $a$ and $b$, then it is evident from Theorem 85 that: the aggregate consisting of all the elements in $c$ and in all separation lines intersecting $a$ and $b$ which are parallel to $c$ must be identical with the aggregate consisting of all the elements in $d$ and in all separation lines intersecting $a$ and $b$ which are parallel to $d$.

This follows since each element in the one set of separation lines must also lie in the other set.

Thus the aggregate which we obtain in this way is independent of the particular set of parallel separation lines intersecting $a$ and $b$ which we may select and so we have the following definition.

*Definition.* The aggregate of all elements of all mutually parallel separation lines which intersect two neutral-parallel optical lines will be called an *optical plane.**

It is evident that through any element of an optical plane there is *one single optical line* lying in the optical plane.

For if $a$ and $b$ be two neutral-parallel optical lines which are intersected by a separation line $d$ in the elements $A$ and $B$ respectively, and if $C$ be any other element in $d$, then there is a neutral-parallel to $a$ and $b$ through $C$ which we may call $c$.

But through each element of $c$ other than $C$ there is a separation line parallel to $d$ which, by Theorem 82 ($a$ and $b$), must intersect both $a$ and $b$, and so every element of the optical line $c$ lies in the optical plane defined by $a$ and $b$.

An optical plane differs in this respect from an inertia plane, since the latter contains two optical lines passing through any element of it.

*Definition.* In analogy with the case of an inertia plane, an optical line which lies in any optical plane will be called a *generator* of the optical plane.

## THEOREM 86

*If two distinct elements of a general line lie in an optical plane, then every element of the general line lies in the optical plane.* ·

Let the optical plane be determined by the two neutral-parallel optical lines $a$ and $b$.

If the two elements lie in a general line which is known to intersect both $a$ and $b$, no proof is required.

Let $C$ be any element in any separation line $AB$ which intersects $a$ in $A$ and $b$ in $B$, and let $D'$ be any element in any separation line $A'B'$ parallel to $AB$ and intersecting $a$ in $A'$ and $b$ in $B'$.

We have to show that every element of the general line $CD'$ lies in the optical plane.

By Theorem 82 ($a$ or $b$) an optical line through $C$ parallel to $a$ or $b$ will intersect $A'B'$ in some element, say $C'$.

If $C'$ should coincide with $D'$, then $CD'$ would be an optical line which would be neutrally parallel to $a$ or $b$ and we already know that

---

* The name "optical plane" has been adopted because of certain analogies with an optical line.

each element of it must lie either in a separation line parallel to $AB$ and intersecting both $a$ and $b$, or in $AB$ itself.

Thus if $C'$ should coincide with $D'$, the general line $CD'$ is such that every element of it lies in the optical plane.

If $C'$ does not coincide with $D'$, then an optical line through $D'$ parallel to $CC'$ will intersect $AB$ in some element, say $D$ (Theorem 82 ($a$ or $b$)).

Now $DD'$ must be a neutral-parallel of $CC'$ and either of the optical lines $a$ or $b$ must be either parallel to $CC'$ and $DD'$ or identical with one of them.

If $a$ is identical with $CC'$ or $DD'$, then $a$ intersects $CD'$, while if $b$ is identical with $CC'$ or $DD'$, then $b$ intersects $CD'$.

If $a$ is not identical with $CC'$ or $DD'$, then, by Theorem 84, $a$ must intersect $CD'$, and similarly if $b$ is not identical with $CC'$ or $DD'$, then $b$ must intersect $CD'$.

Thus in all these cases $CD'$ intersects both $a$ and $b$ and therefore every element of $CD'$ lies in the optical plane determined by $a$ and $b$.

## THEOREM 87

*If $e$ be a general line in an optical plane and $A$ be any element of the optical plane which does not lie in $e$, then there is one single general line through $A$ in the optical plane which does not intersect $e$.*

We saw in the course of proving Theorem 86 that if an optical plane be determined by two neutral-parallel optical lines $a$ and $b$, then any general line containing two elements in the optical plane and therefore any general line lying in the optical plane, must either be a neutral-parallel of $a$ or $b$, or else must intersect both $a$ and $b$.

Suppose first that $e$ is a separation line in the optical plane determined by $a$ and $b$, then $e$ must intersect both $a$ and $b$.

Since $A$ does not lie in $e$ it must lie in a separation line $f$ parallel to $e$ and intersecting both $a$ and $b$.

Now through $A$ there is an optical line, say $c$, which is a neutral-parallel of $a$ or $b$ and which by Theorem 82 ($a$ and $b$) must intersect $e$ and must lie in the optical plane, while any other general line $f'$ through $A$ and lying in the optical plane must intersect both $a$ and $b$.

But $f'$ is not parallel to $e$ and therefore by Theorem 83 it must intersect it.

Suppose next that $e$ is an optical line.

Then $e$ must either be parallel to $a$ and $b$ or be identical with one of them.

Through $A$ there is an optical line parallel to $a$ or $b$ and therefore parallel to $e$, and this optical line must lie in the optical plane.

Any other general line through $A$ in the optical plane intersects both $a$ and $b$ and so by Theorem 84 it must also intersect $e$.

Thus there is in all cases one single general line through $A$ in the optical plane which does not intersect $e$.

## THEOREM 88

*If $A$, $B$ and $C$ be three elements in an optical plane which do not all lie in one general line and if $D$ be an element linearly between $A$ and $B$, while $E$ is an element linearly between $A$ and $C$, there exists an element which lies both linearly between $B$ and $E$ and linearly between $C$ and $D$.*

The proof of this theorem is quite analogous to that of Theorem 76, the only difference being that $V$ is here an optical plane instead of an inertia plane and, as such, it cannot contain any inertia line.

Thus the words "which does not lie in $V$" may be omitted from the first sentence of the proof.

## THEOREM 89

*If $A$, $B$ and $C$ be three elements in an optical plane which do not all lie in one general line and if $D$ be an element linearly between $A$ and $B$ while $F$ is an element linearly between $C$ and $D$, there exists an element, say $E$, which is linearly between $A$ and $C$ and such that $F$ is linearly between $B$ and $E$.*

The proof of this theorem is quite analogous to that of Theorem 77, the only difference being that $V$ is here an optical plane instead of an inertia plane and, as such, it cannot contain any inertia line.

Thus the words "which does not lie in $V$" may be omitted from the first sentence of the proof.

## REMARKS

It will be observed that Theorem 88 is the analogue of Peano's axiom (14) for the case of elements in an optical plane, while Theorem 89 is the corresponding analogue of his axiom (13).

Further, Theorem 87 corresponds to the Euclidean axiom of parallels for the case of general lines in an optical plane.

## THEOREM 90

*If $A$, $B$ and $C$ be three elements in an optical plane which do not all lie in one general line and if $D$ be an element linearly between $A$ and $B$ while $DE$ is a general line through $D$ parallel to $BC$ and intersecting $AC$ in the element $E$, then $E$ is linearly between $A$ and $C$.*

In the first place $E$ cannot be identical with $A$ for then the general line $DE$ would be identical with the general line $BA$ and would therefore intersect $BC$.

Again $E$ cannot be identical with $C$ for once more $BC$ and $DE$ would intersect.

Thus we must either have $C$ linearly between $A$ and $E$, or $A$ linearly between $C$ and $E$, or $E$ linearly between $A$ and $C$.

If $C$ were linearly between $A$ and $E$, then since $D$ is linearly between $A$ and $B$ it would follow by Theorem 88 that there existed an element which was both linearly between $B$ and $C$ and linearly between $E$ and $D$.

Thus in this case also $BC$ and $DE$ would intersect.

Next if $A$ were linearly between $C$ and $E$, then since $D$ is linearly between $A$ and $B$ it would follow similarly by Theorem 89 that $BC$ and $DE$ must intersect.

Thus the only possibility is that $E$ is linearly between $A$ and $C$.

## THEOREM 91

*If three parallel general lines $a$, $b$ and $c$ in one optical plane intersect a general line $d_1$ in $A_1$, $B_1$ and $C_1$ respectively and intersect a second general line $d_2$ in $A_2$, $B_2$ and $C_2$ respectively, then if $B_1$ is linearly between $A_1$ and $C_1$ we shall also have $B_2$ linearly between $A_2$ and $C_2$.*

If $A_1$ should be identical with $A_2$ the result follows directly from Theorem 90.

Similarly it follows directly if $C_1$ should be identical with $C_2$.

If $B_1$ should be identical with $B_2$ the following method is still valid.

The general line $C_1 A_2$ cannot be identical with the general line $c$ and therefore $C_1 A_2$ must intersect the general line $b$ (which is parallel to $c$) in some element, say $B'$.

Then, since $B_1$ is linearly between $A_1$ and $C_1$, it follows by Theorem 90 that $B'$ is linearly between $C_1$ and $A_2$.

Similarly, since $B'$ is linearly between $A_2$ and $C_1$, it follows that $B_2$ is linearly between $A_2$ and $C_2$.

THEOREM 92

*If two elements A and B lie in one optical line and if two other elements C and D lie in a neutral-parallel optical line, and if A be after B, then:*

(1) *If C be after D the general lines AD and BC intersect in an element which is both linearly between A and D and linearly between B and C.*

(2) *If the general lines AD and BC intersect in an element E which is either linearly between A and D or linearly between B and C, we shall also have C after D.*

Let $A$ and $B$ lie in an optical line $a$ and let $C$ and $D$ lie in a neutral-parallel optical line $c$.

Let $a_1$ be any inertia line through $A$ and let $b_1$ be a parallel inertia line through $B$.

Then $a_1$ and $b_1$ lie in an inertia plane, say $P$.

Let $B'$ be any element in $b_1$ which is *after* $B$ and let $a'$ be an optical line through $B'$ parallel to $a$.

Then $a'$ will intersect $a_1$ in some element, say $A'$, and, by Theorem 57, since $A$ is *after* $B$, we must have $A'$ *after* $B'$.

But, since $B'$ is not an element of $a$ but is *after* $B$, an element of $a$, it follows that $a'$ is an after-parallel of $a$.

Since further $a$ and $c$ are neutral-parallels, it follows by Theorem 26 (*b*) that $a'$ is an after-parallel of $c$.

Thus $a'$ and $c$ lie in an inertia plane, say $Q$.

Proceeding now to prove the first part of the theorem, we have $A'$ after $B'$ and $C$ after $D$ and so it follows by Theorem 68 and the definition of "linearly between" that $A'D$ and $B'C$ intersect in an element, say $E'$, which is linearly between $A'$ and $D$ and also linearly between $B'$ and $C$.

But since $a_1$ is an inertia line there is an inertia plane containing $a_1$ and the element $E'$ which we may call $R$, and similarly there is an inertia plane containing $b_1$ and the element $E'$ which we may call $S$.

Now since $a_1$ and $b_1$ are parallel general lines in the inertia plane $P$ it follows, by Theorem 51, that the inertia planes $R$ and $S$ have a general line, say $e_1$, in common, which is parallel to $a_1$ and $b_1$ and must therefore be an inertia line.

Since $e_1$ lies both in $R$ and $S$ it must intersect $BC$ and $AD$ which lie respectively in $S$ and $R$ and are separation lines.

Suppose $e_1$ intersects $BC$ in $E$ and $AD$ in $\bar{E}$, then $E$ and $\bar{E}$ lie in the inertia line $e_1$ and so, if they were distinct, they could not lie in one optical plane.

But $E$ and $\bar{E}$ each lie in the optical plane determined by the neutral-parallel optical lines $a$ and $c$ and so $\bar{E}$ is identical with $E$.

But since $D$, $A$ and $A'$ are elements in the inertia plane $R$ which do not all lie in one general line, and since $E'$ is linearly between $A'$ and $D$

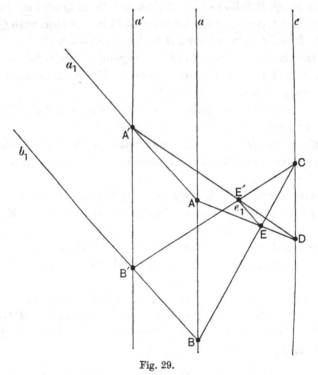

Fig. 29.

and $E'E$ is parallel to $A'A$, it follows, by Theorem 72, that $E$ is linearly between $A$ and $D$.

Similarly since $E'$ is linearly between $B'$ and $C$, and $C$, $B$ and $B'$ lie in the inertia plane $S$ and are not all in one general line and since $E'E$ is parallel to $B'B$ it follows that $E$ is linearly between $B$ and $C$.

Thus the first part of the theorem is proved.

Proceeding now to prove the second part of the theorem; since $AD$ and $BC$ intersect in the element $E$ and since $a_1$ and $b_1$ are inertia lines it follows that there is an inertia plane, say $R$, which contains $a_1$ and the element $E$, and another inertia plane, say $S$, containing $b_1$ and the element $E$.

It follows, since $a_1$ and $b_1$ are parallel and lie in the inertia plane $P$, that $R$ and $S$ have a general line, say $e_1$, in common, which is parallel to $a_1$ and $b_1$ (Theorem 51) and must therefore be an inertia line.

Now the element $E$ could not lie in the optical line $c$, since then it would have to coincide with both $C$ and $D$ and could not therefore be linearly between $A$ and $D$ or linearly between $B$ and $C$.

Thus, since $E$ and $c$ lie in one optical plane and $c$ also lies in the inertia plane $Q$, it follows that $E$ does not lie in $Q$ and so the inertia line $e_1$ cannot have more than one element in common with $Q$.

If now $E$ be linearly between $A$ and $D$, then since $D$, $A$ and $A'$ lie in the inertia plane $R$ and are not all in one general line, it follows, since $e_1$ is parallel to $AA'$, that $e_1$ must intersect $A'D$ in an element, say $E'$, which is linearly between $A'$ and $D$.

Also, since $B'C$ is not parallel to $e_1$ and lies in the inertia plane $S$ with it, it follows that $B'C$ must intersect $e_1$.

But $B'C$ lies in $Q$ and we have seen that $e_1$ and $Q$ cannot have more than one element in common and therefore $A'D$ and $B'C$ intersect $e_1$ in the same element $E'$.

If we suppose instead that $E$ is linearly between $B$ and $C$, we find in a similar way that $B'C$ and $A'D$ intersect $e_1$ in an element $E'$ which is linearly between $B'$ and $C$.

But by the definition of "linearly between" the element $E'$ must in either case be between the parallel optical lines $a'$ and $c$ in the inertia plane $Q$.

Thus, since $a'$ and $c$ are parallel optical lines in the inertia plane $Q$ and since $A'$ is *after* $B'$ and the element of intersection of $A'D$ and $B'C$ lies between $a'$ and $c$, it follows by Theorem 69 that $C$ is *after* $D$, as was to be proved.

### Theorem 93

(a) *If $A_0$ and $A_x$ be two elements in a general line $a$ which lies in the same optical plane with another general line $b$ which intersects $a$ in the element $C$ such that either $A_0$ is linearly between $C$ and $A_x$, or $A_x$ is linearly between $C$ and $A_0$ and if an optical line through $A_0$ intersects $b$ in $B_0$ so that $B_0$ is after $A_0$, then a parallel optical line through $A_x$ will intersect $b$ in an element which is after $A_x$.*

The proof of this theorem is exactly analogous to that of Theorem 71, using Theorem 92 (1) in place of Theorem 68.

There is also a (b) form of this theorem which may be proved in an analogous manner.

## THEOREM 94

*If $A$, $B$ and $C$ be three elements in an optical plane which do not all lie in one general line and if $D$ be the mean of $A$ and $B$, then a general line through $D$ parallel to $BC$ intersects $AC$ in an element which is the mean of $A$ and $C$.*

Let $a_1$ be any inertia line through $A$ while $b_1$ and $c_1$ are parallel inertia lines through $B$ and $C$ respectively.

Then $b_1$ and $c_1$ lie in one inertia plane, say $P$, $c_1$ and $a_1$ in a second inertia plane, say $Q$, and $a_1$ and $b_1$ in a third inertia plane, say $R$.

Let one of the optical lines through $A$ in the inertia plane $Q$ intersect $c_1$ in $C'$ and let one of the optical lines through $A$ in the inertia plane $R$ intersect $b_1$ in $B'$.

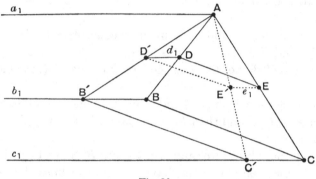

Fig. 30.

Then $AC'$ and $AB'$ may be taken as generators of opposite sets of an inertia plane, say $S$, containing $A$, $B'$ and $C'$.

Let $d_1$ be an inertia line through $D$ parallel to $a_1$, $b_1$ and $c_1$.

Then $d_1$ will lie in $R$ and will intersect the optical line $AB'$ in some element, say $D'$.

If now a general line be taken through $D$ parallel to $BC$, it will lie in the optical plane, and since the general line $AC$ is distinct from the general line $BC$ it follows from Theorem 87 that this general line through $D$ parallel to $BC$ must intersect $AC$ in some element, say $E$.

Let $e_1$ be an inertia line through $E$ parallel to $a_1$, $b_1$, $c_1$, $d_1$.

Then $e_1$ will lie in the inertia plane $Q$ and will intersect the optical line $AC'$ in some element, say $E'$.

Now $d_1$ and $e_1$ being parallel inertia lines will lie in an inertia plane, say $T$, which contains the two intersecting general lines $DE$ and $d_1$ which are respectively parallel to $BC$ and $b_1$ in $P$.

Thus by Theorem 52 the inertia planes $T$ and $P$ are parallel.

But the inertia plane $S$ has the general line $D'E'$ in common with $T$ and the general line $B'C'$ in common with $P$ and so $D'E'$ is parallel to $B'C'$.

Now since $A$, $B$ and $B'$ lie in the inertia plane $R$ and since $D$ is the mean of $A$ and $B$ and since $DD'$ is parallel to $BB'$, it follows by Theorem 78 provided that $A$, $B$ and $B'$ do not lie in one general line, that $D'$ is the mean of $A$ and $B'$.

The only case in which $A$, $B$ and $B'$ do lie in one general line is when $B'$ coincides with $B$ and then $D'$ is identical with $D$ so that $D'$ is still the mean of $A$ and $B'$.

Again, since $A$, $B'$ and $C'$ lie in one inertia plane $S$ and do not all lie in one general line and since $D'$ is the mean of $A$ and $B'$, and $D'E'$ is parallel to $B'C'$, it follows by Theorem 78 that $E'$ is the mean of $A$ and $C'$.

Further, since $A$, $C$ and $C'$ lie in one inertia plane $Q$, since $E'$ is the mean of $A$ and $C'$ and since $E'E$ is parallel to $C'C$, it follows, provided that $A$, $C$ and $C'$ do not lie in one general line, that $E$ is the mean of $A$ and $C$.

The only case in which $A$, $C$ and $C'$ do lie in one general line is when $C'$ coincides with $C$ and then $E'$ coincides with $E$ so that $E$ is still the mean of $A$ and $C$.

(It is to be noted that we cannot have both $B'$ coinciding with $B$ and $C'$ with $C$, for then we should have two optical lines $AB'$ and $AC'$ passing through the same element $A$ and lying in an optical plane, which is impossible.) Thus the theorem is proved.

Since there is only one general line through $D$ parallel to $BC$ and this must pass through the mean of $A$ and $C$, it follows directly that if $E$ be the mean of $A$ and $C$, then the general line $DE$ is parallel to $BC$.

*Definition.* If a pair of parallel general lines in an optical plane be intersected by another pair of parallel general lines, then the four general lines will be said to form a *general parallelogram in the optical plane.*

The terms *corner*, *side line*, *diagonal line*, *adjacent* and *opposite* will be used in a similar sense for the case of a general parallelogram in an optical plane as for one in an inertia plane.

## THEOREM 95

*If we have a general parallelogram in an optical plane, then:*

(1) *The two diagonal lines intersect in an element which is the mean of either pair of opposite corners.*

(2) *A general line through the element of intersection of the diagonal lines and parallel to either pair of opposite side lines intersects either of the other side lines in an element which is the mean of the pair of corners through which that side line passes.*

The proof of this theorem is exactly analogous to that of Theorem 79, using Theorem 94 in place of Theorem 78.

## THEOREM 96

*If $A$, $B$, $C$, $D$ be the corners of a general parallelogram in an optical plane, $AB$ and $DC$ being one pair of parallel side lines and $BC$ and $AD$ the other pair of parallel side lines, then if $E$ be the mean of $A$ and $B$ while $F$ is the mean of $D$ and $C$, the general lines $AF$ and $EC$ are parallel to one another.*

The proof of this theorem is exactly analogous to that of Theorem 80, using Theorem 95 in place of Theorem 79 and Theorem 94 in place of Theorem 78.

## THEOREM 97

*If three parallel general lines $a$, $b$ and $c$ in one optical plane intersect a general line $d_1$ in $A_1$, $B_1$ and $C_1$ respectively, and intersect a second general line $d_2$ in $A_2$, $B_2$ and $C_2$ respectively, and if $B_1$ be the mean of $A_1$ and $C_1$, then $B_2$ will be the mean of $A_2$ and $C_2$.*

The proof of this theorem is exactly analogous to that of Theorem 81, using Theorem 94 in place of Theorem 78, and Theorem 95 in place of Theorem 79.

## REMARKS

If $P$ and $P'$ be parallel inertia planes and if $a$ be any generator of $P$, there is one single generator of $P'$ which is a neutral-parallel of $a$.

This is easily seen, for if $b_1$ be any generator of $P'$ belonging to the set not parallel to $a$ and if $B$ be any element in $b_1$, then either:

(1) $B$ is *before* an element of $a$,

or  (2) $B$ is *after* an element of $a$,

or  (3) $B$ is neither *before* nor *after* any element of $a$.

In cases (1) and (2), since $B$ does not lie in $a$ and, since $b_1$ neither intersects $a$ nor is parallel to it, it follows by Post. XII ($a$ and $b$) that there is one single element of $b_1$ which is neither *before* nor *after* any element of $a$.

Thus there is always an element of $b_1$ which is neither *before* nor *after* any element of $a$.

Let $B_0$ be such an element and let $a'$ be the generator of $P'$ parallel to $a$ and passing through $B_0$.

Then $a'$ is a neutral-parallel of $a$.

Again, there can be no other generator of $P'$ besides $a'$ which is a neutral-parallel of $a$, for any other generator of $P'$ parallel to $a'$ must be either a before- or after-parallel of $a'$ and therefore by Theorem 26 ($a$ or $b$) such a generator must be a before- or after-parallel of $a$.

Again, if $P$ and $P'$ be parallel inertia planes and if $A$ be any element of $P$, while $a$ and $b$ are the two generators of $P$ which pass through $A$, then there is one single generator of $P'$, say $a'$, which is neutrally parallel to $a$ and there is one single generator of $P'$, say $b'$, which is neutrally parallel to $b$.

The optical lines $a'$ and $b'$ being generators of opposite sets must intersect in some element, say $A'$.

Then $A'$ is neither *before* nor *after* any element of $a$ and also neither *before* nor *after* any element of $b$.

Similarly $A$ is neither *before* nor *after* any element of $a'$ and also neither *before* nor *after* any element of $b'$.

The elements $A$ and $A'$ will be spoken of as *representatives* of one another in the parallel inertia planes $P$ and $P'$.

Thus we have the following definition.

*Definition.* If $P$ and $P'$ be parallel inertia planes and if $A$ and $A'$ be elements in $P$ and $P'$ respectively such that the two generators of $P'$ passing through $A'$ are respectively neutral-parallels of the generators of $P$ which pass through $A$, then the elements $A$ and $A'$ will be called *representatives* of one another in the parallel inertia planes $P$ and $P'$.

It is evident that the elements $A$ and $A'$ must lie in a separation line.

## Theorem 98

*If $P_1$ and $P_2$ be two parallel inertia planes and if $A_1$ be any element in $P_1$ while $A_2$ is its representative in $P_2$, then if $A_1'$ be any other element in $P_1$ and $A_2'$ its representative in $P_2$ the separation lines $A_1 A_2$ and $A_1' A_2'$ are parallel to one another.*

Let $a_1$ and $b_1$ be the generators of $P_1$ which pass through $A_1$ and let $a_2$ and $b_2$ be the generators of $P_2$ which pass through $A_2$, the optical lines

$a_1$ and $a_2$ being neutrally parallel to one another and the optical lines $b_1$ and $b_2$ being also neutrally parallel to one another.

Consider first the case where $A_1'$ lies in one of the generators $a_1$ or $b_1$ which pass through $A_1$.

It will be sufficient if we consider the case where $A_1'$ lies in $a_1$.

Then $A_2'$ will lie in $a_2$.

Let $b_1'$ be the second generator of $P_1$ which passes through $A_1'$ and let $b_2'$ be the second generator of $P_2$ which passes through $A_2'$.

Then $b_1'$ will be parallel to $b_1$ while $b_2'$ will be parallel to $b_2$ and the optical lines $b_1'$ and $b_2'$ will be neutrally parallel to one another by the definition of representative elements.

Now since $a_1$ and $a_2$ are neutral-parallel optical lines they determine an optical plane which contains the separation lines $A_1 A_2$ and $A_1' A_2'$ which must therefore either intersect or be parallel to one another.

Now, by Theorem 45, no element of the general line $A_1 A_2$ with the exception of $A_1$ is either *before* or *after* any element of $b_1$, and similarly, no element of the general line $A_1' A_2'$ with the exception of $A_1'$ is either *before* or *after* any element of $b_1'$.

Now suppose, if possible, that $A_1 A_2$ and $A_1' A_2'$ intersect in some element $A_0$.

Then $A_0$ could not coincide with either $A_1$ or $A_1'$ and so would require to be neither *before* nor *after* any element of $b_1$ and also neither *before* nor *after* any element of $b_1'$.

If then $b_0$ were an optical line through $A_0$ parallel to $b_1$ and $b_1'$, we should have $b_0$ neutrally parallel to both $b_1$ and $b_1'$.

Thus by Theorem 28, $b_1$ would require to be neutrally parallel to $b_1'$.

But $b_1$ and $b_1'$ are parallel generators of the inertia plane $P_1$ and so one must be an after-parallel of the other.

Thus the supposition that $A_1 A_2$ and $A_1' A_2'$ intersect leads to a contradiction and therefore is not true.

It follows that $A_1 A_2$ and $A_1' A_2'$ are parallel, which proves the theorem in this special case.

Next consider the case where $A_1'$ does not lie either in $a_1$ or $b_1$.

Let $b_1'$ be the generator of $P_1$ through $A_1'$ parallel to $b_1$ and let $b_2'$ be the generator of $P_2$ through $A_2'$ parallel to $b_2$.

Let $b_1'$ and $a_1$ intersect in $B_1$ and let $b_2'$ and $a_2$ intersect in $B_2$.

Then since $a_1$ and $a_2$ are neutrally parallel and also $b_1'$ and $b_2'$ are neutrally parallel, it follows by the case already proved that $A_1 A_2$ and $B_1 B_2$ are parallel to one another.

Similarly $A_1'A_2'$ and $B_1B_2$ are parallel to one another.

Thus by Theorem 50 $A_1'A_2'$ and $A_1A_2$ are parallel to one another.

### SETS OF THREE ELEMENTS WHICH DETERMINE OPTICAL PLANES

If $A_1$, $A_2$ and $A_3$ be three distinct elements which do not all lie in one general line and do not all lie in one inertia plane, they either may or may not all lie in one optical plane.

In those cases in which they do lie in an optical plane they determine the optical plane containing them.

We have the following criteria by which we may say that the three elements do lie in one optical plane.

CASE I.   Three elements $A_1$, $A_2$, $A_3$ lie in one optical plane if $A_1$ and $A_2$ lie in an optical line while $A_3$ is an element which is neither *before* nor *after* any element of the optical line.

This is clearly true since if $a$ be the optical line containing $A_1$ and $A_2$, there is an optical line, say $b$, through $A_3$ and neutrally parallel to $a$.

These two optical lines may be taken as generators of an optical plane which will contain $A_1$, $A_2$ and $A_3$.

Now if $P$ be this optical plane it is the only one which contains $A_1$, $A_2$ and $A_3$, for suppose that $A_1$, $A_2$ and $A_3$ also lie in an optical plane $P'$ determined by the two generators $a'$ and $b'$.

Then, since $P'$ contains $A_2$ and $A_3$, it follows by Theorem 86 that $P'$ contains every element of the general line $A_2A_3$ and since $A_2A_3$ is a separation line it cannot be parallel to either $a'$ or $b'$ and must therefore intersect both $a'$ and $b'$.

Again since $P'$ contains $A_1$ and $A_2$ it follows that $P'$ contains every element of $A_1A_2$: that is, it contains the optical line $a$.

Also since $P'$ contains $A_3$ it must contain the optical line through $A_3$ parallel to $a$: that is, it contains $b$.

Further $a$ cannot intersect either $a'$ or $b'$ and so must be either parallel to both or identical with one.

Similarly $b$ cannot intersect either $a'$ or $b'$ and so must be either parallel to both or identical with one.

Now every element in the optical plane $P$ must either lie in the separation line $A_2A_3$ or in a separation line parallel to $A_2A_3$ and intersecting $a$ and $b$.

But such a separation line must also intersect $a'$ and $b'$ and will therefore lie in the optical plane $P'$.

Similarly every element in the optical plane $P'$ must either lie in the separation line $A_2 A_3$ or in a separation line parallel to $A_2 A_3$ and intersecting $a'$ and $b'$.

But such a separation line must also intersect $a$ and $b$ and will therefore lie in the optical plane $P$.

Thus every element in $P$ lies also in $P'$ and every element in $P'$ lies also in $P$.

Thus the optical planes $P$ and $P'$ are identical and so there is only one optical plane containing the three elements $A_1, A_2, A_3$.

CASE II. Three elements $A_1, A_2, A_3$ lie in one optical plane if $A_1$ and $A_2$ lie in a separation line while $A_3$ is an element which is *before one single element* of $A_1 A_2$, or is *after one single element* of $A_1 A_2$.

This may be shown as follows:

Let $A_3$ be *before* the one single element $A_4$ of the separation line $A_1 A_2$ and let $A_1 A_2$ be denoted by $a$.

Then $A_3 A_4$ cannot be an inertia line, for, if it were, we know that it would lie in an inertia plane containing $a$.

Thus the three elements $A_1, A_2, A_3$ would lie in one inertia plane, contrary to what was proved on pp. 72–73.

Thus $A_3 A_4$ cannot be an inertia line and so, since $A_3$ is *before* $A_4$, it must be an optical line.

Now $A_4$ must be distinct from at least one of the two elements $A_1$, $A_2$, and without loss of generality we may suppose $A_4$ distinct from $A_1$.

Then $A_1$ is neither *before* nor *after* $A_4$ since they are both elements of the separation line $A_1 A_2$.

Further, $A_1$ cannot be *before* any element of the optical line $A_3 A_4$ which is *before* $A_4$, for then $A_1$ would be *before* $A_4$, which is impossible.

Similarly $A_1$ cannot be *after* any element of the optical line $A_3 A_4$ which is *after* $A_4$.

Again $A_1$ cannot be *after* any element of the optical line $A_3 A_4$ which is *before* $A_4$; for if $A_5$ were such an element of $A_3 A_4$ we should have $A_5$ *before* two distinct elements of $a$ and so $A_5$, $A_1$ and $A_4$ would lie in one inertia plane which would also contain $A_3$, contrary to what has already been shown.

Similarly $A_1$ cannot be *before* any element of the optical line $A_3 A_4$ which is *after* $A_4$.

Thus $A_1$ is neither *before* nor *after* any element of the optical line $A_3 A_4$ and so through $A_1$ there is one single optical line which is neutrally parallel to $A_3 A_4$.

Thus these two optical lines may be taken as generators of an optical plane and, since the separation line $a$ intersects both these optical lines and contains the elements $A_1$ and $A_2$, it follows that the three elements $A_1$, $A_2$, $A_3$ lie in an optical plane.

Further, there is only one optical plane containing $A_1$, $A_2$ and $A_3$; for any optical plane containing $A_1$ and $A_2$ must also contain $A_4$, and since, by Case I, there is only one optical plane containing $A_3$, $A_4$ and $A_1$, it follows that there is only one optical plane containing $A_1$, $A_2$ and $A_3$.

Similarly, if $A_3$ be *after one single element* of the separation line $A_1A_2$, there is one single optical plane containing the three elements $A_1$, $A_2$ and $A_3$.

## THEOREM 99

*If two optical parallelograms have a pair of opposite corners in common lying in an inertia line, then their separation diagonal lines are such that no element of the one is either* before *or* after *any element of the other*.

Let $A$ and $B$ be the two common opposite corners lying in the inertia line $a$, and let $B$ be *after $A$*.

Let $C$ and $D$ be the two remaining corners of the one optical parallelogram which we shall suppose to lie in the inertia plane $P$, and let $E$ and $F$ be the two remaining corners of the other optical parallelogram which we shall suppose to lie in the inertia plane $Q$.

Then by Theorem 60 the two optical parallelograms have a common centre, say $O$, which is *after $A$* and *before $B$*.

Then the general lines $CD$ and $EF$ are separation lines and so their common element is neither *before* nor *after* any element of either of them.

Let $CD$ be denoted by $c$ and $EF$ by $e$.

Now, since $C$ and $E$ are two distinct elements in the $\alpha$ sub-set of $A$ which do not lie in one optical line, it follows by Theorem 13 that $C$ is neither *before* nor *after $E$*, and similarly $C$ is neither *before* nor *after $F$*.

Let $E_1$ be any element in $e$ such that $E$ is linearly between $O$ and $E_1$ and let the optical line through $E_1$ parallel to $EA$ intersect $a$ in $A_1$, while the optical line through $E_1$ parallel to $EB$ intersects $a$ in $B_1$.

Then by Theorem 72 $A$ is linearly between $A_1$ and $O$, while $B$ is linearly between $O$ and $B_1$.

Thus $A_1$ must be *before $A$* and $B_1$ must be *after $B$*.

Again, since $A_1$ is *before $O$*, and $O$ and $E_1$ lie in a separation line, we must have $A_1$ *before $E_1$*.

Similarly, since $B_1$ is *after* $O$, and $O$ and $E_1$ lie in a separation line, we must have $B_1$ *after* $E_1$.

But $A$ is *after* $A_1$ and $C$ is *after* $A$ and therefore $C$ is *after* $A_1$, and since $A$ is the only element common to $a$ and the $\beta$ sub-set of $C$, it follows that $A_1 C$ is an inertia line.

If then $C$ were *before* $E_1$ it would follow by Theorem 12 that $C$ should

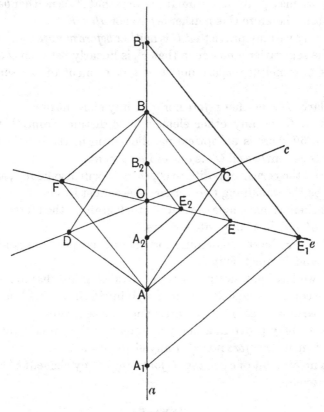

Fig. 31.

lie in the optical line $A_1 E_1$ which it clearly cannot do since $A_1 C$ is an inertia line.

Thus $C$ is not *before* $E_1$.

Further, $B$ is *after* $C$ and $B_1$ *after* $B$ and therefore $B_1$ is *after* $C$ and $B_1 C$ is an inertia line.

If then $C$ were *after* $E_1$ it would follow by Theorem 12 that $C$ should lie in the optical line $B_1 E_1$ which it clearly cannot do since $B_1 C$ is an inertia line.

Thus $C$ is not *after* $E_1$.

In a similar manner we may prove that $C$ is neither *before* nor *after* any element $F_1$ of the separation line $e$ such that $F$ is linearly between $O$ and $F_1$.

Again let $E_2$ be any element of $e$ which is linearly between $O$ and $E$ and let the optical line through $E_2$ parallel to $EA$ intersect $a$ in $A_2$ while the optical line through $E_2$ parallel to $EB$ intersects $a$ in $B_2$.

Then we may prove in a similar manner that $E_2$ is neither *before* nor *after* $C$ and therefore $C$ is neither *before* nor *after* $E_2$.

Similarly we may prove that $C$ is neither *before* nor *after* any element $F_2$ of the separation line $e$ such that $F_2$ is linearly between $O$ and $F$.

Thus $C$ is neither *before* nor *after* any element of the separation line $e$.

Similarly $D$ is neither *before* nor *after* any element of $e$.

Again if $C'$ be any other element in $c$ distinct from $O$, then by Theorem 59 there is an optical parallelogram in the inertia plane $P$ having $O$ as centre and $C'$ as one of its corners.

If $D'$ be the corner opposite to $C'$, then $D'$ will also lie in $c$, and if $A'$ and $B'$ be the remaining two corners these must lie in $a$.

Then there is one single optical parallelogram in the inertia plane $Q$ having $A'$ and $B'$ as opposite corners.

If $E'$ and $F'$ be the remaining corners of this optical parallelogram, then $E'$ and $F'$ must lie in $e$.

Thus we have got two new optical parallelograms having a pair of opposite corners $A'$ and $B'$ in common, lying in the inertia line $a$, while their separation diagonal lines are $c$ and $e$ respectively.

Thus we may prove in a manner similar to that already employed that $C'$ is neither *before* nor *after* any element of $e$.

Thus no element of $c$ is either *before* or *after* any element of $e$, as was to be proved.

## REMARKS

It is evident from the above that any general line which intersects the separation lines $c$ and $e$ in distinct elements must itself be a separation line.

It also appears from this theorem that it is possible to have an element which does not lie in a separation line and which is neither *before* nor *after* any elements of the separation line.

If two distinct elements $A_1$ and $A_2$ lie in a separation line $a$, while $A_3$ is an element which does not lie in $a$ and is neither *before* nor *after* any element of $a$, then we have already seen (p. 73) that $A_1$, $A_2$ and $A_3$

cannot lie in an inertia plane and it is also evident that they cannot lie in an optical plane.

For suppose, if possible, that $A_3$ does lie in an optical plane containing the separation line $a$; then there would be a generator of the optical plane passing through $A_3$ and intersecting $a$ in some element, say $A_4$.

Since $A_3$ is supposed not to lie in $a$, the elements $A_3$ and $A_4$ would require to be distinct and since they would then lie in an optical line we should have $A_3$ either *before* or *after* $A_4$: an element of $a$, contrary to hypothesis.

Thus $A_1$, $A_2$ and $A_3$ cannot lie in an optical plane.

Again if two distinct elements $A_1$ and $A_2$ lie in a separation line while $A_3$ is an element which does not lie in $A_1 A_2$ and is neither *before* nor *after* any element of $A_1 A_2$, then the element $A_2$ is neither *before* nor *after* any element of $A_3 A_1$.

For if $A_2$ were either *before* or *after* any element of $A_3 A_1$, then the three elements $A_1$, $A_2$, $A_3$ would lie either in an inertia or optical plane contrary to what we have just shown.

Similarly $A_1$ is neither *before* nor *after* any element of $A_2 A_3$.

Again if $a$ be a separation line and $A$ be an element which is not an element of $a$ and is neither *before* nor *after* any element of $a$, then if $B$ be any element of $a$, no element of the general line $AB$ is either *before* or *after* any element of $a$.

This is easily seen, for suppose, if possible, that $C$ is some element of $AB$ which is either *before* or *after* some element of $a$.

Then $C$ could not lie in $a$ and would lie either in an inertia or optical plane containing $a$.

But such inertia or optical plane would contain the element $A$ and so the separation line $a$ and the element $A$ would lie in an inertia or optical plane, contrary to what we have already proved.

Thus no element of $AB$ is either *before* or *after* any element of $a$.

*Definition.* An inertia line and a separation line which are diagonal lines of the same optical parallelogram will be said to be *conjugate* to one another.

It is evident that if an inertia line and a separation line are conjugate they lie in one inertia plane and intersect one another.

It is also evident that if $A$ be an element lying in an inertia or separation line $a$ in an inertia plane $P$, then there is only one separation or inertia line through $A$ and lying in $P$ which is conjugate to $a$; since, if two optical parallelograms lie in $P$ and have $a$ as a common diagonal line, then their other diagonal lines do not intersect (Post. XVI).

From this it also follows that if two intersecting separation lines $b$ and $c$ be both conjugate to the same inertia line $a$, then $a$, $b$ and $c$ cannot lie in the same inertia plane and we shall have $a$ and $b$ in one inertia plane, say $P$, while $a$ and $c$ lie in another, say $Q$.

If $O$ be the element of intersection of $b$ and $c$, then $O$ must lie both in $P$ and $Q$ and therefore in the inertia line $a$.

If $A_1$ be any element in $a$ distinct from $O$, there is one optical parallelogram in the inertia plane $P$ having $O$ as centre and $A_1$ as one of its corners.

If $A_2$ be the corner opposite $A_1$, then there is an optical parallelogram in $Q$ also having $A_1$ and $A_2$ as a pair of opposite corners and therefore having the same centre $O$.

The separation lines $b$ and $c$ will be the separation diagonal lines of the optical parallelograms in $P$ and $Q$ respectively, and so it follows by Theorem 99 that no element of $b$ is either *before* or *after* any element of $c$.

By considerations similar to the above, we can see that if two intersecting inertia lines $b$ and $c$ be both conjugate to the same separation line $a$, then $a$ and $b$ must lie in one inertia plane while $a$ and $c$ lie in another distinct inertia plane.

Further if $O$ be the element of intersection of $b$ and $c$, then $O$ lies in $a$.

In this case however, since $b$ and $c$ are two intersecting inertia lines, they must lie in one inertia plane which must be distinct from both the others.

Again it is clear that if $a$ be an inertia or separation line lying in an inertia plane $P$ with a separation or inertia line $b$ which is conjugate to $a$, then any general line $c$ lying in $P$ and parallel to $b$ is also conjugate to $a$.

Also conversely it is clear that if $a$ be an inertia or separation line lying in an inertia plane $P$ with two distinct separation or inertia lines $b$ and $c$ which are each conjugate to $a$, then $b$ and $c$ must be parallel to one another.

## THEOREM 100

*If an inertia line $a$ be conjugate to a separation line $b$, and if an inertia line $a'$ be co-directional with $a$ while a separation line $b'$ is co-directional with $b$, and if $a'$ and $b'$ intersect one another, then $a'$ is conjugate to $b'$.*

Let $P$ be the inertia plane containing $a$ and $b$ and let $O$ be the element of intersection of $a$ and $b$, while $O'$ is the element of intersection of $a'$ and $b'$.

Two cases have to be considered:

(1) $O'$ lies in the inertia plane $P$.

(2) $O'$ does not lie in the inertia plane $P$.

Consider first case (1).

Here both $a'$ and $b'$ must lie in $P$.

Then since $a'$ is co-directional with $a$ and $a$ is conjugate to $b$ and since $a$, $b$ and $a'$ lie in one inertia plane, it follows that $a'$ is conjugate to $b$.

Also since $b'$ is co-directional with $b$ and $a'$ is conjugate to $b$, and since $a'$, $b$ and $b'$ lie in one inertia plane, it follows that $a'$ is conjugate to $b'$.

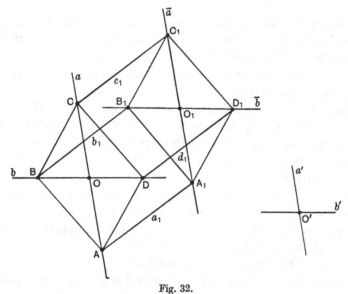

Fig. 32.

Consider next case (2).

Here $a'$ and $b'$ lie in an inertia plane $P'$ which must be distinct from $P$, since the element $O'$ does not lie in $P$ and therefore, by Theorem 52, $P'$ must be parallel to $P$.

Now let $A$ be any element of $a$ which is *before* $O$.

Then we know that there is one single optical parallelogram lying in $P$ which has $O$ as centre and $A$ as one of its corners.

Let $C$ be the corner opposite to $A$ and let $B$ and $D$ be the remaining pair of corners, which must both lie in $b$, since $b$ is conjugate to $a$ and intersects it in $O$.

Now, since $P$ and $P'$ are parallel inertia planes and since $b$ is a general line in $P$, then, as we showed on p. 76, there is at least one inertia plane, say $Q$, containing $b$ and another general line, say $\bar{b}$ in $P'$.

Then $\bar{b}$ must be parallel to $b$ and since $b$ is a separation line, $\bar{b}$ must also be a separation line.

Let $b_1$ and $d_1$ be parallel optical lines in $Q$ which pass through $B$ and $D$ respectively and let them intersect the separation line $\bar{b}$ in $B_1$ and $D_1$ respectively.

Let an optical line be taken through $B_1$ parallel to $BC$ and let an optical line be taken through $D_1$ parallel to $DC$.

Then these two optical lines will be generators of opposite sets of the inertia plane $P'$ and consequently will intersect in some element, say $C_1$.

Similarly if an optical line be taken through $B_1$ parallel to $BA$ and an optical line be taken through $D_1$ parallel to $DA$ these two optical lines will also lie in $P'$ and will intersect in some element, say $A_1$.

Now let optical lines $a_1$ and $c_1$ be taken through $A$ and $C$ respectively and parallel to $b_1$ and $d_1$.

Then $C$ is *after* $O$ and therefore also *after* both $B$ and $D$ and consequently $c_1$ is an after-parallel of $b_1$ and also an after-parallel of $d_1$.

Thus $B_1C$ and $D_1C_1$ must both intersect $c_1$ and this latter optical line cannot lie in $P'$ and so cannot have more than one element in common with $P'$.

But $C_1$ lies in $P'$ and is the one element common to $B_1C_1$ and $D_1C_1$ and so the optical line $c_1$ must pass through $C_1$.

In an analogous way we find that $a_1$ is a before-parallel of both $b_1$ and $d_1$ and must pass through the element $A_1$.

Further, $c_1$ must be an after-parallel of $a_1$.

But now, by hypothesis, $a$ is an inertia line so that $a$ and $a_1$ lie in an inertia plane, which must also contain $c_1$ since $c_1$ is parallel to $a_1$ and passes through the element $C$ of $a$.

Thus $A_1C_1$ must be an inertia line parallel to $a$ and we may denote it by $\bar{a}$.

Then $\bar{a}$ and $\bar{b}$ are diagonal lines of the optical parallelogram whose corners are $A_1, B_1, C_1$ and $D_1$ and so $\bar{a}$ is conjugate to $\bar{b}$; which intersects it in some element, say $O_1$.

But $a'$ and $\bar{a}$ are each parallel to $a$ and therefore $a'$ and $\bar{a}$ are co-directional while $b'$ and $\bar{b}$ are each parallel to $b$ and so $b'$ and $\bar{b}$ are co-directional.

Thus, by case (1), $a'$ is conjugate to $b'$, as was to be proved.

Thus the theorem holds in all cases.

*Definitions.* If $A$ be any element and $a$ be an inertia line not containing $A$, while $B$ is the element common to $a$ and the $\alpha$ sub-set of $A$, then we shall speak of $B$ as *the first element of a which is* after $A$.

Similarly if $C$ be the element common to $a$ and the $\beta$ sub-set of $A$, we shall speak of $C$ as *the last element of a which is* before $A$.

POSTULATE XVIII. **If $a$, $b$ and $c$ be three parallel inertia lines which do not all lie in one inertia plane* and $A_1$ be any element in $a$ and if**

> **$B_1$ be the first element in $b$ which is after $A_1$,**
>
> **$C_1$ be the first element in $c$ which is after $A_1$,**
>
> **$B_2$ be the first element in $b$ which is after $C_1$,**
>
> **$C_2$ be the first element in $c$ which is after $B_1$,**

**then the first element in $a$ which is after $B_2$ and the first element in $a$ which is after $C_2$ are identical.**

It is evident that there is a $(b)$ form of this postulate in which the word *last* is substituted for the word *first* and the word *before* for the word *after*, but this is not independent, as may be readily seen.

Thus let $A_1$ be any element in $a$ and let $B_1$ be the last element in $b$ which is *before* $A_1$ and let $C_2$ be the last element in $c$ which is *before* $B_1$ while $A_2$ is the last element in $a$ which is *before* $C_2$.

Then $C_2$ is the first element in $c$ which is *after* $A_2$,

> $B_1$ is the first element in $b$ which is *after* $C_2$,
>
> $A_1$ is the first element in $a$ which is *after* $B_1$.

Thus if $B_2$ be the first element in $b$ which is *after* $A_2$ and if $C_1$ be the first element in $c$ which is *after* $B_2$, it follows by Post. XVIII that the first element in $a$ which is *after* $C_1$ is identical with the first element in $a$ which is *after* $B_1$: that is, with the element $A_1$.

But $C_1$ is the last element in $c$ which is *before* $A_1$ and $B_2$ is the last element in $b$ which is *before* $C_1$ while $A_2$ is the last element in $a$ which is *before* $B_2$.

Thus the last element in $a$ which is *before* $B_2$ and the last element in $a$ which is *before* $C_2$ are identical.

---

* If $a$, $b$ and $c$ do all lie in one inertia plane, the same result may easily be deduced from the other postulates.

## Theorem 101

*If an inertia line c be conjugate to two intersecting separation lines d and e, then if A be any element of d and B be any distinct element of e, the general line AB is conjugate to a set of inertia lines which are parallel to c.*

Let $C_1$ be the element of intersection of the separation lines $d$ and $e$.

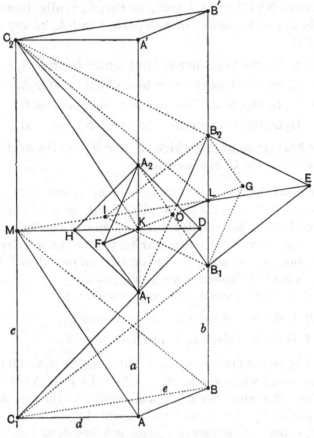

Fig. 33.

Then we know that $c$ and $d$ lie in one inertia plane, say $P$, while $c$ and $e$ lie in another distinct inertia plane, say $Q$, and the element $C_1$ lies in $c$.

If $A$ or $B$ should coincide with $C_1$, then the general line $AB$ must coincide with $e$ or $d$ and the result follows directly.

We shall suppose therefore that neither $A$ nor $B$ coincides with $C_1$.

Then since by Theorem 99 $A$ is neither *before* nor *after* $B$ and since $A$ and $B$ are distinct it follows that $AB$ is a separation line.

Let $a$ be an inertia line through $A$ parallel to $c$ while $b$ is an inertia line through $B$ parallel to $c$.

Then since $A$ and $B$ lie in a separation line it follows that $a$ and $b$ must be distinct and therefore are parallel to one another.

Thus $a$ and $b$ must lie in an inertia plane which we shall call $R$.

Further, $a$ must lie in the inertia plane $P$ while $b$ must lie in the inertia plane $Q$.

Now let $A_1$ be the first element in $a$ which is *after* $C_1$,

    let $B_1$ be the first element in $b$ which is *after* $C_1$,

    let $A_2$ be the first element in $a$ which is *after* $B_1$,

    let $B_2$ be the first element in $b$ which is *after* $A_1$.

If now $C_2$ be the first element in $c$ which is *after* $A_2$, it follows by Post. XVIII that $C_2$ is also the first element in $c$ which is *after* $B_2$.

Now the optical lines $C_1A_1$ and $C_2A_2$ cannot be parallel; for since $A_1$ is *after* $C_1$ and $c$ and $a$ are parallel inertia lines in the inertia plane $P$, it would then follow by Theorem 57 (*b*) that $A_2$ was *after* $C_2$.

But we know that $C_2$ is *after* $A_2$ and so $C_1A_1$ and $C_2A_2$ are not parallel, and, since they lie in one inertia plane, it follows that they must intersect in some element, say $D$.

Similarly $C_1B_1$ and $C_2B_2$ must intersect in some element, say $E$.

Also since $a$ and $b$ are parallel inertia lines in the inertia plane $R$ and since $B_2$ is *after* $A_1$ and $A_2$ *after* $B_1$, it follows in a similar manner that the optical lines $A_1B_2$ and $A_2B_1$ must intersect in some element, say $O$.

Now $A_2$ cannot be identical with $A_1$ for then we should have the three elements $C_1$, $B_1$ and $A_1$ lying in pairs in three distinct optical lines, which is impossible by Theorem 14.

Further, since $B_1$ is *after* $C_1$ and $A_2$ is *after* $B_1$, it follows that $A_2$ is *after* $C_1$.

But $A_2$ cannot be *before* $A_1$ for then we should have $A_2$ *after* one element of the optical line $C_1A_1$ and *before* another element of it which would entail that $A_2$ should lie in the optical line $C_1A_1$, by Theorem 12.

We know however that $A_2$ and $A_1$ are distinct elements in the inertia line $a$ and so $A_2$ cannot be *before* $A_1$.

Thus, since $A_1$ and $A_2$ are distinct elements in an inertia line and $A_2$ is not *before* $A_1$, it follows that $A_2$ is *after* $A_1$.

Similarly $B_2$ is *after* $B_1$.

Let an optical line through $A_1$ parallel to $DA_2$ be taken and an optical line through $A_2$ parallel to $DA_1$ and let these intersect in $H$.

Then $A_1$, $D$, $A_2$, $H$ form the corners of an optical parallelogram in the inertia plane $P$, having its centre, say $K$, in the inertia line $a$.

Again let an optical line through $B_1$ parallel to $EB_2$ be taken and an optical line through $B_2$ parallel to $EB_1$ and let these intersect in $I$.

Then $B_1$, $E$, $B_2$, $I$ form the corners of an optical parallelogram in the inertia plane $Q$, having its centre, say $L$, in the inertia line $b$.

If now we take optical parallelograms having $C_1$ and $C_2$ as opposite corners in each of the inertia planes $P$ and $Q$ then, by Theorem 60, these have a common centre, say $M$, lying in the inertia line $c$.

Also $D$ will be one of the remaining corners of the optical parallelogram in $P$ while $E$ will be one of the remaining corners of the optical parallelogram in $Q$.

Thus $MD$ and $ME$ will each be conjugate to $c$.

Further, since the general lines $MD$ and $d$ are both conjugate to $c$ and lie in the same inertia plane $P$, they must be parallel to one another.

Similarly the general lines $ME$ and $e$ must also be parallel to one another.

But now the optical parallelogram in the inertia plane $P$ having $C_1$ and $C_2$ as a pair of opposite corners, and the optical parallelogram whose corners are $A_1$, $D$, $A_2$, $H$ have diagonal lines $c$ and $a$ respectively which do not intersect, and so since they both lie in $P$ their other diagonal lines do not intersect.

But these other diagonal lines have the element $D$ in common and so must be identical.

Thus the general lines $MD$ and $KD$ are identical and so $K$ lies in $MD$.

Similarly $L$ lies in $ME$.

Now let an optical line through $A_1$ parallel to $OA_2$ be taken and an optical line through $A_2$ parallel to $OA_1$ and let these intersect in $F$.

Then $A_1$, $F$, $A_2$, $O$ form the corners of an optical parallelogram in the inertia plane $R$, and by Theorem 60 this must have the same centre $K$ as the optical parallelogram whose corners are $A_1$, $D$, $A_2$, $H$.

Again let an optical line through $B_1$ parallel to $OB_2$ be taken and an optical line through $B_2$ parallel to $OB_1$ and let these intersect in $G$.

Then $B_1$, $G$, $B_2$, $O$ form the corners of an optical parallelogram in the inertia plane $R$, and by Theorem 60 this must have the same centre $L$ as the optical parallelogram whose corners are $B_1$, $E$, $B_2$, $I$.

But now the optical parallelograms whose corners are $A_1$, $F$, $A_2$, $O$ and $B_1$, $G$, $B_2$, $O$ have the diagonal lines $a$ and $b$ which do not intersect and so, since both lie in the same inertia plane $R$, their other diagonal lines do not intersect.

That is, $FO$ and $GO$ do not intersect and so since they have the element $O$ in common they must be identical.

Thus $O$ lies in the general line $FG$; that is, in the general line $KL$.

Thus $KL$ is conjugate to both $a$ and $b$.

Now let a general line through $C_2$ parallel to $C_1A$ intersect $a$ in $A'$, and let a general line through $C_2$ parallel to $C_1B$ intersect $b$ in $B'$.

Then $A, A', C_2, C_1$ form the corners of a general parallelogram in the inertia plane $P$, while $B, B', C_2, C_1$ form the corners of a general parallelogram in the inertia plane $Q$.

Also, since $MK$ and $C_2A'$ are both parallel to $C_1A$, and since $M$ is the mean of $C_1$ and $C_2$, it follows by Theorem 81 that $K$ must be the mean of $A$ and $A'$.

Similarly $L$ is the mean of $B$ and $B'$.

Thus by Theorem 80 the general lines $AM$ and $KC_2$ are parallel to one another and similarly the general lines $BM$ and $LC_2$ are parallel to one another.

But now, since $A_2$ is *after* $A_1$ and since $K$ is the centre of optical parallelograms having $A_1$ and $A_2$ as opposite corners, it follows that $K$ is *after* $A_1$ and *before* $A_2$.

But, since $A_2$ is *before* $C_2$, it follows that $K$ is *before* $C_2$.

But $A_2$ is the only element common to $a$ and the $\beta$ sub-set of $C_2$ and $K$ is distinct from $A_2$.

Thus since $K$ is *before* $C_2$ and does not lie in the $\beta$ sub-set of $C_2$, it follows that $KC_2$ must be an inertia line.

Similarly $LC_2$ is an inertia line.

Thus $KC_2$ and $LC_2$ lie in an inertia plane, say $S$, while $MA$ and $MB$ being respectively parallel to these must, by Theorem 52, lie in an inertia plane, say $S'$, parallel to $S$.

But now the general lines $KL$ and $AB$ lie in $S$ and $S'$ respectively and also both lie in the inertia plane $R$.

Thus $AB$ is parallel to $KL$ and so, since $KL$ is conjugate to $a$ and $b$, we must also have $AB$ conjugate to $a$ and $b$, and therefore also conjugate to every inertia line in $R$ parallel to $a$ and $b$.

But since $a$ and $b$ are parallel to $c$, therefore all these inertia lines are parallel to $c$ and so the theorem is proved.

## Theorem 102

*If $P_1$ and $P_2$ be parallel inertia planes and if $A_1$ be any element in $P_1$ while $A_2$ is its representative in $P_2$, then the separation line $A_1A_2$ is conjugate to every inertia line in $P_1$ which passes through $A_1$, and similarly $A_1A_2$ is conjugate to every inertia line in $P_2$ which passes through $A_2$.*

Let $a_1$ and $b_1$ be the two generators of the inertia plane $P_1$ which pass through the element $A_1$, and let $a_2$ and $b_2$ be the two generators of the inertia plane $P_2$ which pass through $A_2$, and let $a_2$ be neutrally parallel to $a_1$ while $b_2$ is neutrally parallel to $b_1$.

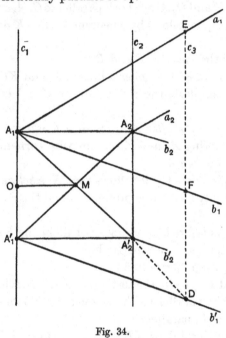

Fig. 34.

Let $c_1$ be any arbitrary inertia line in $P_1$ which passes through $A_1$.

Then $c_1$ and the element $A_2$ lie in an inertia plane, say $Q$, which contains the inertia line $c_1$ in common with $P_1$ and the element $A_2$ in common with $P_2$ and therefore, by Theorem 46, must have a general line in common with $P_2$ which must be parallel to $c_1$ and pass through $A_2$.

Let this parallel to $c_1$ through $A_2$ be denoted by $c_2$.

Then $c_2$ must also be an inertia line.

Let $A_1'$ be the one single element common to $c_1$ and the $\beta$ sub-set of $A_2$, while $A_2'$ is the one single element common to $c_2$ and the $\beta$ sub-set of $A_1$, so that $A_2A_1'$ and $A_1A_2'$ are optical lines.

Also $A_1'$ is *before* $A_2$ while $A_2'$ is *before* $A_1$.

But, since $A_1A_2$ is a separation line while $A_1A_1'$ is an inertia line and $A_1'$ is *before* $A_2$, it follows that $A_1'$ must also be *before* $A_1$.

Similarly $A_2'$ must be *before* $A_2$.

Now $A_1'$ cannot be *after* $A_2'$, for then, since $A_1'$ is before $A_1$, it would require to lie in the optical line $A_1A_2'$, which is impossible, since $A_1A_1'$ is an inertia line.

Similarly $A_2'$ cannot be *after* $A_1'$, and accordingly $A_1'A_2'$ must be a separation line.

Now let $b_1'$ and $b_2'$ be optical lines through $A_1'$ and $A_2'$ respectively parallel to $b_1$ and accordingly parallel to one another.

We shall presently show that $b_1'$ and $b_2'$ must be neutral parallels, but let us first consider any element $D$ which lies in $b_1'$ and *before* $A_1'$ and let $c_3$ be an inertia line through $D$ parallel to $c_1$ and $c_2$. Let $c_3$ intersect $a_1$ in $E$ and $b_1$ in $F$.

Then, since $D$ is *before* $A_1'$ and since $b_1'$ and $b_1$ are parallel optical lines, it follows that $F$ is *before* $A_1$, so that $F$ lies in the $\beta$ sub-set of $A_1$.

Thus $E$ must lie in the $\alpha$ sub-set of $A_1$ and, since $D$ is *before* $A_1'$, it follows that $A_1'$ is in the $\alpha$ sub-set of $D$.

Suppose now, if possible, that $b_2'$ is an after-parallel of $b_1'$.

Then $A_2'$ would be *after* some element of $b_1'$ and so there would be one single element common to $b_1'$ and the $\beta$ sub-set of $A_2'$.

This hypothetical element would therefore be *before* $A_2'$ and would also have to be *before* $A_1'$, since $A_1'A_2'$ is a separation line.

If now, we try to identify this hypothetical element with $D$, we shall find it impossible, for, if we suppose $D$ to be in the $\beta$ sub-set of $A_2'$ we should have $A_2'$ in the $\alpha$ sub-set of $D$ and accordingly we should have:

> $A_1'$ the first element in $c_1$ which is *after* $D$;
>
> $A_2'$ the first element in $c_2$ which is *after* $D$;
>
> $A_1$ the first element in $c_1$ which is *after* $A_2'$;
>
> $A_2$ the first element in $c_2$ which is *after* $A'_1$;
>
> $E$   the first element in $c_3$ which is *after* $A_1$;

and so, by Post. XVIII, $E$ should be the first element in $c_3$ which is *after* $A_2$.

But $A_2$ is neither *before* nor *after* any element of $a_1$, and so $E$ could not be *after* $A_2$.

Thus the assumption of the existence of an element common to $b_1'$ and the $\beta$ sub-set of $A_2'$ leads to a contradiction, and so $b_2'$ cannot be an after-parallel of $b_1'$.

Similarly, if $b_2'$ were supposed to be a before-parallel of $b_1'$, we should require $b_1'$ to be an after-parallel of $b_2'$, and a similar method would show this also to be impossible.

Thus, since $b_2'$ is parallel to $b_1'$, and cannot be either an after- or before-parallel of $b_1'$, it follows that $b_2'$ is a neutral-parallel of $b_1'$.

But now the separation lines $A_1A_2$ and $A_1'A_2'$ cannot intersect, for, since $A_2$ is neither *before* nor *after* any element of $b_1$, while $A_2'$ is neither *before* nor *after* any element of $b_1'$, it would follow, by Theorem 45, that such an element of intersection, if it existed, would be neither *before* nor *after* any element of either $b_1$ or $b_1'$, and so there would be an optical line through it which would be neutrally parallel to both $b_1$ and $b_1'$.

But, if this were so, it would follow, by Theorem 28, that $b_1'$ was a neutral-parallel of $b_1$, contrary to what we have already seen, that the element $A_1'$ of $b_1'$ is *before* the element $A_1$ of $b_1$.

Thus the separation lines $A_1A_2$ and $A_1'A_2'$ cannot intersect and so, since they both lie in the inertia plane $Q$ and are distinct, it follows that they are parallel.

Thus $A_1$, $A_2$, $A_2'$, $A_1'$ form the corners of a parallelogram in the inertia plane $Q$ and its diagonal lines are $A_1A_2'$ and $A_2A_1'$ which are both optical lines which must intersect in some element, say $M$.

If then a general line be taken through $M$ parallel to $A_1A_2$ and meeting $A_1A_1'$ in an element $O$, it follows, by Theorem 79, that $O$ is the mean of $A_1$ and $A_1'$.

Thus an optical parallelogram in the inertia plane $Q$ having $A_1$ and $A_1'$ as a pair of opposite corners will have $OM$ as its separation diagonal line.

Thus $OM$ is conjugate to $c_1$ and, since $A_1A_2$ is parallel to $OM$ and in the same inertia plane $Q$ as are $OM$ and $c_1$, it follows that $A_1A_2$ is also conjugate to $c_1$, and therefore conjugate to every inertia line in $P_1$ which passes through $A_1$.

Similarly $A_1A_2$ is conjugate to every inertia line in $P_2$ which passes through $A_2$ and so the theorem is proved.

### Theorem 103

*If two inertia lines b and c intersect in an element $A_1$ and are both conjugate to a separation line a, then a is conjugate to every inertia line in the inertia plane containing b and c which passes through the element $A_1$.*

We have already seen that $a$ cannot lie in the inertia plane containing $b$ and $c$ and also that it passes through the element of intersection of $b$ and $c$.

Let $P_1$ be the inertia plane containing $b$ and $c$ and let $A_2$ be any element of $a$ distinct from $A_1$.

Let $P_2$ be an inertia plane through $A_2$ and parallel to $P_1$.

Let $A_2'$ be the representative of $A_1$ in the inertia plane $P_2$.

We shall show that $A_2'$ must be identical with $A_2$.

Since the inertia line $b$ and the separation line $a$ intersect in the element $A_1$ they must lie in one inertia plane which contains the inertia line $b$ in common with the inertia plane $P_1$ and the element $A_2$ in common with the parallel inertia plane $P_2$.

Thus the inertia plane containing $b$ and $a$ has a general line, say $b'$, in common with $P_2$, and $b'$ is parallel to $b$ and is therefore also an inertia line.

Similarly the inertia plane containing $c$ and $a$ has an inertia line, say $c'$, in common with $P_2$, and $c'$ is parallel to $c$.

Further $b'$ and $c'$ must both pass through $A_2$ and must be distinct since $b$ and $c$ are distinct.

Now since $A_1$ and $A_2'$ are representatives of one another in the parallel inertia planes $P_1$ and $P_2$, it follows, by Theorem 102, that the separation line $A_1 A_2'$ is conjugate to any inertia line in $P_1$ which passes through $A_1$.

Thus $A_1 A_2'$ must be conjugate to both $b$ and $c$.

Suppose now, if possible, that $A_2'$ is distinct from $A_2$.

Then $b$ is conjugate to both $A_1 A_2$ and $A_1 A_2'$ and so, by Theorem 101, $b'$ would be conjugate to $A_2 A_2'$.

Similarly $c$ is conjugate to both $A_1 A_2$ and $A_1 A_2'$ and so $c'$ would be conjugate to $A_2 A_2'$.

But then we should have two distinct inertia lines $b'$ and $c'$ both passing through $A_2$ and conjugate to the same general line $A_2 A_2'$ in the inertia plane $P_2$ which contains $b'$ and $c'$, and this we know is impossible.

Thus $A_2'$ cannot be distinct from $A_2$ and so $A_2$ must be the representative of $A_1$ in the inertia plane $P_2$.

It follows accordingly that the separation line $a$ is conjugate to every inertia line in $P_1$ which passes through $A_1$, and so the theorem is proved.

It is to be noted that in proving the above theorem we have also incidentally proved the following important result:

If two inertia lines $b$ and $c$ intersect in an element $A_1$ and are both conjugate to a separation line $a$, then $a$ is such that no element of it, with the exception of $A_1$, is either *before* or *after* any element of either of the generators of the inertia plane containing $b$ and $c$ which pass through $A_1$.

THEOREM 104

*If an optical line $b$ and an inertia line $c$ intersect in an element $A_1$ and if a separation line $a$ passing through $A_1$ be such that no element of $a$ except $A_1$ is either* before *or* after *any element of $b$ and if further $a$ be conjugate to $c$, then $a$ is conjugate to every inertia line which passes through $A_1$ and lies in the inertia plane containing $b$ and $c$.*

Let $A_2$ be any element of $a$ distinct from $A_1$ and let $b'$ be an optical line through $A_2$ parallel to $b$ while $c'$ is an inertia line through $A_2$ parallel to $c$.

Then $b'$ must be a neutral-parallel of $b$.

Let $P_1$ be the inertia plane containing $b$ and $c$ and let $P_2$ be the inertia plane containing $b'$ and $c'$.

Then, since $A_2$ is neither *before* nor *after* any element of the optical line $b$, it follows that $A_2$ does not lie in $P_1$ and so the inertia planes $P_1$ and $P_2$ are parallel to one another.

Let $A_2'$ be the representative of $A_1$ in $P_2$; then $A_2'$ must lie in $b'$ by the definition of representative elements and by Theorem 102 $A_1 A_2'$ is conjugate to $c$.

But $A_1 A_2$ is conjugate to $c$ and so if $A_2'$ were distinct from $A_2$ we should have $c$ conjugate to two intersecting separation lines and so by Theorem 99 no element of $A_1 A_2$ could be either *before* or *after* any element of $A_1 A_2'$.

But if $A_2'$ were distinct from $A_2$, then, since they each lie in the optical line $b'$, it would follow that the one must be *after* the other.

Thus the supposition that $A_2'$ is distinct from $A_2$ leads to a contradiction and so $A_2'$ must be identical with $A_2$.

Thus it follows by Theorem 102 that $A_1 A_2$ (that is $a$) is conjugate to every inertia line which passes through $A_1$ and lies in $P_1$.

Thus the theorem is proved.

It follows directly from the above that no element of $a$ with the exception of $A_1$ is either *before* or *after* any element of the second generator of $P$ which passes through $A_1$.

## THEOREM 105

*If a separation line a be conjugate to two intersecting inertia lines b and c, then any inertia line in the inertia plane containing b and c is conjugate to a set of separation lines which are parallel to a.*

Let the inertia lines $b$ and $c$ intersect in the element $A_1$.

Then we know that $a$ must also pass through $A_1$, but does not lie in the inertia plane containing $b$ and $c$.

Let $P_1$ be the inertia plane containing $b$ and $c$; let $A_2$ be any element in $a$ distinct from $A_1$ and let $P_2$ be an inertia plane through $A_2$ parallel to $P_1$.

Then we have seen in the course of proving Theorem 103 that $A_1$ and $A_2$ are representatives of one another in the parallel inertia planes $P_1$ and $P_2$ respectively, and further every inertia line in $P_1$ which passes through $A_1$ is conjugate to $a$.

Let $d$ be any inertia line in the inertia plane $P_1$ and let $A_1'$ be any element in $d$ while $A_2'$ is the representative of $A_1'$ in $P_2$.

Then by Theorem 102 the separation line $A_1'A_2'$ is conjugate to $d$.

But, provided $A_1'$ be distinct from $A_1$, it follows by Theorem 98 that $A_1'A_2'$ is parallel to $A_1A_2$: that is to $a$, and, since there are an infinite number of elements in $d$, it follows that $d$ is conjugate to a set of separation lines which are parallel to $a$.

Thus the theorem is proved.

## THEOREM 106

*If b and c be any two intersecting inertia lines, there is at least one separation line which is conjugate to both b and c.*

Let the inertia lines $b$ and $c$ intersect in the element $A_1$ and let $P_1$ be the inertia plane containing $b$ and $c$.

Let any element be taken which does not lie in $P_1$ and through it let an inertia plane $P_2$ be taken parallel to $P_1$.

Let $A_2$ be the element in $P_2$ which is the representative of $A_1$.

Then by Theorem 102 the separation line $A_1A_2$ is conjugate to any inertia line in $P_1$ which passes through $A_1$.

Thus the separation line $A_1A_2$ is conjugate to both $b$ and $c$, and so the theorem is proved.

THEOREM 107

*If $b$ and $c$ be any two intersecting separation lines such that no element of the one is either before or after any element of the other, there is at least one inertia line which is conjugate to both $b$ and $c$.*

Let the separation lines $b$ and $c$ intersect in the element $A_1$ and let $Q$ be any inertia plane containing $b$.

Now, since no element of $c$ is either *before* or *after* any element of $b$, it follows that $c$ and $b$ do not lie in one inertia plane and therefore $c$ does not lie in $Q$.

Let $d_1$ be the inertia line through $A_1$ in the inertia plane $Q$ which is conjugate to $b$.

Then $d_1$ being an inertia line which intersects $c$, it follows that $d_1$ and $c$ lie in an inertia plane, say $R$, which must be distinct from $Q$.

Let $e$ be any other inertia line in $R$ distinct from $d_1$ and passing through the element $A_1$.

Then $e$ being an inertia line which intersects $b$, it follows that $e$ and $b$ lie in an inertia plane, say $Q'$, which must also be distinct from $R$.

Let $d_1'$ be the inertia line through $A_1$ in the inertia plane $Q'$ which is conjugate to $b$.

Then $d_1'$ may either coincide with $e$ or be distinct from it.

Consider first the case where $d_1'$ coincides with $e$.

Since then both $d_1$ and $d_1'$ will lie in the inertia plane $R$ and since the separation line $b$ is conjugate to both $d_1$ and $d_1'$, it follows by Theorem 103 that $b$ is conjugate to every inertia line in the inertia plane $R$ which passes through the element $A_1$.

Let $a$ be the inertia line through $A_1$ in the inertia plane $R$ which is conjugate to $c$.

Then $a$ must also be conjugate to $b$ and so the theorem will hold in this case.

Consider next the case where $d_1'$ is distinct from $e$.

Since $d_1$ and $d_1'$ are intersecting inertia lines, they will lie in an inertia plane, say $P_1$, which will be distinct from both $Q$ and $Q'$.

Also, since $d_1'$ does not lie in $R$ in this case, it follows that $P_1$ is distinct from $R$, which has the inertia line $d_1$ in common with $P_1$.

Since $A_1$ is the only element of $c$ which is also an element of $d_1$, it follows that $A_1$ is the only element of $c$ which lies in $P_1$.

Let $A_0$ be any element of $c$ distinct from $A_1$ and let $d_0$ and $d_0'$ be inertia lines through $A_0$ parallel to $d_1$ and $d_1'$ respectively.

Then, by Theorem 52, $d_0$ and $d_0'$ lie in an inertia plane, say $P_2$, which is parallel to $P_1$.

Again, since $A_0$ is an element of the inertia plane $R$ and since $d_0$ is parallel to $d_1$, it follows that $d_0$ lies in $R$.

Further, since the inertia line $e$ is distinct from $d_1$, it cannot be parallel to $d_0$ and must therefore intersect $d_0$ in some element, say $B$.

But the inertia line $e$ also lies in $Q'$ and so $Q'$ contains the element $B$ in common with $P_2$.

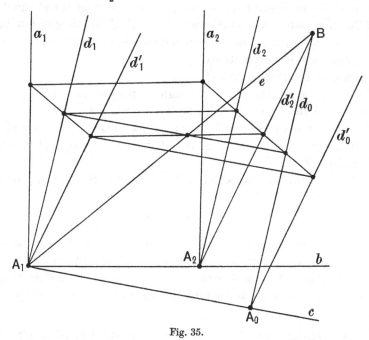

Fig. 35.

Since however $Q'$ has the inertia line $d_1'$ in common with $P_1$, it follows by Theorem 46 that $Q'$ and $P_2$ have a general line, say $d_2'$, in common which must be parallel to $d_1'$ and is therefore an inertia line.

Now, since $b$ is a separation line and $d_2'$ an inertia line in the inertia plane $Q'$, it follows that $b$ and $d_2'$ intersect in some element, say $A_2$.

But $A_2$ being an element of $b$ is an element of the inertia plane $Q$, and accordingly $Q$ has the element $A_2$ in common with the inertia plane $P_2$.

Since however $Q$ has the inertia line $d_1$ in common with $P_1$, it follows by Theorem 46 that $Q$ and $P_2$ have a general line, say $d_2$, in common which must be parallel to $d_1$ and is therefore an inertia line.

But now since $d_1$ and $d_2$ are parallel inertia lines in the inertia plane $Q$ and since $d_1$ is conjugate to $b$, it follows that $d_2$ is conjugate to $b$.

Also since $d_1'$ and $d_2'$ are parallel inertia lines in the inertia plane $Q'$ and since $d_1'$ is conjugate to $b$, it follows that $d_2'$ is conjugate to $b$.

Thus the separation line $b$ is conjugate to the two intersecting inertia lines $d_2$ and $d_2'$ in the inertia plane $P_2$ and so, by Theorem 103, $b$ is conjugate to every inertia line in $P_2$ which passes through the element $A_2$.

Now since no element of $c$ is either *before* or *after* any element of $b$ it follows that the element $A_0$ is neither *before* nor *after* the element $A_2$ and therefore, since $A_2$ and $A_0$ are distinct, $A_2A_0$ is a separation line.

Now let $a_2$ be the inertia line in the inertia plane $P_2$ which passes through $A_2$ and is conjugate to $A_2A_0$.

Then $a_2$ is also conjugate to $b$.

Thus if $a_1$ be an inertia line through $A_1$ parallel to $a_2$ it follows by Theorem 101 that $a_1$ is conjugate to $A_1A_0$: that is to $c$.

But $a_1$ and $a_2$ being parallel inertia lines through elements of the separation line $b$ and $a_2$ being conjugate to $b$, it follows that $a_1$ is also conjugate to $b$.

Thus $a_1$ is conjugate to both $b$ and $c$ and so the theorem is proved.

### THEOREM 108

*If $a$ be a separation line and $B$ be any element which is not an element of $a$ and is neither* before *nor* after *any element of $a$ while $c$ is a general line passing through $B$ and parallel to $a$, then if $A$ be any element of $a$, while $C$ is an element of $c$ distinct from $B$, a general line through $C$ parallel to $BA$ will intersect $a$.*

Let the general line $BA$ be denoted by $b$ and let the general line through $C$ parallel to $b$ be denoted by $d$.

Then, as was pointed out in the remarks at the end of Theorem 99, no element of $b$ is either *before* or *after* any element of $a$ and so, since $a$ and $b$ intersect in $A$, it follows by Theorem 107 that there is at least one inertia line which is conjugate to both $a$ and $b$, and must therefore pass through $A$.

Let $a_1$ be such an inertia line and let $b_1$ be an inertia line through $B$ parallel to $a_1$, while $c_1$ is an inertia line through $C$ parallel to $a_1$ and $b_1$.

Then $a$ and $a_1$ lie in an inertia plane which we may call $P_a$, while $a_1$, $b$ and $b_1$ lie in an inertia plane which we may call $P_b$, and $b_1$, $c$ and $c_1$ lie in an inertia plane which we may call $P_c$.

Then, since $B$ and $a$ do not lie in one inertia plane, it follows that $B$

is not an element of $P_a$ and so, since $b_1$ and $c$ are respectively parallel to $a_1$ and $a$, it follows that $P_c$ is parallel to $P_a$.

Now, since $b_1$ and $a_1$ both lie in $P_0$ and are parallel to one another, and since $a_1$ is conjugate to $b$, it follows that $b_1$ is also conjugate to $b$.

But since $c$ is parallel to $a$ and $b_1$ is parallel to $a_1$, while $c$ and $b_1$ intersect in $B$, it follows, by Theorem 100, that since $a_1$ is conjugate to $a$, therefore $b_1$ is conjugate to $c$.

Thus $b_1$ is conjugate to both $b$ and $c$, which are two distinct and intersecting separation lines and therefore cannot lie in one inertia plane.

Thus $C$ is not an element of $P_b$ and so, if $P_d$ be an inertia plane containing $c_1$ and $d$, then, since $c_1$ is parallel to $b_1$ while $d$ is parallel to $b$, it follows that the inertia plane $P_d$ is parallel to $P_b$.

Then, by Theorem 54, $P_a$ and $P_d$ have a general line in common and, if we call this general line $d_1$, then, since $P_d$ and $P_b$ are parallel, $d_1$ must be parallel to $a_1$ and must be an inertia line.

Now since $c_1$ is parallel to $b_1$ and $d$ is parallel to $b$, and $c_1$ and $d$ intersect, it follows, by Theorem 100, that, since $b_1$ is conjugate to $b$, therefore $c_1$ is conjugate to $d$.

Again since $b_1$ is conjugate to both $b$ and $c$ and since $A$ is an element in $b$ while $C$ is a distinct element in $c$, it follows, by Theorem 101, that the general line $c_1$ is conjugate to $CA$.

Thus $c_1$ is conjugate to both $d$ and $CA$.

Now since $d$ is a separation line while $d_1$ is an inertia line and both lie in the inertia plane $P_d$, it follows that $d$ and $d_1$ must intersect in some element, say $D$.

Thus, since $A$ is an element in $CA$ while $D$ is a distinct element in $d$, it follows, by Theorem 101, that $a_1$ is conjugate to $DA$.

But $a_1$ is conjugate to the separation line $a$ which also passes through $A$, and so, since both $DA$ and $a$ lie in the inertia plane $P_a$ which contains $a_1$, it follows that the general lines $DA$ and $a$ are identical.

Thus $D$ lies in $a$ and therefore the general line $d$ intersects $a$.

Thus the theorem is proved.

## REMARKS

If $a$ be a separation line and $B$ be any element which is not an element of $a$ and is neither *before* nor *after* any element of $a$, then if $b$ be a separation line through $B$ parallel to $a$, no element of $b$ is either *before* or *after* any element of $a$.

This is easily seen: for if $C$ were an element of $b$ which was either *before* or *after* an element of $a$, then the separation line $a$ and the element $C$ would lie either in one inertia plane, or in one optical plane.

Such inertia or optical plane would contain the general line through $C$ parallel to $a$: that is to say it would contain $b$.

Thus the separation line $a$ and the element $B$ would lie in one inertia or optical plane, which we already know is impossible.

Thus no element of $b$ is either *before* or *after* any element of $a$ and therefore any general line which intersects both $a$ and $b$ must be a separation line.

Again, if $AB$ and $DC$ be two parallel separation lines such that no element of the one is either *before* or *after* any element of the other and if $CB$ is parallel to $DA$, then no element of $DA$ is either *before* or *after* any element of $CB$.

This is easily seen: for we know that no element of $CB$ is either *before* or *after* any element of $AB$ and therefore the element $A$ is neither *after* nor *before* any element of $CB$.

Thus since $DA$ is parallel to $CB$ it follows that no element of $DA$ is either *before* or *after* any element of $CB$.

### THEOREM 109

*If $A$ and $B$ be two elements lying respectively in two parallel separation lines $a$ and $b$ which are such that no element of the one is either* before *or* after *any element of the other, and if $A'$ be a second and distinct element in $a$, there is only one general line through $A'$ and intersecting $b$ which does not intersect the general line $AB$.*

We have seen by Theorem 108 that the general line through $A'$ parallel to $AB$ must intersect $b$.

Let $B'$ be the element of intersection. Then the general lines $AB$ and $A'B'$, being parallel, cannot intersect.

Let any other general line through $A'$ and intersecting $b$ intersect it in the element $C$.

Then if $C$ should coincide with $B$ the general lines $A'C$ and $AB$ have the element $B$ in common and therefore intersect.

Suppose next that $C$ does not coincide with $B$.

Since $B$ is neither *before* nor *after* any element of $a$ and since therefore no element of $AB$ is either *before* or *after* any element of $a$, it follows, by Theorem 107, that there is at least one inertia line, say $a_1$, which is conjugate to both $AB$ and $a$ and therefore passes through $A$.

Let $b_1$ be an inertia line through $B$ parallel to $a_1$, and let $a_1'$ and $b_1'$ be inertia lines through $A'$ and $B'$ respectively and also parallel to $a_1$.

Then $a_1$ and $a_1'$ lie in one inertia plane, say $P_1$, which contains also the separation line $a$; while $b_1$ and $b_1'$ lie in an inertia plane, say $P_2$, containing $b$.

Since the elements $B$, $A$ and $A'$ cannot lie in one inertia plane and since $b_1$ and $b$ are respectively parallel to $a_1$ and $a$, it follows that $P_2$ is parallel to $P_1$.

Again $a_1$ and $b_1$ lie in an inertia plane, say $Q$, containing $AB$, while $a_1'$ and $b_1'$ lie in an inertia plane, say $Q'$, containing $A'B'$.

Since the elements $B$, $A$ and $A'$ cannot lie in one inertia plane and since $a_1'$ and $A'B'$ are respectively parallel to $a_1$ and $AB$, it follows that $Q'$ is parallel to $Q$.

Now the inertia line $a_1'$ and the element $C$ lie in an inertia plane, say $R$, and so $R$ has the element $C$ in common with $P_2$.

Thus, by Theorem 51, $R$ and $P_2$ have a general line in common, say $c_1$, which is parallel to $a_1'$ and $b_1'$.

But now $Q$ is an inertia plane through $B$, which is an element of $P_2$ not lying in $b_1'$, and $Q$ is parallel to $Q'$ and therefore, by Theorem 53, the inertia planes $R$ and $Q$ have a general line in common, say $f_1$, which is parallel to $a_1'$.

Now $f_1$ must be an inertia line and therefore will intersect the separation line $AB$ in some element, say $F$, which must be distinct from $A$, since otherwise $R$ would coincide with $P_1$ and could therefore have no element in common with $P_2$, contrary to hypothesis.

Now, since $a_1$ is parallel to $b_1'$ while $a$ is parallel to $b$ and, since $a_1$ is conjugate to $a$ while $b_1'$ and $b$ intersect, therefore $b_1'$ is conjugate to $b$.

Similarly, since $AB$ is parallel to $A'B'$ and, since $a_1$ is conjugate to $AB$ while $b_1'$ and $A'B'$ intersect, therefore $b_1'$ is conjugate to $A'B'$.

But now since $a_1$ is conjugate to the two intersecting separation lines $AB$ and $a$, and since $F$ is an element in $AB$, while $A'$ is a distinct element in $a$, it follows by Theorem 101 that $a_1'$ must be conjugate to $A'F$.

Again since $b_1'$ is conjugate to the two intersecting separation lines $A'B'$ and $b$, and since $A'$ is an element in $A'B'$ while $C$ is a distinct element in $b$, it follows in a similar manner that $a_1'$ must be conjugate to $A'C$.

Thus $a_1'$ is conjugate to $A'F$ and to $A'C$ and since $A'F$ and $A'C$ each lie in the inertia plane $R$ and have an element in common, it follows that they must be identical.

Thus $F$ lies in $A'C$ and also in $AB$ and so $A'C$ intersects $AB$.

Thus there is only one general line through $A'$ and intersecting $b$ which does not intersect the general line $AB$.

## Theorem 110

*If a and b be two parallel separation lines such that no element of the one is either* before *or* after *any element of the other, and if one general line intersects a in A and b in B, while a second general line intersects a in A' and b in B', then a general line through any element of AB and parallel to a or b intersects A'B'.*

Let $D$ be any element of $AB$ and let $d$ be a general line through $D$ parallel to $a$ or $b$.

We have to show that $d$ intersects $A'B'$.

If $D$ should coincide with $A$ or $B$ no proof is required and so we shall suppose it distinct from both.

If $A'B'$ be parallel to $AB$, then no element of $AB$ is either *before* or *after* any element of $A'B'$ and the result follows directly by Theorem 108.

If $A'B'$ be not parallel to $AB$, then by Theorem 109 the general lines $AB$ and $A'B'$ must intersect in some element, say $C$.

Now the general lines $AB$ and $A'B'$ being supposed distinct, $C$ must be distinct from at least one of the elements $A$ and $B$ and, without limitation of generality, we may suppose that $C$ is distinct from $B$.

We shall then have $B'$ distinct from $B$ and so $B'$ will not be an element of $AB$.

Thus through $B'$ there is a parallel to $AB$ and by Theorem 108 this parallel must intersect $d$ in some element, say $D'$.

But now $D'B'$ and $DB$ are parallel separation lines such that no element of the one is either *before* or *after* any element of the other and both are intersected by the general lines $D'D$ and $B'C$.

Further since we have supposed $D$ to be distinct from $B$ therefore $D'$ is distinct from $B'$ and so by Theorem 109 there is only one general line through $B'$ and intersecting $DB$ which does not intersect $D'D$.

But $B'B$ being parallel to $D'D$ must be this one general line, and so, since $B'C$ (that is $A'B'$) is distinct from $B'B$, it follows that $A'B'$ intersects $D'D$.

Thus in all cases a general line through any element of $AB$ and parallel to $a$ or $b$ intersects $A'B'$.

## Theorem 111

*If a and b be two parallel separation lines such that no element of the one is either* before *or* after *any element of the other, and if E be any element in a separation line AB which intersects a in A and b in B, and if A'B' be any other separation line intersecting a in A' and b in B' but not parallel to AB, then E either lies in A'B' or in a separation line parallel to A'B' which intersects both a and b.*

If $E$ does not lie in $A'B'$, then by Theorem 110 a separation line through $E$ parallel to $a$ or $b$ intersects $A'B'$ in an element which is neither *before* nor *after* any element of $a$ or $b$ and so, by Theorem 108, a general line through $E$ parallel to $A'B'$ intersects $a$ and also $b$.

Thus $E$ must lie in a separation line parallel to $A'B'$ and intersecting both $a$ and $b$ when it does not lie in $A'B'$ itself.

## Remarks

If $a$ and $b$ be two parallel separation lines such that no element of the one is either *before* or *after* any element of the other and if $c$ and $d$ be any two non-parallel separation lines intersecting both $a$ and $b$, then it is evident from Theorem 111 that the aggregate consisting of all the elements in $c$ and in all separation lines intersecting $a$ and $b$ which are parallel to $c$ must be identical with the aggregate consisting of all the elements in $d$ and in all separation lines intersecting $a$ and $b$ which are parallel to $d$.

This follows since each element in the one set of separation lines must also lie in the other set.

Thus the aggregate which we obtain in this way is independent of the particular set of separation lines intersecting $a$ and $b$ which we may select and so we have the following definition.

*Definition.* If $a$ and $b$ be two parallel separation lines such that no element of the one is either *before* or *after* any element of the other, then the aggregate of all elements of all mutually parallel separation lines which intersect both $a$ and $b$ will be called a *separation plane.**

If a separation plane $P$ be determined by the two parallel separation lines $a$ and $b$, then any element $C$ in $P$ must lie in a separation line, say $c$, which intersects both $a$ and $b$.

Any other element $D$ in $P$ must either lie in $c$ or in a separation line, say $d$, parallel to $c$ and intersecting both $a$ and $b$.

---

* The name "separation plane" has been adopted from its analogy to a separation line.

If $D$ lies in $c$, then $D$ is neither *before* nor *after* $C$.

If $D$ lies in a separation line $d$ parallel to $c$, we know that no element of $d$ is either *before* or *after* any element of $c$ and so again $D$ is neither *before* nor *after* $C$.

Thus we have the general result that: *no element of a separation plane is either* before *or* after *any other element of it.*

### THEOREM 112

*If two distinct elements of a general line lie in a separation plane, then every element of the general line lies in the separation plane.*

Let the separation plane be determined by the two parallel separation lines $a$ and $b$ which are such that no element of the one is either *before* or *after* any element of the other.

If the two given elements lie in a separation line which is known to intersect both $a$ and $b$ no proof is required.

Otherwise let $C$ be any element in any separation line $AB$ which intersects $a$ in $A$ and $b$ in $B$ and let $D'$ be any element in any separation line $A'B'$ parallel to $AB$ and intersecting $a$ in $A'$ and $b$ in $B'$.

We have to show that every element of the general line $CD'$ lies in the separation plane.

Now no element of $AB$ is either *before* or *after* any element of $A'B'$ and so by Theorem 108 a general line through $C$ parallel to $a$ or $b$ will intersect $A'B'$ in some element, say $C'$.

If $D'$ should coincide with $C'$, then $CD'$ would be parallel to $a$ or $b$ and, since $C$ cannot be either *before* or *after* any element of $a$ or $b$, it follows that no element of $CD'$ could be either *before* or *after* any element of $a$ or $b$.

Thus in this case, by Theorem 108, a general line through any element of $CD'$ distinct from $C$ taken parallel to $AB$ will intersect both $a$ and $b$.

Thus every element of $CD'$ will in this case lie in the separation plane.

If $D'$ should not coincide with $C'$, then since $CD'$ is distinct from $CC'$ and intersects $A'B'$ it follows by Theorem 109 that $CD'$ must intersect both $a$ and $b$.

Thus again every element of $CD'$ lies in the separation plane determined by $a$ and $b$.

## Theorem 113

*If e be a general line in a separation plane and if A be any element of the separation plane which does not lie in e, then there is one single general line through A in the separation plane which does not intersect e.*

We saw in the course of proving Theorem 112 that if a separation plane be determined by two parallel separation lines $a$ and $b$ such that no element of the one is either *before* or *after* any element of the other, then any general line containing two elements in the separation plane and therefore any general line lying in the separation plane must either be parallel to $a$ or $b$, or else must intersect both $a$ and $b$.

Suppose first that $e$ intersects both $a$ and $b$.

Since $A$ does not lie in $e$ it must lie in a separation line $d$ parallel to $e$ and intersecting both $a$ and $b$.

Now through $A$ there is a separation line, say $c$, parallel to $a$ or $b$ and which, by Theorem 108, must intersect $e$ and must lie in the separation plane, while any other general line $f$ through $A$ and lying in the separation plane must intersect both $a$ and $b$.

Thus, by Theorem 109, $f$ being supposed distinct from $d$ must intersect $e$.

Suppose next that $e$ is parallel to $a$ or $b$.

Through $A$ there is a separation line parallel to $a$ or $b$ and therefore parallel to $e$ and which, as we know, lies in the separation plane.

Any other general line through $A$ in the separation plane must intersect both $a$ and $b$ and so, by Theorem 110, it must be intersected by $e$.

Thus there is in all cases one single general line through $A$ in the separation plane which does not intersect $e$.

## Theorem 114

*If A, B and C be three elements in a separation plane which do not all lie in one general line and if D be an element linearly between A and B, while E is an element linearly between A and C, there exists an element which lies both linearly between B and E and linearly between C and D.*

The proof of this theorem is quite analogous to that of Theorem 76, the only difference being that $V$ is here a separation plane instead of an inertia plane and, as such, it cannot contain any inertia line.

Thus the words "which does not lie in $V$" may be omitted from the first sentence of the proof.

### THEOREM 115

*If A, B and C be three elements in a separation plane which do not all lie in one general line and if D be an element linearly between A and B while F is an element linearly between C and D, there exists an element, say E, which is linearly between A and C and such that F is linearly between B and E.*

The proof of this theorem is quite analogous to that of Theorem 77, the only difference being that $V$ is here a separation plane instead of an inertia plane and, as such, it cannot contain any inertia line.

Thus the words "which does not lie in $V$" may be omitted from the first sentence of the proof.

### REMARKS

It will be observed that Theorem 114 is the analogue of Peano's axiom (14) for the case of elements in a separation plane, while Theorem 115 is the corresponding analogue of his axiom (13).

Further, Theorem 113 corresponds to the Euclidean axiom of parallels for the case of general lines in a separation plane.

### THEOREM 116

*If A, B and C be three elements in a separation plane which do not all lie in one general line and if D be an element linearly between A and B while DE is a general line through D parallel to BC and intersecting AC in the element E, then E is linearly between A and C.*

The proof of this theorem is exactly analogous to that of Theorem 90, using Theorem 114 in place of Theorem 88, and Theorem 115 in place of Theorem 89.

### THEOREM 117

*If three parallel general lines a, b and c in one separation plane intersect a general line $d_1$ in $A_1$, $B_1$ and $C_1$ respectively and intersect a second general line $d_2$ in $A_2$, $B_2$ and $C_2$ respectively, then if $B_1$ is linearly between $A_1$ and $C_1$ we shall also have $B_2$ linearly between $A_2$ and $C_2$.*

The proof of this theorem is exactly analogous to that of Theorem 91, using Theorem 116 in place of Theorem 90.

## THEOREM 118

*If A, B and C be three elements in a separation plane which do not all lie in one general line and if D be the mean of A and B, then a general line through D parallel to BC intersects AC in an element which is the mean of A and C.*

The proof of this theorem is exactly analogous to that of Theorem 94.

It is to be noted however that for the case of a separation plane we can never have $B'$ coinciding with $B$ or $C'$ coinciding with $C$, since a separation plane cannot contain an optical line.

Since there is only one general line through $D$ parallel to $BC$ and this must pass through the mean of $A$ and $C$, it follows directly that if $E$ be the mean of $A$ and $C$, then the general line $DE$ is parallel to $BC$.

*Definition.* If a pair of parallel general lines in a separation plane be intersected by another pair of parallel general lines, then the four general lines will be said to form a *general parallelogram in the separation plane.*

The terms *corner, side line, diagonal line, adjacent* and *opposite* will be used in a similar sense for the case of a general parallelogram in a separation plane as for one in an inertia or optical plane.

## THEOREM 119

*If we have a general parallelogram in a separation plane, then:*

(1) *The two diagonal lines intersect in an element which is the mean of either pair of opposite corners.*

(2) *A general line through the element of intersection of the diagonal lines and parallel to either pair of opposite side lines intersects either of the other side lines in an element which is the mean of the pair of corners through which that side line passes.*

The proof of this theorem is exactly analogous to that of Theorem 79, using Theorem 118 in place of Theorem 78.

## THEOREM 120

*If A, B, C, D be the corners of a general parallelogram in a separation plane; AB and DC being one pair of parallel side lines and BC and AD the other pair of parallel side lines, then if E be the mean of A and B, while F is the mean of D and C, the general lines AF and EC are parallel to one another.*

The proof of this theorem is exactly analogous to that of Theorem 80, using Theorem 119 in place of Theorem 79, and Theorem 118 in place of Theorem 78.

## Theorem 121

*If three parallel general lines a, b and c in one separation plane intersect a general line $d_1$ in $A_1$, $B_1$ and $C_1$ respectively and intersect a second general line $d_2$ in $A_2$, $B_2$ and $C_2$ respectively, and if $B_1$ be the mean of $A_1$ and $C_1$, then $B_2$ will be the mean of $A_2$ and $C_2$.*

The proof of this theorem is exactly analogous to that of Theorem 81, using Theorem 118 in place of Theorem 78, and Theorem 119 in place of Theorem 79.

### Sets of three elements which determine separation planes

If $A_1$, $A_2$ and $A_3$ be three distinct elements which do not all lie in one general line and do not all lie in one inertia plane or in one optical plane, then they must all lie in one separation plane, as we shall shortly show.

In those cases in which they do all lie in one separation plane they determine the separation plane containing them.

We have the following criterion by which we may say that the three elements do lie in one separation plane.

Three elements $A_1$, $A_2$, $A_3$ lie in one separation plane if $A_1$ and $A_2$ lie in a separation line while $A_3$ is an element which is not an element of the separation line and is neither *before* nor *after* any element of the separation line.

This is clearly true since if $a$ be the separation line containing $A_1$ and $A_2$, there is a separation line $b$ through $A_3$ and parallel to $a$ which is such that no element of $b$ is either *before* or *after* any element of $a$.

The separation lines $a$ and $b$ then determine a separation plane which will contain $A_1$, $A_2$ and $A_3$.

If $P$ be this separation plane it is the only one which contains $A_1$, $A_2$ and $A_3$, for suppose $A_1$, $A_2$ and $A_3$ also lie in a separation plane $P'$ determined by the two parallel separation lines $a'$ and $b'$, which are such that no element of $b'$ is either *before* or *after* any element of $a'$.

Now since $P'$ contains $A_1$, $A_2$ and $A_3$ it must contain the three general lines $A_1 A_2$, $A_2 A_3$ and $A_3 A_1$, by Theorem 112.

At most only one of these general lines could be parallel to $a'$ or $b'$.

Suppose first that $A_1 A_2$ or $a$ is not parallel to $a'$ or $b'$.

Then $a$ must intersect both $a'$ and $b'$, and since $A_3$ is an element of $P'$ the separation line $b$ through $A_3$ parallel to $a$ must lie in $P'$ and must intersect both $a'$ and $b'$.

Then every element in $P$ must lie in a separation line intersecting both $a$ and $b$ and parallel to $a'$ or $b'$.

But we know that every element of any such separation line $c$ must lie in $P'$, for by Theorem 110 a general line through any element of $c$ parallel to $a$ or $b$ must intersect $a'$ and $b'$.

Similarly every element in $P'$ must lie in $P$ and so $P'$ must be identical with $P$.

Next suppose that $a$ is parallel to $a'$ or $b'$.

Then $A_1A_3$ cannot be parallel to $a'$ or $b'$ and so must intersect both $a'$ and $b'$.

Then any element in $P$ must lie either in $A_3A_1$ or in a general line parallel to $A_3A_1$ and intersecting both $a$ and $b$.

But any such general line must also intersect both $a'$ and $b'$, and so every element in $P$ must also lie in $P'$, and similarly every element in $P'$ must also lie in $P$.

Thus again $P'$ must be identical with $P$.

Thus there is only one separation plane containing the three elements $A_1$, $A_2$ and $A_3$.

Any three distinct elements $A_1$, $A_2$ and $A_3$ which do not all lie in one general line must all lie either in an inertia plane, an optical plane, or a separation plane.

This is easily seen; for $A_1$ and $A_2$ must lie either in an optical line, an inertia line, or a separation line.

If $A_1A_2$ be an optical line we must have either

(1) $A_3$ *after* an element of $A_1A_2$,

or (2) $A_3$ *before* an element of $A_1A_2$,

or (3) $A_3$ neither *before* nor *after* any element of $A_1A_2$.

We cannot have $A_3$ *after* one element of $A_1A_2$ and *before* another element of it, since $A_3$ is not an element of $A_1A_2$ (Theorem 12).

In cases (1) and (2), as we have seen, $A_1$, $A_2$ and $A_3$ lie in an inertia plane.

In case (3) we have seen that $A_1$, $A_2$ and $A_3$ lie in an optical plane.

If $A_1A_2$ be an inertia line we know that the three elements must always lie in an inertia plane.

If $A_1A_2$ be a separation line we must have either

(1) $A_3$ *after* at least two distinct elements of $A_1A_2$,

or (2) $A_3$ *before* at least two distinct elements of $A_1A_2$,

or (3) $A_3$ *after* one single element of $A_1A_2$,

or (4) $A_3$ *before* one single element of $A_1A_2$,

or (5) $A_3$ neither *before* nor *after* any element of $A_1A_2$.

We cannot have $A_3$ *after* one element of $A_1A_2$ and *before* another element of it for then we should have one element of $A_1 A_2$ *after* another element of it, contrary to the hypothesis that $A_1 A_2$ is a separation line.

We have already seen that in cases (1) and (2) $A_1$, $A_2$ and $A_3$ lie in an inertia plane.

Also in cases (3) and (4) we have seen that $A_1$, $A_2$ and $A_3$ lie in an optical plane.

Finally in case (5) we have seen that $A_1$, $A_2$ and $A_3$ lie in a separation plane.

This exhausts all the possibilities which are logically open and so we see that $A_1$, $A_2$ and $A_3$ must always lie either in an inertia plane, an optical plane, or a separation plane.

It follows directly that any two intersecting general lines $a$ and $b$ must lie either in an inertia plane, an optical plane, or a separation plane, which we may call $P$.

Any element in $P$ must lie either in $b$ or in a general line parallel to $b$ and intersecting $a$.

Also, conversely, any element in $b$ or in any general line which intersects $a$ and is parallel to $b$, must lie in $P$.

Thus we have the following definition:

*Definition.* If $a$ and $b$ be any two intersecting general lines, then the aggregate of all elements of the general line $b$ and of all general lines parallel to $b$ which intersect $a$ will be called a *general plane*.

Thus a general plane is a common designation for an inertia plane, an optical plane, or a separation plane.

By combining Theorems 76, 88 and 114 we now see that the analogue of Peano's axiom (14) holds in general for our geometry; while by combining Theorems 77, 89 and 115 we see that the analogue of his axiom (13) also holds in general.

Again by combining Theorems 47, 87 and 113 we get what corresponds to the Euclidean axiom of parallels for the case of general lines in a general plane.

Peano's fifteenth axiom is as follows:

A point can be found external to any plane.

It is evident in our geometry that, since there is more than one general plane, there is an element external to any general plane, and so the analogue of Peano's axiom (15) also holds.

If $a$ and $b$ be two intersecting general lines in a general plane $P$ and if through any element $A$ not lying in $P$ two general lines $a'$ and $b'$ be taken respectively parallel to $a$ and $b$, then if $P'$ be the general plane determined by $a'$ and $b'$, the two general planes $P$ and $P'$ can have no element in common.

This is easily seen, for in the first place the general line $a'$ can have no element in common with $P$, for then, since it is parallel to $a$, every element of $a'$ would have to lie in $P$, contrary to the hypothesis that the element $A$ does not lie in $P$.

Similarly $b'$ can have no element in common with $P$.

If now $B$ be any element in $a'$ distinct from $A$ and if $b''$ be a general line through $B$ parallel to $b'$, then $b''$ must also be parallel to $b$ and, since $B$ does not lie in $P$, it follows that $b''$ can have no element in common with $P$.

But any element in $P'$ must lie either in $b'$ or in a general line parallel to $b'$ which intersects $a'$ and therefore the general plane $P'$ can have no element in common with $P$.

### THEOREM 122

*If $a$ and $b$ be any two intersecting general lines in a general plane $P$ and if through any element $O'$ not lying in $P$ two general lines $a'$ and $b'$ respectively parallel to $a$ and $b$ be taken determining a general plane $P'$, then there is a general line through $O'$ and lying in $P'$ which is parallel to any general line in $P$.*

Let the general lines $a$ and $b$ intersect in the element $O$ and let $A$ and $B$ be any two elements distinct from $O$ and lying in $a$ and $b$ respectively.

Then the general lines $OO'$ and $a$ determine a general plane which must contain $a'$, since $a'$ is parallel to $a$ and intersects $OO'$.

Thus a general line through $A$ parallel to $OO'$ will intersect $a'$ in some element, say $A'$.

Similarly a general line through $B$ parallel to $OO'$ will intersect $b'$ in some element, say $B'$.

Then $BB'$ will be parallel to $AA'$.

But $AB$ and $AA'$ determine a general plane which must contain $BB'$ and so the general lines $AB$ and $A'B'$ must lie in one general plane.

But $AB$ lies in $P$ while $A'B'$ lies in $P'$, and so $A'B'$ can have no element in common with $AB$ and must therefore be parallel to it.

Let the general line $AB$ be denoted by $c$ and the general line $A'B'$ by $c'$.

Let $c_1$ be a general line through $O$ parallel to $c$ while $c_1'$ is a general line through $O'$ parallel to $c'$.

Then $c_1$ will lie in $P$ and $c_1'$ will lie in $P'$, and since $c'$ is parallel to $c$ we must also have $c_1'$ parallel to $c_1$.

R                                                                    13

Now any general line in $P$ and passing through $O$, with the exception of $c_1$, must intersect $c$ in some element, say $X$.

If $X$ should coincide with either $A$ or $B$, we know that $O'A'$ and $O'B'$ are respectively parallel to $OA$ and $OB$, so that we shall suppose $X$ distinct from $A$ and $B$.

If now a general line be taken through $X$ parallel to $AA'$, such general line will lie in the general plane determined by $AB$ and $AA'$ and will therefore intersect $A'B'$ in some element, say $X'$.

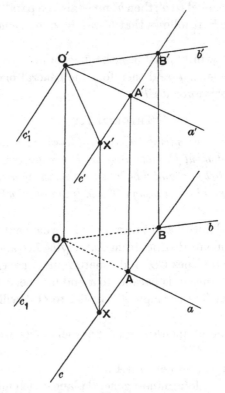

Fig. 36.

Now $XX'$ must be parallel to $OO'$ and so $XX'$ must lie in the general plane determined by $OX$ and $OO'$.

Thus $OX$ and $O'X'$ lie in one general plane.

But $OX$ lies in $P$ while $O'X'$ must lie in $P'$ and, since $P$ and $P'$ have no element in common, it follows that $O'X'$ is parallel to $OX$.

Thus through $O'$ there is a general line in $P'$ which is parallel to any general line in $P$ which passes through $O$, and since any general line in $P$ which does not pass through $O$ is parallel to one which does pass through $O$, it follows that there is a general line through $O'$ and lying in $P'$ which is parallel to any general line in $P$.

It also follows directly from the above that through *any* element of $P'$ there is a general line in $P'$ which is parallel to any general line in $P$.

### Remarks

We have already given a definition of the parallelism of inertia planes and are now in a position to give a definition of the parallelism of general planes which will include that of inertia planes as a special case.

*Definition.* If $P$ be a general plane and if through any element $A$ outside $P$ two general lines be taken respectively parallel to two intersecting general lines in $P$, then the two general lines through $A$ determine a general plane which will be said to be *parallel* to $P$.

Theorem 52 shows that this definition agrees with that given for the case of inertia planes.

If $P$ be a general plane and $A$ be any element outside it, while $P'$ is a general plane through $A$ parallel to $P$, then it is evident from Theorem 122 that, since $P'$ contains the general line through $A$ parallel to any general line in $P$, the general plane $P'$ must be uniquely determined when we know $P$ and $A$.

Thus *through any element outside a general plane $P$ there is one single general plane parallel to $P$.*

Also it is clear that this general plane must be of the same kind as $P$.

Again, since two distinct general lines which are parallel to a third general line are parallel to one another, it follows that: *two distinct general planes which are parallel to a third general plane are parallel to one another.*

*Definition.* If $P$ be a general plane and if through any element $A$ outside $P$ a general line $a$ be taken parallel to any general line in $P$, then the general line $a$ will be said to be *parallel* to the general plane $P$.

### Theorem 123

*If a general plane $P$ have one element in common with each of a pair of parallel general planes $Q$ and $R$, then, if $P$ have a second element in common with $Q$ it also has a second element in common with $R$.*

Let the general plane $P$ have the element $A$ in common with $Q$ and the element $A'$ in common with $R$.

Further let $P$ and $Q$ have a second element $B$ in common.

Then, as was observed at the end of Theorem 122, there is a general line, say $c$, through $A'$ in the general plane $R$ which is parallel to $AB$.

But $c$ must also lie in $P$, and so any element of $c$ distinct from $A'$ is a second element common to $P$ and $R$.

Thus the theorem is proved.

## THEOREM 124

*If two parallel general lines $a$ and $b$ lie in one general plane $R$ and if two other distinct general planes $P$ and $Q$ containing $a$ and $b$ respectively have an element $A_1$ in common, then $P$ and $Q$ have a general line in common which is parallel to $a$ and $b$.*

The proof of this theorem is exactly analogous to that of Theorem 51, using Theorem 123 in place of Theorem 46.

## THEOREM 125

*If three distinct general planes $P$, $Q$ and $R$ and three parallel general lines $a$, $b$ and $c$ be such that $a$ lies in $P$ and $R$, $b$ in $Q$ and $P$ and $c$ in $R$ and $Q$, then if $Q'$ be a general plane parallel to $Q$ through some element of $P$ which does not lie in $b$ the general planes $R$ and $Q'$ have a general line in common which is parallel to $c$.*

The proof of this theorem is analogous to that of Theorem 53, using Theorem 123 in place of Theorem 46.

Since however a general plane does not always contain an optical line, we take any general line through the element $A$ distinct from $a$, which lies in the general plane $P$, and such general line must intersect $b$ in an element which we shall call $B$.

We then take any general line through $B$ distinct from $b$, which lies in the general plane $Q$ and this general line must intersect $c$ in some element which we shall call $C$.

Then $BA$ and $BC$ lie in a general plane which we shall call $S$.

The demonstration from this point on is similar to that of Theorem 53, once more using Theorem 123 in place of Theorem 46.

If a pair of parallel general lines be both intersected by another pair of parallel general lines then the four general lines will form a *general parallelogram* either in an *acceleration plane*, an *optical plane*, or a *separation plane*.

Thus a *general parallelogram* may now be defined in this way without specifying which type of general plane it lies in.

## THEOREM 126

*If two general parallelograms have a pair of adjacent corners $A$ and $B$ in common, their remaining corners either lie in one general line or else form the corners of another general parallelogram.*

Let $A$, $B$, $C$, $D$ be the corners of the one general parallelogram and $A$, $B$, $C'$, $D'$ the corners of the other, and let $AC$ and $BD$ be a pair of opposite side lines of the first general parallelogram while $AC'$ and $BD'$ are a pair of opposite side lines of the second.

Then $CD$ and $C'D'$ being each parallel to $AB$ must either be parallel to one another or else must be identical.

In the latter case the corners $C$, $D$, $C'$, $D'$ lie in one general line.

Suppose now that $CD$ and $C'D'$ are distinct and therefore parallel; we have to prove that $CC'$ is parallel to $DD'$.

Two cases have to be considered:

    (1) The two general parallelograms lie in distinct general planes,

or   (2) The two general parallelograms lie in the same general plane.

We shall first consider case (1).

Since $CD$ and $C'D'$ are parallel they must lie in a general plane, say $P$.

Again $AC$ and $AC'$ must lie in a general plane, say $Q$, distinct from the general planes of either of the general parallelograms, since by hypothesis $C'$ does not lie in the general plane containing $A$, $B$, $C$ and $D$.

Similarly $BD$ and $BD'$ must lie in a general plane, say $R$, distinct from the general planes of either of the general parallelograms.

Further, the element $A$ cannot lie in $R$, since otherwise $A$, $B$, $C$, $D$, $C'$ and $D'$ would all lie in one general plane, contrary to hypothesis.

But $AC$ is parallel to $BD$, while $AC'$ is parallel to $BD'$ and therefore $Q$ is parallel to $R$.

Thus the general lines $CC'$ and $DD'$ can have no element in common, and since they both lie in $P$, it follows that they are parallel.

Thus $C$, $C'$, $D'$, $D$ form the corners of another general parallelogram.

We have next to consider case (2).

Let $P$ be the general plane containing the two given general parallelograms, and let $Q$ be any other general plane distinct from $P$ and containing the general line $AB$.

Let $AC_1$ be any general line distinct from $AB$ which passes through $A$ and lies in $Q$.

Through any element $C_1$ of $AC_1$ distinct from $A$ let a general line be taken parallel to $AB$ and let it meet the general line through $B$ parallel to $AC_1$ in the element $D_1$.

Then by the case already proved the general lines $CC_1$ and $DD_1$ are parallel.

Similarly $C'C_1$ and $D'D_1$ are parallel to one another.

But now the general parallelograms whose corners are $C_1$, $D_1$, $D$, $C$ and $C_1$, $D_1$, $D'$, $C'$ cannot lie in one general plane; for the general lines $CD$ and $C'D'$ both lie in $P$, while $C_1D_1$ does not lie in $P$.

Thus again by case (1) $CC'$ is parallel to $DD'$ and so $C$, $C'$, $D'$, $D$ form the corners of a general parallelogram.

Thus the theorem holds in all cases.

## THEOREM 127

(1) *If three distinct elements $A$, $B$ and $C$ in a general plane $P$ do not all lie in one general line and if $D$ be any element linearly between $B$ and $C$, then any general line passing through $D$ and lying in $P$ and which is distinct from $BC$ and $AD$ must either intersect $AC$ in an element linearly between $A$ and $C$, or else intersect $AB$ in an element linearly between $A$ and $B$.*

(2) *If further $E$ be an element linearly between $C$ and $A$ and if $F$ be an element linearly between $A$ and $B$, then $D$, $E$ and $F$ cannot lie in one general line.*

In order to prove the first part of the theorem let $a$ be any general line passing through $D$ and lying in $P$.

Then $a$ must either be parallel to $AC$ or else intersect $AC$ in some element, say $E$.

If $a$ be parallel to $AC$, then it follows by Theorems 72, 90 and 116 that $a$ must intersect $AB$ in an element which is linearly between $A$ and $B$.

If $a$ intersects $AC$ in an element $E$, then provided $a$ be distinct from $BC$ and $AD$ we must either have:

     (i)  $E$ linearly between $A$ and $C$,

or   (ii)  $C$ linearly between $A$ and $E$,

or  (iii)  $A$ linearly between $C$ and $E$.

In case (ii) it follows by the analogue of Peano's axiom (13) that $a$ intersects $AB$ in an element linearly between $A$ and $B$, while in case (iii) it follows by the analogue of Peano's axiom (14) that $a$ intersects $AB$ in an element linearly between $A$ and $B$.

Thus the first part of the theorem is proved.

In order now to prove the second part of the theorem it is to be observed in the first place that since the elements $D$, $E$ and $F$ lie in three distinct general lines $BC$, $CA$ and $AB$ and are distinct from the elements of intersection of these, therefore the elements $D$, $E$ and $F$ are all distinct.

If then $D$, $E$ and $F$ lay in one general line, we should require to have either:

$E$ linearly between $D$ and $F$,

or $F$ linearly between $E$ and $D$,

or $D$ linearly between $F$ and $E$.

Now the elements $F$, $C$ and $B$ do not lie in one general line and we have $D$ linearly between $B$ and $C$.

If then we had also $E$ linearly between $D$ and $F$ it would follow that $A$ must be linearly between $B$ and $F$, contrary to the hypothesis that $F$ is linearly between $A$ and $B$.

Thus $E$ cannot be linearly between $D$ and $F$.

Similarly $F$ cannot be linearly between $E$ and $D$, and further $D$ cannot be linearly between $F$ and $E$.

It follows therefore that $D$, $E$ and $F$ cannot lie in one general line and so the second part of the theorem is proved.

## Theorem 128

*If an inertia line $a$ be conjugate to two intersecting separation lines $b$ and $c$, then $b$ and $c$ lie in a separation plane such that any separation line in it is conjugate to a set of inertia lines which are parallel to $a$.*

Let the separation lines $b$ and $c$ intersect in the element $A$.

Then we know that $a$ must also pass through $A$ and that the separation lines $b$ and $c$ must be such that no element of the one is either *before* or *after* any element of the other, and so there must be a separation plane, say $P$, which contains them.

Let $B$ and $C$ be elements in $b$ and $c$ respectively and let them both be distinct from $A$.

Then $BC$ is a separation line which we may call $d$ and which lies in the separation plane $P$.

Let $e$ be an inertia line through $B$ parallel to $a$.

Then $e$ is conjugate to $b$ and, by Theorem 101, it must also be conjugate to $d$.

Now we know that there is only one general line in $P$ and passing through $A$ which does not intersect $d$.

Let $AF$ be any general line passing through $A$ and intersecting $d$ in $F$.

Then, by Theorem 101, since $e$ is conjugate to $b$ and $d$, it follows that $a$ must be conjugate to $AF$.

Again, if $d'$ be the general line through $A$ parallel to $d$, it must lie in the separation plane $P$, and, since $e$ is conjugate to $d$, while $a$ and $d'$ are respectively parallel to $e$ and $d$ and, since $a$ and $d'$ intersect one another, it follows by Theorem 100 that $a$ must be conjugate to $d'$.

Thus every separation line passing through $A$ in the separation plane $P$ is conjugate to $a$ and therefore also conjugate to any inertia line which intersects it and is parallel to $a$.

Consider now any separation line $f$ in $P$ which does not pass through $A$.

Then there is a separation line $f'$ passing through $A$ and parallel to $f$, and $a$ must be conjugate to $f'$.

Thus by Theorem 100 any inertia line intersecting $f$ and parallel to $a$ must be conjugate to $f$.

Thus any separation line in $P$, whether it pass through $A$ or not, must be conjugate to a set of inertia lines which are parallel to $a$, and so the theorem is proved.

## THEOREM 129

*If $O$ be any element in a separation line $b$ lying in a separation plane $P$ and if $a$ be an inertia line through $O$ which is conjugate to every separation line in $P$ which passes through $O$, then there is one and only one such separation line which is conjugate to every inertia line passing through $O$ and lying in the inertia plane containing $a$ and $b$.*

Let $Q$ be the inertia plane containing $a$ and $b$ and let $Q'$ be an inertia plane parallel to $Q$ through any element of $P$ which does not lie in $b$.

Then by Theorem 123 $P$ and $Q'$ will have a general line, say $b'$, in common which must be parallel to $b$ and must be a separation line.

Let $c$ be one of the generators of $Q$ which pass through $O$.

Then since $Q'$ is parallel to $Q$ there is one single generator of $Q'$, say $c'$, which is neutrally parallel to $c$.

Let $c'$ intersect $b'$ in $O'$.

Then $O'$ is neither *before* nor *after* any element of $c$ and so no element of the general line $OO'$ with the exception of $O$ is either *before* or *after* any element of $c$.

But $OO'$ lies in $P$ and therefore is conjugate to $a$, and so, by Theorem 104, $OO'$ is conjugate to every inertia line in $Q$ which passes through $O$.

Thus, as in Theorem 104, $O$ and $O'$ are representatives of one another in the parallel inertia planes $Q$ and $Q'$, and further, we may show as in

Theorem 103 that if $O''$ be any element of $Q'$ distinct from $O'$ the general line $OO''$ cannot be conjugate to two distinct inertia lines in $Q$ which pass through $O$.

But now any separation line in $P$ which passes through $O$ must either be identical with $b$ or else intersect $b'$ in some element.

If it should intersect $b'$ in any element other than $O'$ it cannot be conjugate to more than one inertia line in $Q$ which passes through $O$.

Also if it be identical with $b$ it cannot be conjugate to more than one inertia line in $Q$ which passes through $O$.

Thus there is one and only one separation line in $P$ which passes through $O$ and is conjugate to every inertia line passing through $O$ and lying in the inertia plane $Q$.

THEOREM 130

*If a separation line $a$ have an element $O$ in common with an inertia plane $P$ and be conjugate to every inertia line in $P$ which passes through $O$, and if $c$ be any such inertia line and $b$ be the separation line in $P$ which passes through $O$ and is conjugate to $c$, then $b$ is conjugate to every inertia line in the inertia plane containing $a$ and $c$ which passes through the element $O$.*

Let $A_1$ be any element in $a$ distinct from $O$, and let $d$ be any inertia line in $P$ which passes through $O$ and is distinct from $c$.

Let $B_1$ be the one single element common to $d$ and the $\alpha$ sub-set of $A_1$.

Then $A_1 B_1$ is an optical line and $B_1$ is *after* $A_1$ and so, since $A_1 O$ is a separation line while $B_1 O$ is an inertia line, we must have $B_1$ *after* $O$.

Let $D$ be the one single element common to $c$ and the $\alpha$ sub-set of $B_1$ and let $E$ be the one single element common to $c$ and the $\beta$ sub-set of $B_1$.

Then $B_1 D$ and $B_1 E$ are optical lines lying in $P$.

Also $D$ is *after* $B_1$ while $B_1$ is *after* both $A_1$ and $O$ and so $D$ is *after* both $A_1$ and $O$.

Let the optical line through $O$ parallel to $B_1 D$ intersect the optical line through $D$ parallel to $B_1 E$ in $F$ and let the optical line through $E$ parallel to $B_1 D$ intersect $DF$ in $B_2$.

Then $B_1$, $E$, $B_2$, $D$ are the corners of an optical parallelogram lying in $P$. Let $C$ be its centre.

Then, since $DF$ and $OF$ are both optical lines and since $D$ is *after* $O$ but is not in an optical line with it, it follows that $F$ is *after* $O$.

But now since $A_1$ is not an element of the optical line $DF$ but is *before* an element of it, it follows that there is one single element common to the optical line $DF$ and the $\alpha$ sub-set of $A_1$.

Let $B_2'$ be this element, which we shall prove must be identical with $B_2$.

Then, since $D$ is *after* $A_1$ but is not in an optical line with it, it follows that $B_2'$ cannot be either identical with $D$ or *after* $D$ and therefore, since $B_2'$ and $D$ lie in an optical line, it follows that $B_2'$ is *before* $D$.

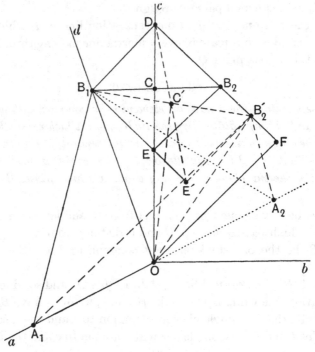

Fig. 37.

But $B_1$ and $D$ lie in another optical line and $B_1$ is also *before* $D$ and therefore $B_2'$ is neither *before* nor *after* $B_1$, so that $B_1 B_2'$ is a separation line.

Again, since $OF$ is one of the generators of $P$ which pass through $O$ and since by hypothesis $A_1 O$ is conjugate to every inertia line in $P$ which passes through $O$ and $A_1$ is distinct from $O$, it follows that $A_1$ is neither *before* nor *after* any element of $OF$.

Thus $A_1$ is not *before* $F$ and so $B_2'$ can neither be *before* $F$ nor identical with it and so, since $B_2'$ and $F$ lie in an optical line, it follows that $B_2'$ is *after* $F$.

But, since $F$ is *after* $O$, it follows that $B_2'$ is *after* $O$ and, since $B_2'$ and $O$ do not lie in one optical line, it follows that $B_2'O$ must be an inertia line.

Let the optical line through $B_2'$ parallel to $DB_1$ intersect $B_1E$ in $E'$. Then $B_1$, $E'$, $B_2'$, $D$ form the corners of an optical parallelogram of which $B_1B_2'$ is the separation diagonal line, and accordingly, $E'D$ is the inertia diagonal line.

Let $C'$ be the centre of this optical parallelogram.

Then $C'$ is linearly between $B_1$ and $B_2'$ and so, since $O$ is *before* both $B_1$ and $B_2'$ and is not in the general line $B_1B_2'$, it follows by Theorem 73 that $C'O$ is an inertia line.

Now in the inertia plane containing $a$ and $d$ take the second optical line which passes through $B_1$ and let it intersect $a$ in the element $A_2$.

Then, since $OB_1$ is conjugate to $a$, it follows that $A_1$, $B_1$ and $A_2$ are three corners of an optical parallelogram having $O$ as its centre.

But we showed that $OB_2'$ must be an inertia line and, as it lies in $P$ and passes through $O$, it must also be conjugate to $a$.

But $O$ is the mean of $A_1$ and $A_2$ while $A_1B_2'$ is an optical line and so $A_2B_2'$ must also be an optical line.

But now, since $C'$ is the mean of $B_1$ and $B_2'$, it follows that $B_1$, $A_1$ and $B_2'$ are three corners of an optical parallelogram of which $C'$ is the centre and so $B_1B_2'$ is conjugate to $C'A_1$.

Similarly $B_1$, $A_2$ and $B_2'$ are three corners of an optical parallelogram of which $C'$ is the centre and so $B_1B_2'$ is conjugate to $C'A_2$. Further, since $B_1B_2'$ is a separation line, it follows that $C'A_1$ and $C'A_2$ are both inertia lines.

Thus $B_1B_2'$ is conjugate to two inertia lines passing through the element $C'$ and therefore it must be conjugate to every inertia line passing through $C'$ and lying in the inertia plane containing $C'A_1$ and $C'A_2$.

But the element $O$ lies in $A_1A_2$ and $C'O$ must therefore lie in this inertia plane and, moreover, we showed that $C'O$ must be an inertia line.

Thus $B_1B_2'$ must be conjugate to $C'O$.

But $B_1B_2'$ is conjugate to $C'D$, which lies in $P$ as does also $C'O$ and therefore the inertia lines $C'O$ and $C'D$ must be identical; so that $C'$ lies in $OD$.

It follows that $E'$ must be identical with $E$ and $B_2'$ must be identical with $B_2$.

Further, $C'$ must be identical with $C$ and so $B_1B_2$ is conjugate to both $CA_1$ and $CA_2$ and therefore is also conjugate to every inertia line in the inertia plane containing $CA_1$ and $CA_2$ which passes through $C$.

But this is the inertia plane which contains $a$ and $c$, while the separation line $b$ which lies in $P$, passes through $O$ and is conjugate to $c$, must be parallel to $B_1B_2$, since $B_1B_2$ also lies in $P$.

Thus, by Theorem 100, $b$ is conjugate to every inertia line in the inertia plane containing $a$ and $c$ which passes through the element $O$.

Thus the theorem is proved.

### Remarks

All the postulates which have hitherto been introduced may be represented by ordinary geometric figures involving not more than three dimensions.

This may be done in the manner described in the introduction: the $\alpha$ and $\beta$ sub-sets being represented by cones.

We have now however to introduce a new postulate which cannot be represented along with the others in a three-dimensional figure and which therefore gives our geometry a sort of four-dimensional character.

The new postulate is as follows:

**POSTULATE XIX. If P be any optical plane, there is at least one element which is neither before nor after any element of P.**

Since any element in an optical plane must lie in a generator, it will be *after* certain elements and *before* certain other elements of that optical plane.

It follows that any element such as is postulated in Post. XIX must lie outside $P$.

Again if $P$ be an optical plane and $A$ be any element which is neither *before* nor *after* any element of $P$, then an optical line through $A$ parallel to any generator of $P$ will be a neutral-parallel and accordingly any generator of an optical plane lies in at least one other distinct optical plane.

Since we already know that any optical line lies in at least one optical plane, it follows that *there are at least two distinct optical planes containing any optical line.*

This might be taken as an alternative form of the postulate.

If $P$ and $Q$ be two distinct optical planes having an optical line $a$ in common, then any element of $Q$ which does not lie in $a$ must lie in a generator of $Q$, say $b$, which is a neutral-parallel of $a$.

Since any generator of $P$ which is distinct from $a$ is also a neutral-parallel of $a$, it follows by Theorem 28 that $b$ is a neutral-parallel of every generator of $P$.

Since every element of $P$ lies in a generator it follows that no element of $Q$ lying outside $a$ is either *before* or *after* any element of $P$.

Although Post. XIX is required in order to prove that there are at least two distinct optical planes containing any optical line, it is possible, without using this postulate, to prove that there are at least two distinct optical planes containing any separation line.

This may be done in the following manner:

Let $b$ be the separation line and $O$ be any element in it.

We already know that if we take any two inertia planes containing $b$, then $b$ is conjugate to one single inertia line in each of them which passes through $O$.

If $a_1$ and $a_2$ be two such inertia lines, then, as was shown in Theorem 103, $b$ is conjugate to every inertia line in the inertia plane containing $a_1$ and $a_2$ which passes through $O$.

Further, if $c_1$ and $c_2$ be the two generators of this inertia plane which pass through $O$ it was also shown in the course of proving Theorem 103 that if we take any element $O'$ of $b$ distinct from $O$ such element is neither *before* nor *after* any element of either $c_1$ or $c_2$.

Thus if we take an optical line through $O'$ parallel to $c_1$ it will be a neutral-parallel and so $b$ and $c_1$ lie in an optical plane.

Similarly $b$ and $c_2$ lie in an optical plane.

These optical planes must be distinct since $c_1$ and $c_2$ are distinct optical lines which both pass through $O$.

## Theorem 131

*If $b$ be any separation line and $O$ be any element in it, there are at least two inertia planes containing $O$ and such that $b$ is conjugate to every inertia line in each of them which passes through $O$.*

Let $Q$ be an optical plane containing $b$ and let $c$ be the generator of $Q$ which passes through $O$.

Then by Post. XIX it follows, as we have already shown, that there is at least one other optical plane, say $R$, containing the optical line $c$.

Let $d$ be any separation line in $R$ and passing through $O$.

Then no element of $d$ except $O$ is either *before* or *after* any element of $Q$ and $O$ itself is neither *before* nor *after* any element of $Q$ which lies outside $c$.

Thus no element of $d$ is either *before* or *after* any element of $b$ and so, by Theorem 107, there is at least one inertia line, say $a$, which is conjugate to both $b$ and $d$.

Thus, as was shown in Theorem 128, $a$ must be conjugate to every separation line which lies in the separation plane containing $b$ and $d$ and which passes through $O$.

Let $S$ be the separation plane containing $b$ and $d$, and let $P$ be the inertia plane containing $a$ and $c$.

Then since $b$ is conjugate to $a$ and since no element of $b$ with the exception of $O$ is either *before* or *after* any element of $c$ it follows, by Theorem 104, that $b$ is conjugate to every inertia line in $P$ which passes through $O$.

Similarly, since $d$ is conjugate to $a$ and since no element of $d$ with the exception of $O$ is either *before* or *after* any element of $c$, it follows that $d$ is conjugate to every inertia line in $P$ which passes through $O$.

Thus any inertia line in $P$ which passes through $O$ is conjugate to both $b$ and $d$ and therefore is conjugate to every separation line passing through $O$ and lying in the separation plane $S$.

It follows that $P$ cannot have more than one element in common with $S$, for if it had, it would have a separation line in common with $S$ and every inertia line in $P$ which passed through $O$ would require to be conjugate to one separation line lying in $P$, which is impossible.

Now by Theorem 129 there is one and only one separation line, say $e$, lying in $S$ and passing through $O$ which is conjugate to every inertia line passing through $O$ and lying in the inertia plane containing $a$ and $b$.

Let $T$ be the inertia plane containing $a$ and $e$.

Then, by Theorem 130, since $b$ is conjugate to $a$, it follows that $b$ is conjugate to every inertia line in $T$ which passes through $O$.

But now $b$ is conjugate to every inertia line lying either in $T$ or $P$ which passes through $O$ and, since $T$ contains the separation line $e$ which lies in $S$ while $P$ does not contain any separation line in $S$, it follows that $T$ and $P$ are distinct inertia planes.

Thus the theorem is proved.

### REMARKS

It follows from this that if a separation line $b$ have an element $O$ in common with any inertia plane $U$ and is conjugate to every inertia line in $U$ which passes through $O$, then $b$ is also conjugate to certain other inertia lines passing through $O$ which do not lie in $U$.

It also follows directly that there are certain optical lines passing through $O$, but not lying in $U$, which are such that no element of $b$ with the exception of $O$ is either *before* or *after* any element of them.

*Another important point which arises in the last theorem is that we may have an inertia plane and a separation plane having only one element in common and such that each inertia line through the common element in the former is conjugate to every separation line through it in the latter.*

### THEOREM 132

*If two distinct inertia planes $P$ and $P'$ have a separation line $b$ in common and if another separation line $c$ intersecting $b$ in the element $O$ be conjugate to every inertia line in $P$ which passes through $O$, then if $c$ be conjugate to one inertia line in $P'$ which passes through $O$ it is conjugate to every inertia line in $P'$ which passes through $O$.*

Let $f_1$ and $f_2$ be the two generators of $P$ which pass through $O$ and let $D_1$ be any element in $f_1$ which is *after* $O$.

Let the general line through $D_1$ parallel to $b$ intersect $f_2$ in $D_2$.

Then $D_1 D_2$ is a separation line and so, since $O$ is *before* $D_1$, it must also be *before* $D_2$.

Let $E_1$ be any element linearly between $D_1$ and $D_2$ and let $E_2$ be any element linearly between $E_1$ and $D_2$, while $C$ is any element linearly between $E_1$ and $E_2$.

Then by Theorem 73 $(a)$ $OE_1$ is an inertia line and $E_1$ is *after* $O$.

Similarly, since $O$ is *before* both $E_1$ and $D_2$, it follows that $OE_2$ is an

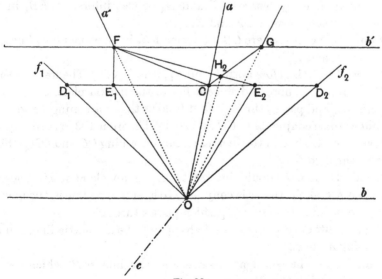

Fig. 38.

inertia line and $E_2$ is *after* $O$, and further since $O$ is *before* both $E_1$ and $E_2$ it follows that $OC$ must be an inertia line and $C$ must be *after* $O$.

Thus $OE_1$, $OE_2$ and $OC$ are three distinct inertia lines in $P$ all passing through $O$ and so the separation line $c$ is conjugate to each of them.

Now if $a'$ be an inertia line in $P'$ which passes through $O$ and to which $c$ is conjugate, it follows by Theorem 103 that $c$ is conjugate to every inertia line passing through $O$ and lying in either of the three inertia planes containing $a'$ and $OE_1$, $a'$ and $OE_2$ or $a'$ and $OC$.

Let $F$ be any element of $a'$ which is *after* $O$ and let $b'$ be the general line through $F$ parallel to $b$.

Then $b'$ must lie in $P'$ and must be parallel to $D_1 D_2$.

Let $Q$ be the general plane containing $b'$ and $FC$.

Then $Q$ contains $D_1 D_2$ and therefore also contains $FE_1$, $FE_2$ and $FC$.

Now any general line in $P'$ which passes through $O$ with the exception of $b$ must intersect $b'$ in some element, say $G$.

If now we consider the general line $CG$, we see that it must lie in $Q$ since $C$ and $G$ are distinct elements in $Q$.

Further, $CG$ must be distinct from $E_1 E_2$ since $E_1 E_2$ is parallel to $b'$ while $CG$ intersects $b'$.

Thus, since $F$, $E_1$ and $E_2$ do not lie in one general line while $C$ is linearly between $E_1$ and $E_2$, it follows by Theorem 127 that, provided $G$ does not coincide with $F$, the general line $CG$ either intersects $FE_2$ in an element linearly between $F$ and $E_2$ or else intersects $FE_1$ in an element linearly between $F$ and $E_1$.

Consider the case where $CG$ intersects $FE_2$ in an element $H_2$ linearly between $F$ and $E_2$.

Then, since $O$ is *before* both $F$ and $E_2$, it follows by Theorem 73 that $OH_2$ is an inertia line, and since it lies in the inertia plane containing $a'$ and $OE_2$ and passes through $O$, it follows that $c$ is conjugate to it.

But $c$ is also conjugate to $OC$ and so, by Theorem 103, $c$ is conjugate to every inertia line in the inertia plane containing $OC$ and $OH_2$ which passes through $O$.

Similarly, if $CG$ should intersect $FE_1$ in an element $H_1$ linearly between $F$ and $E_1$, then $c$ is conjugate to ever inertia line in the inertia plane containing $OC$ and $OH_1$ which passes through $O$.

Thus in either case if $OG$ should happen to be an inertia line, $c$ must be conjugate to it.

Thus $c$ must be conjugate to every inertia line in $P'$ which passes through $O$ and so the theorem is proved.

### REMARKS

Since, in the above theorem, there is one single inertia line through $O$ in the inertia plane $P$ which is conjugate to $b$, such inertia line will be conjugate to both $b$ and $c$ and so it follows, by Theorem 99, that no element of $b$ is either *before* or *after* any element of $c$ and so $b$ and $c$ must lie in a separation plane.

Again if $f_1'$ and $f_2'$ be the two generators of $P'$ which pass through $O$, then no element of $c$ with the exception of $O$ will be either *before* or *after* any element of either $f_1'$ or $f_2'$.

Now let $a_1$ be the one single inertia line through $O$ and lying in $P$ which is conjugate to $b$, and let $a_1'$ be the one single inertia line through $O$ and lying in $P'$ which is conjugate to $b$.

Then $a_1$ and $a_1'$ lie in an inertia plane, say $R$, and both $b$ and $c$ must be conjugate to every inertia line passing through $O$ and lying in $R$.

Thus if $g_1$ and $g_2$ be the two generators of $R$ which pass through $O$, no element of either $b$ or $c$ with the exception of $O$ is either *before* or *after* any element of either $g_1$ or $g_2$.

Thus the optical lines $g_1$ and $g_2$ are such that $g_1$ and $b$ lie in an optical plane and also $g_2$ and $b$ lie in an optical plane.

The optical lines $f_1'$ and $f_2'$ on the other hand are such that both of them lie in an inertia plane containing $b$.

### NORMALITY OF GENERAL LINES HAVING A COMMON ELEMENT

We are now in a position to define what we mean when we say that a general line $a$ is "*normal*" to a general line $b$, which has an element in common with it.

Since $a$ and $b$ are not always general lines of the same kind, it is difficult to give any simple definition which will include all cases; but the introduction of the word "*normal*" is justified by the simplification which is thereby brought about in the statement of many theorems.

Only one case will be found to be strictly analogous to the normality of intersecting straight lines in ordinary geometry: namely the case of two separation lines.

The other cases are so different from our ordinary ideas of lines "*at right angles*" that we do not propose to use this expression in connexion with them.

Thus for instance any optical line is to be regarded as being "normal to itself", and the use of the words "at right angles" would, in this case, clearly be an abuse of language.

The extension of the idea of normality from the cases of general lines having a common element to the cases of general lines which have not a common element is however quite analogous to the corresponding extension in ordinary geometry and will be made subsequently.

We are at present only concerned with the cases of general lines having a common element and shall naturally include among these that of an optical line being "normal to itself".

*Thus the complete definition of the normality of general lines having a common element is to be taken as consisting of the following four particular definitions A, B, C and D.*

*Definition A.* Any optical line will be said to be *normal to itself*.

*Definition B.* If an optical line $a$ and a separation line $b$ have an element $O$ in common and if no element of $b$ with the exception of $O$ be either *before* or *after* any element of $a$, then $b$ will be said to be *normal* to $a$, and $a$ will be said to be *normal* to $b$.

*Definition C.* If an inertia line $a$ and a separation line $b$ be conjugate one to the other, then $a$ will be said to be *normal* to $b$ and $b$ will be said to be *normal* to $a$.

*Definition D.* A separation line $a$ having an element $O$ in common with a separation line $b$ will be said to be *normal* to $b$ provided an inertia plane $P$ exists containing $b$ and such that every inertia line in $P$ which passes through $O$ is conjugate to $a$.

In this last case, since there is one single inertia line in $P$ which passes through $O$ and is conjugate to $b$, it is evident that $a$ and $b$ must lie in a separation plane.

If $c$ be this inertia line then, by Theorem 130, every inertia line which passes through $O$ and lies in the inertia plane containing $c$ and $a$ is conjugate to $b$ and so $b$ satisfies the definition of being normal to $a$.

Let the separation plane containing $a$ and $b$ be denoted by $S$.

Then $c$ is conjugate to both $a$ and $b$ and therefore is conjugate to every separation line in $S$ which passes through $O$.

It follows, by Theorem 129, that there is one and only one separation line in $S$ and passing through $O$ which is conjugate to every inertia line in $P$ which passes through $O$ and the separation line $a$ has this property.

Now it is easy to see that $a$ is the only separation line in $S$ and passing through $O$ which is normal to $b$; for suppose, if possible, that $a'$ is another such separation line.

Then, by the definition, there must exist an inertia plane, say $P'$, containing $b$ and such that every inertia line in $P'$ which passes through $O$ is conjugate to $a'$.

Then there would exist one single inertia line, say $c'$, through $O$ and lying in $P'$ which would be conjugate to $b$.

Thus $c'$ would be conjugate to every separation line in $S$ which passed through $O$ and therefore would be conjugate to $a$.

But now $P'$ could not be identical with $P$, for, as we have seen, $a$ is the only separation line in $S$ and passing through $O$ which is conjugate to every inertia line in $P$ which passes through $O$ and $a'$ has been supposed distinct from $a$.

But, by Theorem 132, it follows that $a$ must be conjugate to every inertia line in $P'$ which passes through $O$.

Thus we should have two distinct separation lines $a$ and $a'$ both lying in $S$ and passing through $O$ and both conjugate to every inertia line in $P'$ which passes through $O$.

But this is impossible by Theorem 129, and so the assumption of the existence of two distinct separation lines in $S$ which pass through $O$ and are normal to $b$ leads to a contradiction and therefore is not true.

Thus there is one and only one separation line in $S$ which passes through $O$ and is normal to $b$.

Again, since $b$ lies in $P$ while $a$ cannot lie in $P$, it follows that if a separation line $a$ be normal to a separation line $b$ having an element in common with it, then $a$ and $b$ must be distinct.

If $b$ be any general line in an inertia plane $P$ and $O$ be any element of $b$, then we know that if $b$ be either an inertia or separation line there is one and only one general line through $O$ and lying in $P$ which is conjugate and therefore *normal* to $b$.

Also, from our definitions, if $b$ be an optical line there is still one and only one general line through $O$ and lying in $P$ which is normal to $b$: namely $b$ itself.

Thus we have the following general result:

*If $P$ be either a separation plane or an inertia plane and if $b$ be any general line in $P$ and $O$ be any element in $b$, then there is one and only one general line lying in $P$ and passing through $O$ which is normal to $b$.*

Now we have seen that if a separation line $a$ be normal to a separation line $b$ having an element in common with it, then $a$ and $b$ lie in a separation plane.

Thus two intersecting separation lines in an optical plane cannot be normal one to another.

Any separation line, however, which lies in an optical plane is normal to every optical line in the optical plane since no element of the separation line except the element of intersection is either *before* or *after* any element of any optical line in the optical plane.

Since there is one and only one optical line which passes through any element of an optical plane and lies in the optical plane we have the following result:

*If P be an optical plane and if b be any separation line in P and O be any element in b, then there is one and only one general line lying in P and passing through O which is normal to b.*

*If on the other hand b be an optical line lying in P, then every general line in P which passes through O (including b itself) is normal to b.*

We have now to prove the general theorem that: *if b and c be two distinct general lines having an element O in common and if a general line a passing through O be normal to both b and c, then a is normal to every general line which passes through O and lies in the general plane containing b and c.*

We have already proved a number of special cases of this general theorem.

(1) If $b$ and $c$ be both optical lines and $a$ be a separation line, then $b$ and $c$ lie in an inertia plane, say $P$, and if $O'$ be any element of $a$ distinct from $O$ there will be an inertia plane, say $P'$, passing through $O'$ and parallel to $P$.

Then $O$ and $O'$ will be representatives of one another in the parallel inertia planes $P$ and $P'$ and so, by Theorem 102, $a$ is conjugate to every inertia line in $P$ which passes through $O$.

Thus $a$ is normal to every separation line in $P$ which passes through $O$, to every inertia line in $P$ which passes through $O$ and to every optical line in $P$ which passes through $O$.

(2) If $b$ and $c$ be both inertia lines and $a$ be a separation line, the same result follows from Theorem 103.

(3) If $b$ be an optical line and $c$ an inertia line while $a$ is a separation line, the same result follows from Theorem 104.

(4) If $b$ be a separation line and $c$ an inertia line while $a$ is a separation line, it follows by Theorem 132 that $a$ must be normal to every inertia line which passes through $O$ and lies in the inertia plane containing $b$ and $c$.

Thus as before, $a$ must be normal to every general line which passes through $O$ and lies in the inertia plane.

(5) If $b$ and $c$ be both separation lines and $a$ an inertia line, then, as we have seen, $b$ and $c$ must lie in a separation plane and, as was shown in

Theorem 128, $a$ is conjugate and therefore normal to every separation line passing through $O$ and lying in this separation plane.

(6) If $b$ be an optical line and $c$ a separation line while $a$ is identical with $b$, then, as we have already seen, $b$ and $c$ lie in an optical plane while $a$ is normal to every general line which passes through $O$ and lies in this optical plane.

Several other cases remain to be considered and these form the subject of Theorems 133 to 135.

We shall postpone the enumeration of the various remaining cases till we have proved these theorems.

## THEOREM 133

*If a separation line c be normal to a separation line b which it intersects in the element O and if further c be normal to an optical line a′ which it also intersects in the element O, then c is normal to every general line passing through O and lying in the general plane containing b and a′.*

By the definition of normality there exists an inertia plane, say $P$, containing $b$ and such that every inertia line in $P$ which passes through $O$ is conjugate to $c$.

In case $a′$ should lie in this particular inertia plane the result follows directly and so we shall suppose that $a′$ does not lie in $P$.

We shall therefore suppose that $a′$ and $b$ lie in a general plane $P′$ distinct from $P$.

From the remarks at the end of Theorem 132 it is evident that $P′$ may be either an inertia plane or an optical plane.

The mode of proof is similar to that employed in Theorem 132 except that $a′$ is here an optical line instead of an inertia line.

Thus the proof that $c$ is conjugate to every inertia line passing through $O$ and lying in either of the three inertia planes containing $a′$ and $OE_1$, $a′$ and $OE_2$, or $a′$ and $OC$, follows in this case from Theorem 104 instead of Theorem 103.

Everything else follows exactly as in Theorem 132 and we find that, if $OG$ be any general line in $P′$ which passes through $O$ and is distinct from $b$, then $OG$ lies in some inertia plane such that every inertia line in the latter which passes through $O$ is conjugate to $c$.

Thus if $OG$ be a separation line it satisfies the condition that $c$ should be normal to it.

Also if $OG$ should be either an optical line or an inertia line $c$ must also be normal to it, and so the theorem is proved.

## REMARKS

From the definition of the normality of intersecting separation lines it is evident that we may have a separation line normal to two (or more) separation lines in an inertia plane.

From the last theorem it is also evident that we may have a separation line normal to two (or more) separation lines in an optical plane.

We may also have a separation line normal to two (or more) separation lines in a separation plane, as may easily be seen from the following considerations:

In the remarks at the end of Theorem 131 it was pointed out that we may have an inertia plane and a separation plane having only one element in common and such that each inertia line through the common element in the former is conjugate to every separation line through it in the latter.

Let $P$ be the inertia plane, $S$ the separation plane and $O$ the common element.

Let $a$ and $b$ be any two separation lines passing through $O$ and lying in $S$, and let $c$ be any separation line passing through $O$ and lying in $P$.

Then $a$ satisfies the definition of being normal to $c$ and therefore $c$ is normal to $a$.

Similarly $c$ must be normal to $b$.

Thus $c$ is normal to the two separation lines $a$ and $b$ which lie in the separation plane $S$.

## THEOREM 134

*If three distinct separation lines $a$, $b$ and $c$ have an element $O$ in common and if $c$ be normal to both $a$ and $b$, then $c$ is normal to every general line which passes through $O$ and lies in the general plane containing $a$ and $b$.*

By the definition of the normality of intersecting separation lines there must exist an inertia plane, say $P$, containing $b$ and such that every inertia line in $P$ which passes through $O$ is conjugate to $c$.

Let $f_1$ and $f_2$ be the two generators of $P$ which pass through $O$ and let $D_1$ be any element in $f_1$ which is *after* $O$.

Let the separation line through $D_1$ parallel to $b$ intersect $f_2$ in $D_2$.

Then $D_2$ must also be *after* $O$.

Let $C$ be any element linearly between $D_1$ and $D_2$.

Then by Theorem 73 $OC$ is an inertia line and $C$ is *after* $O$.

But $c$ is normal to the inertia line $OC$ and to the separation line $a$ and therefore by case (4) on p. 212 $c$ must be normal to every inertia line (and therefore also every general line) which passes through $O$ and lies in the inertia plane containing $OC$ and $a$.

Let $R$ be this inertia plane.

If $R$ should coincide with $P$ the result follows directly and so we shall suppose that $R$ is distinct from $P$.

Let $S$ be the general plane containing $a$ and $b$.

Then $S$ will be distinct from both $P$ and $R$, and, as was pointed out in the remarks at the end of the last theorem, $S$ may be an inertia plane, an optical plane, or a separation plane.

Let one of the generators of $R$ which pass through $C$ intersect $a$ in $G_0$ and let the generator of the opposite set passing through $O$ intersect $CG_0$ in $F$.

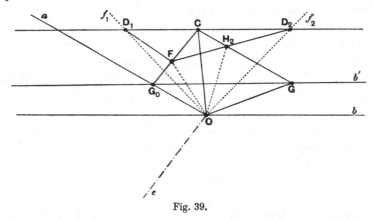

Fig. 39.

Then, since $O$ does not lie in the optical line $CG_0$ but is *before* the element $C$ of it, it follows that $F$ must lie in the $\alpha$ sub-set of $O$ and therefore $F$ is *after* $O$.

Let $b'$ be the general line through $G_0$ parallel to $b$.

Then, since $G_0$ lies in $S$, it follows that $b'$ lies in $S$.

Let $Q$ be the general plane containing $b'$ and $G_0C$.

Then, since $D_1D_2$ is parallel to $b$ and is distinct from $b'$, it follows that it is parallel to $b'$ and, since $D_1D_2$ passes through $C$, it must lie in the general plane $Q$.

Thus $D_1$, $D_2$ and $F$ are three distinct elements in $Q$ which do not all lie in one general line.

Now any general line in $S$ which passes through $O$ and is distinct from $b$ must intersect $b'$ in some element, say $G$.

Then the general line $CG$ lies in $Q$ and is distinct from $D_1D_2$.

If then $G$ does not coincide with $G_0$ it follows, by Theorem 127, that

$CG$ must either intersect $D_1F$ in an element $H_1$ linearly between $D_1$ and $F$, or else must intersect $D_2F$ in an element $H_2$ linearly between $D_2$ and $F$.

But, since $O$ is *before* both $D_1$ and $F$, it follows, by Theorem 73, that $OH_1$ is an inertia line and similarly, since $O$ is *before* both $D_2$ and $F$, it follows that $OH_2$ is an inertia line.

Now $c$ is normal to every general line in $P$ which passes through $O$ and also to every general line in $R$ which passes through $O$ and therefore $c$ is normal to the three optical lines $OD_1$, $OD_2$ and $OF$.

Thus $c$ must be conjugate to every inertia line which passes through $O$ and lies either in the inertia plane containing $OD_1$ and $OF$, or the inertia plane containing $OD_2$ and $OF$.

Thus $c$ is conjugate to $OH_1$ and also to $OH_2$.

But $c$ is conjugate to $OC$ and therefore is conjugate to every inertia line which passes through $O$ and lies in the inertia plane containing $OC$ and $OH_1$ or the inertia plane containing $OC$ and $OH_2$.

Thus, since $OG$ lies in the inertia plane containing $OC$ and $OH_1$ or in the inertia plane containing $OC$ and $OH_2$ as the case may be, it follows that $c$ must be normal to $OG$.

Thus, including the separation lines $a$ and $b$, the separation line $c$ is normal to every general line which passes through $O$ and lies in the general plane $S$.

### Theorem 135

*If two distinct separation lines $a$ and $b$ intersect in an element $O$ and if an optical line $c$ passing through $O$ be normal to both $a$ and $b$, then $c$ is normal to every general line which passses through $O$ and lies in the general plane containing $a$ and $b$.*

From the definition of the normality of an optical line to an intersecting separation line it follows that $c$ and $a$ lie in an optical plane, say $P$, while $c$ and $b$ lie in an optical plane, say $Q$.

If $P$ should be identical with $Q$ we already know that $c$ is normal to every general line in $P$ which passes through $O$ including the optical line $c$ itself.

Let us suppose next that $P$ is distinct from $Q$.

We have already seen that in this case $a$ and $b$ lie in a separation plane, say $S$, and further we have seen that no element of $b$ with the exception of $O$ is either *before* or *after* any element of $P$.

Let $D$ be any element of $b$ distinct from $O$ and let $E$ be any element of $a$ distinct from $O$, while $F$ is any element of $a$ such that $O$ is linearly between $E$ and $F$.

Let $e$ and $f$ be optical lines through $E$ and $F$ respectively and parallel to $c$.

Then $D$ is neither *before* nor *after* any element either of $e$ or of $f$ and so, by Theorem 45, no element of $DE$ with the exception of $E$ is either *before* or *after* any element of $e$, and no element of $DF$ with the exception of $F$ is either *before* or *after* any element of $f$.

But now by Theorem 127 any general line passing through $O$ and lying in $S$ and which is distinct from both $a$ and $b$ must either intersect

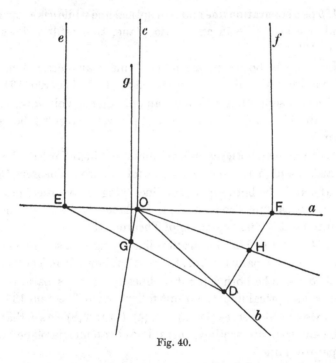

Fig. 40.

$DE$ in some element, say $G$, linearly between $D$ and $E$ or else must intersect $DF$ in some element, say $H$, linearly between $D$ and $F$.

Thus $G$ is neither *before* nor *after* any element of $e$ while $H$ is neither *before* nor *after* any element of $f$.

If then $g$ be an optical line through $G$ parallel to $e$ it will be a neutral-parallel and, since $c$ is a neutral-parallel of $e$ and $G$ does not lie in $c$, it follows by Theorem 28 that $g$ is a neutral-parallel of $c$.

Thus $G$ is neither *before* nor *after* any element of $c$ and therefore, by Theorem 45, no element of $OG$ with the exception of $O$ is either *before* or *after* any element of $c$.

Thus $c$ is normal to $OG$ and similarly it is normal to $OH$.

It follows that $c$ is normal to every general line which passes through $O$ and lies in $S$, and so the theorem is proved.

ENUMERATION OF CASES OF GENERAL THEOREM CONTINUED

We now resume the enumeration of the various cases of the general theorem stated on p. 212 and which was interrupted in order to prove Theorems 133 to 135.

Six cases have already been mentioned and we now proceed with case (7).

(7) If $b$ be a separation line and $c$ an optical line while $a$ is a separation line and if $b$ and $c$ lie in an inertia plane, the result follows from Theorem 133.

(8) If $b$ and $c$ be both separation lines lying in an inertia plane and if $a$ be also a separation line, the result follows from Theorem 134.

(9) If $b$ be a separation line and $c$ an optical line while $a$ is a separation line and if $b$ and $c$ lie in an optical plane, the result follows from Theorem 133.

(10) If $b$ and $c$ be both separation lines lying in an optical plane and if $a$ be also a separation line, the result follows from Theorem 134.

(11) If $b$ and $c$ be both separation lines lying in an optical plane and if $a$ be an optical line also in the optical plane, the result still holds as was pointed out at the beginning of Theorem 135.

(12) If $b$ and $c$ be both separation lines lying in a separation plane and if $a$ be also a separation line, the result follows from Theorem 134.

(13) If $b$ and $c$ be both separation lines lying in a separation plane and if $a$ be an optical line, the result follows from Theorem 135.

If now we combine cases (1), (2), (3), (4), (7) and (8) we see that $b$ and $c$ may be any two intersecting general lines in an inertia plane taking $a$ as a separation line.

If we combine cases (9) and (10) we see that $b$ and $c$ may be any two intersecting general lines in an optical plane taking $a$ as a separation line.

Further, combining cases (6) and (11) we also see that $b$ and $c$ may be any two intersecting general lines in an optical plane taking $a$ as an optical line.

Finally from cases (12), (13) and (5) we see that $b$ and $c$ may be any two intersecting general lines in a separation plane taking $a$ as a separation line, an optical line, or an inertia line.

Thus for all the different possible cases of the normality of general lines having a common element this general result holds.

## THEOREM 136

*If $b$ and $c$ be two separation lines intersecting in an element $O$ and lying in a separation plane $S$ and such that $c$ is normal to $b$, then if $O'$ be any other element of $b$, the normal to $b$ through $O'$ in the separation plane $S$ is parallel to $c$.*

From the definition of the normality of intersecting separation lines it follows that there must exist an inertia plane $P$ containing $b$ and such that every inertia line in $P$ which passes through $O$ is conjugate to $c$.

Let $a_1$ and $a_2$ be any two such inertia lines and let $a_1'$ and $a_2'$ be inertia lines passing through $O'$ and parallel to $a_1$ and $a_2$ respectively.

Let $c'$ be a separation line passing through $O'$ and parallel to $c$.

Then $c'$ will lie in $S$.

But, by Theorem 100, both $a_1'$ and $a_2'$ must be conjugate to $c'$ and so, by Theorem 103, $c'$ is conjugate to every inertia line in the inertia plane containing $a_1'$ and $a_2'$ which passes through the element $O'$.

But this inertia plane is the inertia plane $P$ which contains the separation line $b$ and so $c'$ satisfies the definition of being normal to $b$.

Further, $c'$ passes through $O'$ and lies in $S$ and we have already seen that there is only one normal to $b$ which satisfies these conditions.

Thus the normal to $b$ through $O'$ in the separation plane $S$ is parallel to $c$ as was to be proved.

## THEOREM 137

*If $b$ and $c$ be two separation lines intersecting in an element $O$ and such that $c$ is normal to $b$ and if $b'$ and $c'$ be two other separation lines intersecting in an element $O'$ and respectively parallel to $b$ and $c$, then $c'$ is normal to $b'$.*

Since $c$ is normal to $b$ there must exist an inertia plane $P$ containing $b$ and such that every inertia line in $P$ which passes through $O$ is conjugate to $c$.

Let $a_1$ be one such inertia line which we shall suppose does not also pass through $O'$.

Then through $O'$ there is an inertia line, say $a_1'$, which is parallel to $a_1$.

Thus $b'$ and $a_1'$ determine an inertia plane $P'$ which will be either identical with $P$ or parallel to $P$ according as $O'$ does or does not lie in $P$.

Let $a_2$ be a second inertia line in $P$ and passing through $O$ but not through $O'$.

Then through $O'$ there is an inertia line say $a_2'$ parallel to $a_2$ and lying in $P'$.

Then by Theorem 100 both $a_1'$ and $a_2'$ are conjugate to $c'$ and so, by Theorem 103, $c'$ is conjugate to every inertia line in $P'$ which passes through $O'$.

But $P'$ contains $b'$ and so $c'$ satisfies the definition of being normal to $b'$.

## THEOREM 138

*If an optical line b intersects a separation line c in an element O and if c be normal to b and if further b' and c' be an optical line and a separation line respectively which intersect in an element O' and are respectively parallel to b and c, then c' will be normal to b'.*

From the definition of the normality of a separation line to an optical line it follows that $b$ and $c$ lie in an optical plane, say $P$.

Further, $b'$ and $c'$ lie in a general plane $P'$ which must be either identical with $P$ or parallel to $P$ according as $O'$ does or does not lie in $P$.

In either case $P'$ is an optical plane and accordingly, since $b'$ is an optical line and $c'$ a separation line, it follows that $c'$ must be normal to $b'$.

## REMARKS

By combining Theorems 100, 137 and 138 we obtain the general result that *if b and c be two general lines intersecting in an element O and such that the one is normal to the other and if b' and c' be two other general lines intersecting in an element O' and respectively parallel to b and c, then of these latter two general lines the one is normal to the other.*

If now we remember that an optical line is to be regarded as *normal to itself*, we are in a position to extend the definition of the normality of general lines to the case of general lines which have no element in common, as is done with straight lines in ordinary geometry.

*Definition.* A general line $b$ will be said to be *normal* to a general line $c'$ which has no element in common with it, provided that a general line $b'$ taken through any element of $c'$ parallel to $b$ is normal to $c'$ in the sense already defined.

It is evident from the above considerations that, in these circumstances, if a general line $c$ be taken through any element of $b$ parallel to $c'$ then $c$ will be normal to $b$ and so $c'$ will be normal to $b$.

Further, we have the result that *any two parallel optical lines are to be regarded as normal to one another.*

Again, if $P$ be an inertia plane and if $a$ be any general line in $P$ and $A$ be any element in $P$, then there is one single general line in $P$ and passing through $A$ which is normal to $a$.

If however $a$ be an optical line, the normal to $a$ through $A$ is either identical with $a$ or parallel to it according as $A$ does or does not lie in $a$.

If, on the other hand, $P$ be an optical plane, there is one single general line in $P$ and passing through $A$ which is normal to $a$, except when $a$ is an optical line, in which case every general line in $P$ which passes through $A$ is normal to $a$.

If $P$ be a separation plane there is one single general line in $P$ which passes through $A$ and is normal to $a$ and in this case the normal to $a$ always intersects $a$ as in ordinary geometry.

*Definition.* A general line $a$ will be said to be *normal* to a general plane $P$ provided $a$ be normal to every general line in $P$.

It is evident that if a general line $a$ be normal to two intersecting general lines in a general plane $P$, then $a$ will be normal to $P$.

In case $P$ be an optical plane it is clear that, according to the above definition, any generator of $P$ is normal to $P$.

This is the only case in which a general line can be normal to a general plane which contains it.

In no other case can a general line which is normal to a general plane have more than one element in common with the latter.

As was pointed out in the remarks at the end of Theorem 131 we may have an inertia plane and a separation plane having only one element in common and such that each inertia line through the common element in the former is conjugate to every separation line through it in the latter.

It is evident now that we have here two general planes which are so related that any general line in the one is normal to any general line in the other.

In ordinary *three-dimensional* geometry two planes cannot be so related, and when we speak of one plane being normal to another the normality is not of this complete character.

We shall therefore introduce the following definition:

*Definition.* If two general planes be so related that every general line in the one is normal to every general line in the other, the two general planes will be said to be *completely normal* to one another.

THEOREM 139

*If $P$ be an inertia plane and $O$ be any element in it, there is at least one separation plane passing through $O$ and completely normal to $P$.*

Let $P_1$ be any inertia plane which is parallel to $P$ and let $O_1$ be the representative of $O$ in $P_1$.

Then, by Theorem 102, the separation line $OO_1$ is conjugate to every inertia line in $P$ which passes through $O$ and so $OO_1$ is normal to $P$.

Let $a_1$ be one of the two generators of $P$ which pass through $O$, and let $OO_1$ be denoted by $b_1$.

Then $a_1$ and $b_1$ lie in an optical plane, say $Q_1$.

Now, by Post. XIX, there is at least one element, say $A$, which is neither *before* nor *after* any element of $Q_1$.

Thus through $A$ there is an optical line, say $a_1'$, which is neutrally parallel to $a_1$ and so $a_1$ and $a_1'$ lie in an optical plane, say $R_1$, which is distinct from $Q_1$.

Again if $P_2$ be an inertia plane through $A$ parallel to $P$ it will contain $a_1'$.

Let $O_2$ be the representative of $O$ in $P_2$.

Then $O_2$ must lie in $a_1'$ and so $OO_2$ must lie in the optical plane $R_1$.

But $OO_1$ lies in $Q_1$ while $OO_2$ lies in $R_1$, and $Q_1$ and $R_1$ have only the optical line $a_1$ in common.

Thus since $OO_1$ and $OO_2$ are both separation lines they must be distinct.

Now, by Theorem 102, $OO_2$ is conjugate to every inertia line in $P$ which passes through $O$, and so $OO_2$ is normal to $P$.

Let $OO_2$ be denoted by $b_2$.

Then no element of $b_2$ is either *before* or *after* any element of $b_1$ and, since $b_1$ and $b_2$ have the element $O$ in common, they must lie in a separation plane, say $S$.

Thus any inertia line in $P$ which passes through $O$ is conjugate to both $b_1$ and $b_2$ and therefore also conjugate to every separation line in $S$ which passes through $O$.

Thus every general line in $P$ is normal to every general line in $S$ and so the separation plane $S$ is completely normal to $P$.

Thus, since $S$ passes through $O$, the theorem is proved.

## THEOREM 140

*If P be a separation plane and O be any element in it, there is at least one inertia plane passing through O and completely normal to P.*

If we take any two separation lines in $P$ and passing through $O$ then, by Theorem 107, there is at least one inertia line, say $a_1$, which is conjugate to both of them and therefore is normal to $P$.

Let $b_1$ be any separation line in $P$ which passes through $O$ and let $Q$ be the inertia plane containing $a_1$ and $b_1$.

Then, by Theorem 129, there is one and only one separation line in $P$ and passing through $O$ which is conjugate to every inertia line in $Q$ which passes through $O$.

Let $b_2$ be this separation line.

Then, as was remarked at the end of Theorem 131, $b_2$ is conjugate to certain other inertia lines passing through $O$ which do not lie in $Q$.

Let $a'$ be any such inertia line and let $Q'$ be the inertia plane containing $a'$ and $b_1$.

Then, by Theorem 132, $b_2$ is conjugate to every inertia line in $Q'$ which passes through $O$.

Let $a_2$ be the one single inertia line in $Q'$ and passing through $O$ which is conjugate to $b_1$ and let $R$ be the inertia plane containing $a_1$ and $a_2$.

Then $a_1$ and $a_2$ are each conjugate to both $b_1$ and $b_2$.

Thus both $a_1$ and $a_2$ are conjugate to every separation line in $P$ which passes through $O$ and so every separation line in $P$ which passes through $O$ is conjugate to every inertia line in $R$ which passes through $O$.

Thus every general line in $P$ is normal to every general line in $R$ and so the inertia plane $R$ is completely normal to $P$.

Thus, since $R$ passes through $O$, the theorem is proved.

## THEOREM 141

*If P be an optical plane and O be any element in it, there is at least one optical plane passing through O and completely normal to P.*

Let $a$ be the generator of $P$ which passes through $O$ and let $b$ be any separation line in $P$ which passes through $O$.

Then, by Post. XIX, there is at least one element, say $A$, which is neither *before* nor *after* any element of $P$.

The general line $OA$ is thus a separation line and, by Theorem 45, no element of $OA$ with the exception of $O$ is either *before* or *after* any element of $a$.

Thus $a$ is normal to $OA$ and it is also normal to $b$ and so, since $OA$ and

$b$ must lie in a separation plane, say $S$, it follows that the optical line $a$ is normal to $S$.

But now we know that there is one single separation line, say $c$, which passes through $O$, lies in $S$ and is normal to $b$.

Then $c$ is normal to both $a$ and $b$ and therefore is normal to $P$.

But $c$ and $a$ lie in an optical plane which is distinct from $P$ and which we shall call $R$.

Further, $a$ is an optical line in $P$ and therefore is normal to $P$.

Thus any general line in $P$ is normal to the two intersecting general lines $a$ and $c$ which lie in $R$ and so every general line in $P$ is normal to every general line in $R$.

It follows that $R$ is completely normal to $P$ and, since $R$ passes through $O$, the theorem is proved.

## Remarks

By combining Theorems 139, 140 and 141 we get the general result:

*If P be any general plane and O be any element in it, there is at least one general plane passing through O and completely normal to P.*

If $R$ be this general plane which is completely normal to $P$ and if $O'$ be any element not lying in $P$, then $O'$ either may or may not lie in $R$.

If $O'$ does not lie in $R$, then there is a general plane, say $R'$, passing through $O'$ and parallel to $R$.

It is evident that since $R$ is completely normal to $P$ we must also have $R'$ completely normal to $P$ and so we may generalise the above result and say:

*If P be any general plane and O be any element whatever, there is at least one general plane passing through O and completely normal to P.*

Let $O$ be any element and let $S$ be any separation plane passing through $O$, while $P$ is an inertia plane also passing through $O$ and completely normal to $S$.

Let $a$ be any separation line in $S$ which passes through $O$ and let $b$ be the one single separation line in $S$ passing through $O$ which is normal to $a$.

Let $c$ be any separation line passing through $O$ and lying in $P$ and let $d$ be the one single inertia line in $P$ and passing through $O$ which is normal to $c$.

Then both $c$ and $d$ are normal to both $a$ and $b$ and so *we have the three separation lines a, b and c all passing through O and each of them normal to the other two; while in addition to these we have the inertia line d also passing through O and normal to all three.*

This result marks an important stage in the development of our theory, as it suggests the possibility of setting up a system of normal coordinate axes one of which axes is of a different character from the remaining three.

Another important result is the following:

If $S$ be a separation plane and if $P$ be an inertia plane passing through any element $O$ of $S$ and completely normal to $S$, then there are two generators of $P$ which pass through $O$ and each of them is normal to the separation plane $S$.

*Thus there are at least two optical lines which pass through any element of a separation plane and are normal to it.*

### THEOREM 142

*If $P$ be an inertia or separation plane and $O$ be any element which does not lie in it, there is one single general line passing through $O$ and normal to $P$ which has an element in common with $P$.*

We already know that if $a$ be a separation line and if $O$ be any element which does not lie in it, then, in whatever type of general plane $O$ and $a$ may lie, there is one single general line passing through $O$ and lying in this general plane which is normal to $a$.

Further, if $d$ be this general line normal to $a$, then $d$ must intersect $a$ in some element, say $A$.

Now suppose that $a$ lies in the inertia or separation plane $P$.

Then there is one single general line passing through $A$ and lying in $P$ which is normal to $a$.

Let $b$ be this general line.

Then, since $P$ is an inertia or separation plane and $a$ is a separation line, $b$ must be distinct from $a$ and must be either an inertia or separation line and cannot be an optical line.

Now we know that in whatever type of general plane $O$ and $b$ may lie there is one single general line passing through $O$ and lying in this general plane which is normal to $b$.

Let $c$ be this general line.

Then, since $b$ is not an optical line, this normal to it through $O$ cannot be parallel to $b$ and therefore must intersect $b$ in some element, say $B$.

Now $a$ is normal to the two general lines $d$ and $b$ which intersect in $A$ and accordingly $a$ is normal to every general line in the general plane containing $d$ and $b$ and therefore is normal to $c$.

But $c$ is normal to the two intersecting general lines $a$ and $b$ which lie in $P$ and therefore $c$ is normal to $P$.

Since $c$ has the element $B$ in common with $P$, we have proved that there is at least one general line through $O$ and normal to $P$ which has an element in common with $P$.

It remains to show that there is only one general line having this property.

Consider first the case where $P$ is a separation plane and let $B'$ be any element in $P$ distinct from $B$.

Then $BB'$ is a separation line and so in whatever type of general plane $O$ and $BB'$ may lie there is one single general line passing through $O$, lying in this general plane and normal to $BB'$.

But $OB$ passes through $O$ and is normal to $BB'$ and therefore $OB'$ cannot be normal to $BB'$ and so cannot be normal to $P$.

This proves that $OB$ is the only general line through $O$ and normal to $P$ which has an element in common with $P$ provided $P$ be a separation plane.

This method does not serve if $P$ be an inertia plane, since $BB'$ might, in this case, be an optical line.

If $P$ be an inertia plane, let $P'$ be an inertia plane passing through $O$ and parallel to $P$.

Then $O$ and $B$ must be representatives of one another in the parallel inertia planes $P'$ and $P$.

If $B'$ be any other element in $P$ distinct from $B$ and we suppose that $OB'$ is normal to $P$, then $B'$ would also be the representative of $O$ in $P$, which we know is impossible.

Thus again $OB$ is the only general line through $O$ and normal to $P$ which has an element in common with $P$.

The theorem thus holds for both separation and inertia planes.

## THEOREM 143

*If $P$ be an optical plane and $O$ be any element which does not lie in it, then:*

(1) *If $O$ be neither before nor after any element of $P$ there is one single generator of $P$ such that every general line which passes through $O$ and intersects this generator is normal to $P$.*

(2) *If $O$ be either before or after any element of $P$ there is no general line passing through $O$ and having an element in common with $P$ which is normal to $P$.*

As regards the first part of this theorem, if we carry out the construction of Theorem 142 taking $a$ as a separation line, then, since $P$ is an optical plane, the general line $b$ must be an optical line since it is normal to $a$.

Since $O$ is neither *before* nor *after* any element of $P$, it is neither *before* nor *after* any element of $b$.

If then $OB$ be any general line passing through $O$ and intersecting $b$ in the element $B$, it follows by Theorem 45 that no element of $OB$ with the exception of $B$ is either *before* or *after* any element of $b$.

It follows that $OB$ is normal to $b$.

But, as in Theorem 142, $OB$ is normal to $a$ and thus $OB$ is normal to the two intersecting general lines $a$ and $b$ which lie in $P$ and therefore it is normal to $P$.

Again if $B'$ be any element in $P$ which does not lie in $b$, then $BB'$ is a separation line and so, as in Theorem 142, $OB'$ cannot be normal to $P$.

Thus all general lines through $O$ which have an element in common with $b$ are normal to $P$, and no other general line through $O$ which intersects $P$ can be normal to $P$.

Thus the first part of the theorem is proved.

Suppose next that $O$ is *before* some element, say $E$, in $P$.

Then through $E$ one single generator of $P$ passes which we may denote by $f$.

Since $O$ does not lie in $f$ but is *before* an element of $f$, it follows that through $O$ there is an optical line which is a before-parallel of $f$ and which we shall denote by $c$.

If $f'$ be any other generator of $P$ it will be a neutral-parallel of $f$ and so by Theorem 26(a) $c$ will be a before-parallel of $f'$.

Thus $O$ is *before* elements of every generator of $P$.

Similarly if $O$ be *after* any element of $P$ it is *after* elements of every generator of $P$.

Thus in case $O$ be either *before* or *after* any element of $P$ it will lie in an inertia plane along with any selected generator of $P$.

Let $OB$ be any general line passing through $O$ and having the element $B$ in common with $P$ and let $b$ be the generator of $P$ which passes through $B$.

Then $OB$ and $b$ lie in an inertia plane and intersect in $B$ and so, since $b$ is an optical line, $OB$ cannot be normal to it.

Thus $OB$ cannot be normal to $P$ and therefore there is in this case no general line passing through $O$ and having an element in common with $P$ which is normal to $P$.

## Theorem 144

*If a general line d have an element A in common with a general plane P, there is at least one general line passing through A and lying in P which is normal to d.*

If $d$ lies completely in $P$ we already know that the theorem holds and so we shall suppose that $A$ is the only element common to $d$ and $P$.

We shall first consider the case where $P$ is an inertia or separation plane.

In this case, if $O$ be any element of $d$ distinct from $A$, there is, by Theorem 142, one single general line passing through $O$ and normal to $P$ which has an element in common with $P$.

Let $B$ be this element.

If $B$ should coincide with $A$, then every general line passing through $A$ and lying in $P$ would be normal to $d$.

If $B$ does not coincide with $A$ let $a$ be the one single general line passing through $A$ and lying in $P$ which is normal to $AB$.

Then since $OB$ is normal to $P$ it must be normal to $a$.

Thus $a$ is normal to the two intersecting general lines $AB$ and $OB$ and therefore is normal to the general plane containing them.

Thus the general line $a$ must be normal to $d$ and, since $a$ passes through $A$ and lies in $P$, the theorem is proved for the case where $P$ is an inertia or separation plane.

Suppose next that $P$ is an optical plane and let $b$ be the generator of $P$ which passes through $A$.

Now, since $b$ is an optical line, it follows that the intersecting general lines $b$ and $d$ must lie in a general plane, say $Q$, which must be either an optical plane or an inertia plane.

Suppose first that $Q$ is an optical plane.

Then, since $b$ is an optical line in $Q$ and $d$ intersects $b$, it follows that $d$ must be a separation line and $b$ must be normal to $d$.

But $b$ passes through $A$ and lies in $P$ and so the theorem is proved for this case.

Next consider the case where $Q$ is an inertia plane.

Let $b'$ be any generator of $P$ distinct from $b$.

Then, since $b'$ is a neutral-parallel of $b$, it follows that an inertia plane $Q'$ through any element of $b'$ and parallel to $Q$ will contain $b'$.

If then $A'$ be the representative of $A$ in $Q'$, the general line $AA'$ will be normal to $Q$ and therefore will be normal to $d$.

But, since $b'$ is neutrally parallel to $b$ which contains the element $A$, the element $A'$ must lie in $b'$ and therefore in the optical plane $P$.

Thus the general line $AA'$ must lie in $P$ and, since it passes through $A$ and is normal to $d$, the theorem holds also in this case.

Thus the theorem holds in general.

## THEOREM 145

*If three general lines $a$, $b$ and $c$ have an element $O$ in common, there is at least one general line passing through $O$ which is normal to all three.*

If we take any two of the three given general lines, say $a$ and $b$, it follows, since they have the element $O$ in common, that they lie in a general plane, say $P$.

Then by Theorems 139, 140 and 141 there is at least one general plane passing through $O$ and completely normal to $P$.

Let $Q$ be this general plane.

Then, since $c$ has the element $O$ in common with $Q$, it follows, by Theorem 144, that there is at least one general line, say $d$, passing through $O$ and lying in $Q$ which is normal to $c$.

But, since $d$ lies in $Q$, it is normal to both $a$ and $b$ and thus is normal to all three general lines.

Thus the theorem is proved.

*Definition.* If a general line and a general plane have one single element in common, they will be said to *intersect* in that element.

*Definition.* If a general line $a$ and a general plane $P$ intersect, then the aggregate of all elements of $P$ and of all general planes parallel to $P$ which intersect $a$ will be called a *general threefold*.

It will be found that, just as there are three types of general line and three types of general plane, so there are three types of general threefold.

In the case of general threefolds, however, unlike that of general lines or of general planes, we are able to give a definition which applies to all three types without first considering any of the special cases.

From the definition it is clear that if a general threefold $W$ be determined by a general line $a$ intersecting a general plane $P$, then any other general plane $P_1$ parallel to $P$ and intersecting $a$ may take the place of $P$, so that $a$ and $P_1$ will also serve to determine $W$.

Again if $a$ intersects $P$ in the element $O$ and if $a'$ be a general line parallel to $a$ and intersecting $P$ in another element $O'$, then $a$ and $a'$ will lie in a general plane, say $Q$.

If through any element $O_1$ of $a$ distinct from $O$ the general plane $P_1$ passes parallel to $P$, then, by Theorem 123, the general plane $Q$ must have a second element in common with $P_1$.

Thus $P_1$ and $Q$ have a general line in common which must be parallel to $OO'$ and so the general line $a'$ must intersect $P_1$ in some element $O_1'$.

Thus $a'$ intersects every general plane parallel to $P$ which intersects $a$, and similarly, $a$ intersects every general plane parallel to $P$ which intersects $a'$.

It follows that every element of $a'$ lies in the general threefold determined by $a$ and $P$, and also: that $a'$ and $P$ determine the same general threefold as $a$ and $P$.

## Theorem 146

*If two distinct elements of a general line lie in a general threefold, then every element of the general line lies in the general threefold.*

Let the general threefold $W$ be determined by a general plane $P$ and a general line $a$ which intersects it.

Let $X_1$ and $X_2$ be two distinct elements of a general line $b$ and let them both lie in $W$.

If $X_1$ and $X_2$ should both lie in $P$ or in any one of the general planes which intersect $a$ and are parallel to $P$, then the general line $b$ will lie in that general plane and therefore every element of $b$ must lie in $W$.

We shall next suppose that $X_1$ lies in one of the set of parallel general planes, say $P_1$, while $X_2$ lies in another, say $P_2$.

Then $b$ either may or may not lie in a general plane containing $a$.

Suppose first that $b$ lies in a general plane $Q$ along with $a$.

Then we may have either:

|    | (1) $b$ identical with $a$,   |
|----|-------------------------------|
| or | (2) $b$ parallel to $a$,      |
| or | (3) $b$ intersecting $a$.     |

If $b$ be identical with $a$ the result is obvious.

If $b$ be parallel to $a$ then, as we have already shown, every element of $b$ lies in $W$.

If $b$ intersects $a$, then at least one of the elements $X_1$, $X_2$ must be distinct from the element of intersection of $b$ and $a$.

We may suppose that $X_1$ is distinct from this element of intersection.

Then the element in which $a$ intersects $P_1$ must be distinct from $X_1$ and so the general plane $Q$ has two distinct elements in common with $P_1$.

Further, since the general line $a$ intersects all the general planes parallel to $P_1$ whose elements along with the elements $P_1$ make up $W$, it follows, by Theorem 123, that $Q$ has a general line in common with each of these general planes and all these general lines are parallel to one another.

Now since $b$ does not lie in $P_1$ it follows that $b$ must intersect all these general planes, and similarly a general plane through any element of $b$ distinct from $X_1$ and taken parallel to $P_1$ must intersect $a$.

Thus we see that in this case also every element of $b$ lies in $W$ and further that $b$ and $P_1$ determine the same general threefold as $a$ and $P_1$: namely $W$.

Thus the theorem holds provided $b$ and $a$ lie in one general plane.

Finally suppose as before that $X_1$ lies in $P_1$ and $X_2$ in $P_2$ and that $b$ and $a$ do not lie in one general plane.

Let $a$ intersect $P_1$ in the element $Y_1$ and let $b'$ be a general line through $Y_1$ parallel to $b$.

Then $b$ and $b'$ lie in a general plane, say $R$, which has the two elements $X_1$ and $Y_1$ in common with $P_1$ and has the element $X_2$ in common with the parallel general plane $P_2$.

Thus $R$ has a general line in common with $P_2$ which is parallel to $X_1Y_1$ and so $b'$ must intersect $P_2$ in some element, say $Y_2$.

But now, from what we have already proved, every element of $b'$ must lie in $W$ and also $b'$ and $P_2$ determine the general threefold $W$ equally with $a$ and $P_2$ or $a$ and $P$.

Again, since $b$ is parallel to $b'$, it follows from what we have already proved that every element of $b$ lies in the general threefold determined by $b'$ and $P_2$: that is in $W$; and that $b$ and $P_2$ may also be taken as determining the general threefold $W$.

Thus the theorem holds in general.

### REMARKS

It is evident from the above that if a general threefold $W$ be determined by a general plane $P$ and a general line $a$ which intersects $P$, then a general line $b$ which has two distinct elements in common with $W$, which do not both lie in $P$ or do not both lie in one of the general planes parallel to $P$ and intersecting $a$, will intersect all these general planes including $P$.

Further, $b$ and $P$, or $b$ and any one of these general planes, will also determine $W$.

Again if a general plane $Q$ have two distinct elements $X_1$ and $X_2$ in common with $W$, then $Q$ will have at least one general line in common with $W$: namely the general line $X_1 X_2$ since, by the above theorem, every element of $X_1 X_2$ must lie in $W$, and we already know that every element of it must also lie in $Q$.

It is not however possible from this to prove that $Q$ and $W$ have more than one general line in common.

### THEOREM 147

*If a general plane have three distinct elements in common with a general threefold and if these three elements do not all lie in one general line, then every element of the general plane lies in the general threefold.*

Let the general threefold $W$ be determined by a general plane $P$ and a general line $a$ which intersects $P$.

Let $X_1$, $X_2$ and $X_3$ be three distinct elements of a general plane $Q$ which do not all lie in one general line and suppose that $X_1$, $X_2$ and $X_3$ all lie in $W$.

If all these three elements should lie in $P$ or if they should all lie in one of the general planes parallel to $P$ which intersect $a$, then $Q$ would be identical with the general plane in which they all lie and accordingly every element of $Q$ would lie in $W$.

If $X_1$, $X_2$ and $X_3$ do not all lie in one of this set of general planes, suppose that $X_1$ lies in the general plane $P_1$ of the set while $X_2$ lies in another distinct general plane of the set, say $P_2$.

Then $X_3$ will lie in some general plane $P_3$ of the set which may be either identical with $P_1$ or with $P_2$, or may be distinct from both.

Now, since $X_1$ and $X_2$ lie in two distinct general planes of the set, it follows that the general line $X_1 X_2$ intersects every general plane of the set and therefore must intersect $P_3$ in some element, say $O$.

Further, since $X_1$, $X_2$ and $X_3$ do not all lie in one general line, it follows that $X_3$ and $O$ must be distinct elements.

Thus the general planes $P_3$ and $Q$ have two distinct elements $X_3$ and $O$ in common and therefore have the general line $OX_3$ in common, which accordingly lies in $W$.

Again the general threefold $W$, as we have seen, may be determined by the general plane $P_3$ and the general line $X_1 X_2$ which intersects $P_3$ in $O$.

But now every element of $Q$ lies either in $X_1 X_2$ or in a general line parallel to $X_1 X_2$ and intersecting $OX_3$.

We have however seen that every element of any such general line must lie in $W$.

It follows that every element of $Q$ must lie in $W$.

Thus the theorem holds in all cases.

## THEOREM 148

(1) *If a general line b lies in a general threefold W and if A be any element lying in W but not in b, then the general line through A parallel to b also lies in W.*

(2) *If a general plane P lies in a general threefold W and if A be any element lying in W but not in P, then the general plane through A parallel to P also lies in W.*

The first part of the theorem may be proved as follows:

The general line $b$ and the element $A$ determine a general plane, say $Q$, having three elements in common with $W$ which do not all lie in one general line, and so, by Theorem 147, $Q$ lies in $W$.

But the general line through $A$ parallel to $b$ must lie in $Q$ and therefore must lie in $W$.

This proves the first part of the theorem.

In order to prove the second part let $b$ and $c$ be two intersecting general lines which both lie in $P$ and therefore in $W$.

The element $A$ does not lie in $P$ and therefore cannot lie either in $b$ or $c$.

If then $b'$ and $c'$ be general lines through $A$ parallel to $b$ and $c$ respectively, it follows from the first part of the theorem that $b'$ and $c'$ both lie in $W$.

If then $P'$ be the general plane containing $b'$ and $c'$ it will contain three distinct elements in common with $W$ which do not all lie in one general line and so, by Theorem 147, $P'$ must lie in $W$.

But $P'$ is parallel to $P$ and passes through $A$ and so the theorem is proved.

## THEOREM 149

*If a general threefold W be determined by a general plane P and a general line a which intersects P, then if Q be any general plane lying in W, and if b be any general line lying in W and intersecting Q, the general plane Q and the general line b also determine the same general threefold W.*

It is evident from the remarks at the end of Theorem 146 that the above holds in the special case where $Q$ is one of the set of general planes consisting of $P$ and all general planes parallel to $P$ which intersect $a$.

We shall therefore consider the case where $Q$ is distinct from any one of this set of general planes which we shall for convenience refer to as the *primary set*.

Let $X_1$ be any element in $Q$ and take any two distinct general lines lying in $Q$ and passing through $X_1$.

Then these could not both lie in any general plane of the primary set, for if so $Q$ would require to be identical with that general plane, contrary to hypothesis.

Thus at least one of the two general lines does not lie in any general plane of the primary set.

Suppose $c_1$ be a general line of this character.

Then, since $Q$ lies in $W$, each element of $c_1$ must lie in a distinct general plane of the primary set, and $c_1$ must intersect every general plane of the primary set.

Thus $W$ may be determined by any general plane of the primary set and the general line $c_1$ which intersects it, in place of the general line $a$.

Let $X_2$ be any element of $Q$ which does not lie in $c_1$ and let $c_2$ be a general line through $X_2$ parallel to $c_1$.

Then $c_2$ must also lie in $Q$ and must also intersect every general plane of the primary set.

Further, since $c_1$ and $c_2$ are parallel, they must intersect any general plane of the primary set in distinct elements, and accordingly any general plane of the primary set has a general line in common with $Q$.

Now let $B$ be any element in $b$ other than its element of intersection with $Q$.

Then $B$ must lie in some general plane of the primary set, say $P_1$, since $b$ lies in $W$.

Now, as we have seen, $P_1$ has a general line in common with $Q$ and, since $B$ does not lie in $Q$, it cannot lie in this general line.

If $C$ and $D$ be any two distinct elements in this general line, then $B$, $C$ and $D$ are three distinct elements in $P_1$ which do not all lie in one general line.

But now, if $W'$ be the general threefold determined by $Q$ and $b$, it is evident that $B$, $C$ and $D$ lie in $W'$ and so, by Theorem 147, the general plane $P_1$ must lie in $W'$.

Also since $c_1$ lies in $Q$ it must lie in $W'$, and so, by Theorem 148, every general plane which passes through an element of $c_1$ and is parallel to $P_1$ must lie in $W'$.

But the general threefold $W$ is the aggregate of all elements of $P_1$ and of all general planes parallel to $P_1$ which intersect $c_1$, and so every element of $W$ must lie in $W'$.

But, since $Q$ and $b$ both lie in $W$, it follows by Theorem 148 that every general plane which passes through an element of $b$ and is parallel to $Q$ must lie in $W$.

Since, however, the general threefold $W'$ is the aggregate of all elements of $Q$ and of all general planes parallel to $Q$ which intersect $b$, it follows that every element of $W'$ must lie in $W$.

Thus the general threefolds $W'$ and $W$ consist of the same set of elements and are therefore identical.

Thus $Q$ and $b$ determine $W$, as was to be proved.

### REMARKS

It follows directly from the above theorem that *any four distinct elements which do not all lie in one general plane determine a general threefold containing them.*

For let $A$, $B$, $C$, $D$ be four distinct elements which do not all lie in one general plane.

Then no three of them can lie in one general line.

Let $Q$ be the general plane containing $A$, $B$ and $C$ and let $b$ be the general line $DA$.

Then $b$ cannot have any other element than $A$ in common with $Q$, for then $D$ would have to lie in $Q$ along with $A$, $B$ and $C$ contrary to hypothesis.

Thus $b$ intersects $Q$.

Let $W$ be the general threefold determined by $Q$ and $b$ and let $W'$ be any general threefold containing $A$, $B$, $C$ and $D$.

Then, since $W'$ contains $A$, $B$ and $C$, it follows by Theorem 147 that $W'$ contains $Q$.

Also by Theorem 146 since $W'$ contains $A$ and $D$ it contains $b$.

Thus by Theorem 149 the general threefold $W'$ is identical with $W$: that is to say is identical with one definite general threefold.

Again it is clear that: *any three distinct general lines having a common element and not all lying in one general plane determine a general threefold containing them.*

### THEOREM 150

*If two distinct general planes $P$ and $Q$ lie in a general threefold $W$, then if $P$ and $Q$ have one element in common they have a second element in common.*

Let $A$ be any element in $P$ and let $B$ be any element which lies in $W$ but not in $P$.

Let the general line $AB$ be denoted by $a$.

Then $a$ intersects $P$ and, since it has two distinct elements in common with $W$, it follows that $a$ lies in $W$.

Then by Theorem 149 $P$ and $a$ may be taken as determining $W$ and any element of $W$ lies either in $P$ or in a general plane parallel to $P$ and intersecting $a$.

If now we call this set of mutually parallel general planes the "primary set" we have already seen in proving Theorem 149 that $Q$ must either be identical with some general plane of the primary set or else must have a general line in common with each general plane of the primary set.

But now, since $P$ and $Q$ are supposed to be distinct, $Q$ cannot be identical with $P$, and since $Q$ is supposed to have an element in common with $P$, it follows that $Q$ is not parallel to $P$.

Thus $Q$ cannot be identical with any general plane of the primary set and therefore must have a general line in common with each of them, including $P$.

Thus $P$ and $Q$ must have a second element in common.

## REMARKS

It is further evident from the above considerations that *if two distinct general planes $P$ and $Q$ both lie in a general threefold $W$, then if $P$ and $Q$ have no element in common they must be parallel to one another.*

Now we have already seen that we can have a separation plane $S$ and an inertia plane $P$ having an element $O$ in common and which are completely normal to one another.

We have seen that in this case $P$ and $S$ cannot have a second element in common.

It follows that $P$ and $S$ cannot lie in one general threefold.

Now let $a_1$ and $a_2$ be any two distinct general lines lying in $P$ and passing through $O$.

Then $S$ and $a_1$ determine a general threefold, say $W_1$, while $S$ and $a_2$ determine a general threefold, say $W_2$.

Now $W_1$ and $W_2$ must be distinct, for if $W_2$ were identical with $W_1$, then $W_1$ would contain both $a_1$ and $a_2$ and would therefore contain $P$.

But $W_1$ contains $S$, and so this is impossible.

Thus $W_1$ and $W_2$ are distinct general threefolds each of which contains the separation plane $S$.

Since there are an infinite number of general lines lying in $P$ and passing through $O$, it follows that *there are an infinite number of general threefolds which all contain any separation plane $S$.*

Similarly *there are an infinite number of general threefolds which all contain any inertia plane P.*

Without Post. XIX or some equivalent we cannot from our remaining postulates show that there is more than one general threefold; for the proof of the existence of an inertia plane which is completely normal to a separation plane depends upon Post. XIX.

## THEOREM 151

*If a general plane P and a general line a both lie in a general threefold W and if a does not lie in P, then either a is parallel to a general line in P or else has one single element in common with P.*

Let $B$ be any element lying in $P$ but not in $a$.

Then $a$ and $B$ determine a general plane, say $Q$, which must lie in $W$, since it contains three elements in common with $W$ which do not all lie in one general line.

But since $P$ and $Q$ have the element $B$ in common and both lie in $W$, therefore by Theorem 150 they have a general line in common which we may denote by $b$.

Since then $b$ must pass through the element $B$ which does not lie in $a$, it follows that $a$ and $b$ are two distinct general lines lying in $Q$ and must therefore either be parallel to one another, or else have one element in common, which is also an element of $P$.

Thus $a$ is either parallel to a general line in $P$ or has an element in common with $P$.

Further, $a$ cannot have more than one element in common with $P$, since then it would require to lie in $P$.

## THEOREM 152

*If a, b and c be any three distinct general lines having an element O in common, but not all lying in one general plane, and if a general line d, also passing through O, be normal to a, b and c, then d is normal to every general line in the general threefold containing a, b and c.*

Let $P$ be the general plane containing $b$ and $c$.

Then $a$ intersects $P$ in $O$ and so $P$ and $a$ determine a general threefold, say $W$, containing $a$, $b$ and $c$.

Consider now any general line $e$ in $W$ which passes through $O$ but is distinct from $a$, $b$ and $c$.

Then $a$ and $e$ determine a general plane, say $Q$, which, by Theorem 147, must lie in $W$.

Further, $Q$ cannot be identical with $P$, since $Q$ contains $a$ but $P$ does not contain it.

Again $Q$ and $P$ have the element $O$ in common and therefore, by Theorem 150, they have a general line, say $f$, in common which passes through $O$.

Now, since $d$ is normal to the two intersecting general lines $b$ and $c$, it follows that $d$ is normal to every general line in $P$ and therefore is normal to $f$.

Again, since $d$ is normal to the two intersecting general lines $a$ and $f$, it follows that $d$ is normal to every general line in $Q$ and therefore is normal to $e$.

But $e$ is any general line in $W$ which passes through $O$ but is distinct from $a$, $b$ and $c$, and so $d$ is normal to every general line in $W$ which passes through $O$.

Next let $e$ be any general line in $W$ which does not pass through $O$ and let $e'$ be the general line through $O$ parallel to $e$.

Then, by Theorem 148, $e'$ must also lie in $W$ and so by the first case $d$ is normal to $e'$ and therefore also normal to $e$.

Thus $d$ is normal to every general line in $W$, as was to be proved.

*Definition.* A general line which is normal to every general line in a general threefold will be said to be *normal to the general threefold*.

Since, by Theorem 145, if three distinct general lines not all lying in one general plane have an element $O$ in common there is at least one general line passing through $O$ and normal to all three, it follows that through any element of a general threefold there is always at least one general line which is normal to the general threefold.

### The three types of general threefold

As in the case of general lines and general planes there are three types of each, so too there are three types of general threefold.

This may be shown in the following way:

If $S$ be any separation plane and $O$ be any element in it, there is an inertia plane, say $P$, which passes through $O$ and is completely normal to $S$.

Now if $a$ be any general line in $P$ which passes through $O$, then $a$ must be normal to $S$ and must intersect it.

But $a$ may be either:

|  |  |  |
|---|---|---|
|  | (1) | a separation line, |
| or | (2) | an optical line, |
| or | (3) | an inertia line, |

and if a general threefold be determined by $S$ and $a$, then these three cases give rise to the three different types.

Let $W$ be the general threefold determined by $a$ and $S$ and consider first the case where $a$ is a separation line.

If now $e$ be any general line in $W$ which passes through $O$ and is distinct from $a$, then $a$ and $e$ determine a general plane $Q$ which lies in $W$, and, since $Q$ has the element $O$ in common with $S$, it must have a general line, say $f$, in common with $S$.

Now $f$ must pass through $O$ and since it lies in $S$ therefore $a$ must be normal to $f$.

But $a$ and $f$ are both separation lines and we already know that if two intersecting separation lines are normal to one another they must lie in a separation plane.

Thus $Q$ must be a separation plane and therefore $e$ must be a separation line.

Thus every general line in $W$ which passes through $O$ must be a separation line.

If $e'$ be any other general line in $W$ which does not pass through $O$, then there is a general line through $O$ parallel to $e'$ which, by Theorem 148, must also lie in $W$ and therefore must be a separation line.

But a general line parallel to a separation line must itself be a separation line and so $e'$ is a separation line.

Thus every general line in $W$ is a separation line and so no element of $W$ is either *before* or *after* any other element of it.

It also follows from this that every general plane in $W$ must be a separation plane.

Consider next the case where $a$ is an optical line.

As before let $e$ be any general line in $W$ which passes through $O$ and is distinct from $a$.

Then $a$ and $e$ determine a general plane $Q$ which has a general line $f$ in common with $S$.

As before $a$ is normal to $f$, but in this case $a$ is an optical line while $f$ is a separation line and we know that in these circumstances $a$ and $f$ must lie in an optical plane.

Thus $Q$ must be an optical plane and, since there is only one optical line in an optical plane which passes through any element of it and all other general lines in it which pass through that element are separation lines, it follows that $e$ must be a separation line.

Again let $e'$ be any other general line in $W$ which does not pass through $O$.

Then there is a general line through $O$ parallel to $e'$ and this general line must either be the optical line $a$ or a separation line.

Thus $e'$ must be either an optical line or a separation line.

Again if $O'$ be any element of $W$ distinct from $O$, then $O'$ may or may not lie in $a$.

If $O'$ does not lie in $a$, then $OO'$ is a separation line and there is an optical line through $O'$ parallel to $a$ which, by Theorem 148, must lie in $W$.

Thus there is at least one optical line passing through any element of $W$ and lying in $W$.

Let $e'$ be any general line in $W$ which passes through $O'$ but not through $O$, and which is not parallel to $a$.

Then the general line through $O$ parallel to $e'$ cannot be identical with $a$ and therefore must be a separation line.

Thus $e'$ must be a separation line.

It follows that of all the general lines passing through any given element of $W$ and lying in $W$ one and only one is an optical line and all the others are separation lines.

Further, all the optical lines in $W$ are parallel to one another.

Since there are two optical lines in any inertia plane which pass through any element of it, it follows that no inertia plane can lie in $W$.

Thus every general plane in $W$ must be either a separation plane or an optical plane.

It follows that all the optical lines in $W$ being parallel to one another must be neutral parallels.

Consider finally the case where $a$ is an inertia line.

As before let $e$ be any general line in $W$ which passes through $O$ and is distinct from $a$.

Then $a$ and $e$ determine a general plane $Q$, which lies in $W$ and, since $a$ is an inertia line, $Q$ must be an inertia plane.

Thus $e$ may be either an inertia line, an optical line, or a separation line.

If $O'$ be any element in $W$ which is distinct from $O$ and if $d$ be any general line passing through $O$ and lying in $W$ but distinct from $OO'$, then through $O'$ there is a general line parallel to $d$, which must lie in $W$ and must be of the same type as $d$.

Thus through any element of $W$ there are general lines of all three types lying in $W$.

Again, if $f$ be any general line lying in $S$ and passing through $O$, then, since $a$ is an inertia line, $a$ and $f$ must lie in an inertia plane, say $R$.

Now, since there are an infinite number of general lines such as $f$ which lie in $S$ and pass through $O$, there must be an infinite number of inertia planes such as $R$ which are all distinct but have the inertia line $a$ in common.

In any one of these inertia planes such as $R$ there are two and only two optical lines which pass through $O$.

All these optical lines must be distinct since the inertia planes have only an inertia line in common, and so there are an infinite number of optical lines passing through $O$ and lying in $W$.

Further, any optical line which passes through $O$ and lies in $W$ must clearly lie in one of this set of inertia planes.

Again, if $O'$ be any element of $W$ distinct from $O$ and if $g$ be any optical line passing through $O$ and lying in $W$ but distinct from $OO'$, then there is an optical line through $O'$ parallel to $g$ and lying in $W$.

The general line $OO'$ either may or may not itself be an optical line.

Thus through any element of $W$ there are an infinite number of optical lines which lie in $W$.

Now we have already seen that $W$ contains the separation plane $S$ and also contains inertia planes, and we can easily show that it also contains optical planes.

Thus let $P$ be any inertia plane in $W$ and let $A$ be any element in $W$ but not in $P$.

Then through $A$ there is an inertia plane parallel to $P$ which we may call $P'$.

Let $B$ be the representative of $A$ in $P$ and let $c_1$ and $c_2$ be the two generators of $P$ which pass through $B$.

Then $A$ is neither *before* nor *after* any element of either $c_1$ or $c_2$ and so $A$ and $c_1$ lie in one optical plane, say $T_1$, while $A$ and $c_2$ lie in another optical plane, say $T_2$.

But $T_1$ and $T_2$ each contain three elements in common with $W$ which do not all lie in one general line and so, by Theorem 147, both $T_1$ and $T_2$ lie in $W$.

Thus $W$ contains all three types of general plane.

We thus see that there are at least three types of general threefold and we have investigated a few of their characteristic properties.

We have next to show that any general threefold must belong to one of these three types.

Since any four distinct elements which do not all lie in one general plane lie in one and only one general threefold, it will be sufficient if we examine the nature of any such general threefold.

### SETS OF FOUR ELEMENTS WHICH DETERMINE THE DIFFERENT TYPES OF GENERAL THREEFOLD

Let $A$, $B$, $C$, $D$ be any four distinct elements which do not all lie in one general plane.

Then no three of them can lie in one general line and $A$, $B$ and $C$ must determine a general plane which we shall call $P$.

Now $P$ may be either:

|  | (1) an inertia plane, |
| or | (2) an optical plane, |
| or | (3) a separation plane. |

Suppose first that $P$ is an inertia plane and that $D$ is any element outside it.

Let $W$ be the general threefold containing $A$, $B$, $C$ and $D$ and which must evidently contain $P$.

Then, by Theorem 142, there is one single general line passing through $D$ and normal to $P$ which has an element in common with $P$.

Let this element be denoted by $O$ and let $a$ be any inertia line in $P$ which passes through $O$, while $b$ is the separation line in $P$ and passing through $O$ which is normal to $a$.

Then, since $a$ is an inertia line, the general line $DO$ which is normal to it must be a separation line.

But $DO$ is also normal to $b$ and, since we know that two intersecting separation lines which are normal to one another must lie in a separation plane, it follows that $DO$ and $b$ lie in a separation plane which we shall call $S$.

Now $S$ contains $DO$ and $b$ and therefore contains three elements in common with $W$ which do not all lie in one general line.

It follows, by Theorem 147, that $S$ lies in $W$.

Thus, by Theorem 149, the general threefold determined by $S$ and $a$ is identical with the general threefold determined by $P$ and $DO$.

This latter is however identical with $W$ and so $S$ and $a$ determine $W$.

But $a$ is an inertia line which is normal to the two intersecting separation lines $DO$ and $b$ which lie in $S$ and therefore $a$ is normal to $S$.

Thus the general threefold $W$ is of the third type.

Further, it is evident that if any general threefold contains an inertia plane it must belong to the third type.

Next consider the case where $P$ is an optical plane and $D$ an element outside it.

Two sub-cases arise here: we may have

$D$ before or after some element of $P$,

or $D$ neither before nor after any element of $P$.

We shall suppose first that $D$ is either before or after some element of $P$ and we shall denote the generator of $P$ which passes through this element by $a$.

If, as before, $W$ denote the general threefold containing $A, B, C$ and $D$, then $W$ will contain $P$ and will therefore contain $a$.

But, since $a$ is an optical line and $D$ is an element which does not lie in $a$ but is either before or after some element of $a$, it follows that $a$ and $D$ lie in an inertia plane, say $Q$.

But $Q$ contains three elements in common with $W$ which do not all lie in one general line and so $Q$ must lie in $W$.

But $Q$ is an inertia plane and so it follows that in this case also $W$ is a general threefold of the third type.

We shall next take the case where $P$ is an optical plane and the element $D$ is neither before nor after any element of $P$.

Let $b$ be any separation line in $P$ and $a$ be any optical line in $P$ and let $b$ and $a$ intersect in the element $O$.

If, as before, $W$ denote the general threefold containing $A, B, C$ and $D$, then $W$ will contain $P$ and therefore will contain $a$ and $b$.

Now, since $D$ is neither before nor after any element of $P$, it is neither before nor after any element of $b$ and so $D$ and $b$ lie in a separation plane which we may call $S$.

Further, since $S$ has three elements in common with $W$ which do not all lie in one general line, it follows that $S$ lies in $W$.

Again, $D$ and $a$ must lie in an optical plane and, since $DO$ is a separation line while $a$ is an optical line, it follows that $a$ is normal to $DO$.

But $a$ must also be normal to $b$ for a similar reason and so, since $DO$ and $b$ are intersecting separation lines in $S$, it follows that $a$ is normal to $S$

But, by Theorem 149, the general threefold determined by $S$ and $a$ is identical with that determined by $P$ and $DO$ which again is identical with $W$.

Since however $S$ is a separation plane while $a$ is an optical line normal to it, it follows that $W$ is in this case a general threefold of the second type.

Consider next the case where $P$ is a separation plane and, as in the previous cases, let $W$ denote the general threefold containing $A$, $B$, $C$ and $D$ and therefore also containing $P$.

Three sub-cases occur here; thus we may have:

$D$ neither *before* nor *after* any element of $P$,

or $D$ either *before* or *after* one single element of $P$,

or $D$ either *before* or *after* at least two elements of $P$.

Now, by Theorem 142, there is one single general line passing through $D$ and normal to $P$ which has an element in common with $P$.

Let $O$ be this element.

Then $DO$ may be either a separation line, an optical line, or an inertia line.

Consider first the case where $D$ is neither *before* nor *after* any element of $P$.

Then $D$ is neither *before* nor *after* $O$ and so $DO$ is a separation line and the general threefold $W$ is of the first type.

Next consider the case where $D$ is either *before* or *after* one single element of $P$ and denote this element by $O'$.

Let $b$ and $c$ be two distinct separation lines in $P$ and passing through $O'$.

Then $DO'$ and $b$ lie in an optical plane and $DO'$ and $c$ lie in another optical plane.

Since $D$ is either *before* or *after* $O'$, it follows that $DO'$ is an optical line and therefore is normal to both $b$ and $c$.

Since $b$ and $c$ intersect one another, it follows that $DO'$ is normal to $P$ and therefore $O'$ must be identical with $O$.

Thus in this case the general threefold $W$ is of the second type.

Next let $D$ be either *before* or *after* at least two distinct elements of $P$, say $E$ and $F$.

Then $EF$ is a separation line and $D$ does not lie in it, and so the three elements $D$, $E$ and $F$ lie in an inertia plane, say $Q$.

But $D$, $E$ and $F$ are elements in $W$ and therefore $Q$ must lie in $W$.

Thus, since $Q$ is an inertia plane, it follows that the general threefold $W$ belongs in this case to the third type.

This exhausts all the possibilities which are open and so we see that any general threefold whatever must be of one of the three types which we have considered.

We shall accordingly give special names to these three types.

*Definition.* If a separation line *a* intersects a separation plane *S* and is normal to it, then the aggregate of all elements of *S* and of all separation planes parallel to *S* which intersect *a* will be called a *separation threefold*.

*Definition.* If an optical line *a* intersects a separation plane *S* and is normal to it, then the aggregate of all elements of *S* and of all separation planes parallel to *S* which intersect *a* will be called an *optical threefold*.

*Definition.* If an inertia line *a* intersects a separation plane *S* and is normal to it, then the aggregate of all elements of *S* and of all separation planes parallel to *S* which intersect *a* will be called an *inertia threefold*.\*

We are now in a position to introduce a new postulate which limits the number of dimensions of our set of elements.

POSTULATE XX. **If W be any optical threefold, then any element of the set must be either before or after some element of W.**

If *W* be any optical threefold and *A* be any element of *W*, then through *A* there is one single optical line which lies in *W* and *A* is *before* certain elements of this optical line and is *after* certain others.

Thus in this case *A* is *before* certain elements of *W* and *after* certain other elements of *W*.

If, on the other hand, *A* be any element outside *W*, then, by Post. XX, *A* must be either *before* some element of *W* or *after* some element of *W*.

If *A* be *before* the element *B* of *W*, then there is an optical line, say *b*, passing through *B* and lying in *W*.

If *b'* be the optical line through *A* parallel to *b*, then *b'* will be a before-parallel of *b*.

But any element of *W* which does not lie in *b* must lie in an optical line *c* neutrally parallel to *b* and lying in *W* and so, by Theorem 26, *b'* must be a before-parallel of *c*.

Thus *A* must be *before* certain elements of *c* and, since *A* is not an element of *W* and therefore not an element of *c*, it follows that *A* cannot be *after* any element of *c*.

---

\* In the first edition of this work the term *rotation* threefold was used instead of *inertia* threefold. The change was made in order that the nomenclature might be more systematic.

Thus $A$ is *before* elements of every optical line in $W$ and is not *after* any element of $W$.

Similarly if $A$ be any element outside $W$ and *after* some element of $W$, then $A$ will be *after* elements of every optical line in $W$ and will not be *before* any element of $W$.

*Definition.* An optical line which lies in an optical or inertia threefold will be spoken of as a *generator* of the optical threefold or inertia threefold, as the case may be.

### Theorem 153

*If $P$ be an optical plane and $O$ be any element in it, there is only one general plane passing through $O$ and completely normal to $P$.*

Let $a$ be the generator of $P$ which passes through $O$ and let $b$ be any separation line in $P$ and passing through $O$.

Then we already know that there is at least one optical plane, say $Q$, which passes through $O$ and is completely normal to $P$.

Further this optical plane $Q$ contains $a$.

Now let $c$ be any separation line passing through $O$ and lying in $Q$.

Then $c$ is normal to both $a$ and $b$.

Let $d$ be any other general line which passes through $O$ and is normal to $P$ and let $X$ be any element in $d$ distinct from $O$.

Now, $P$ and $c$ determine an optical threefold, since no element of $c$ with the exception of $O$ is either *before* or *after* any element of $P$.

Let this optical threefold be denoted by $W$.

Then, by Post. XX, the element $X$ is either *before* or *after* some element of $W$.

If $X$ were outside $W$, then, as we have seen, $X$ would be *before* or *after* elements of every generator of $W$ and therefore *before* or *after* elements of $a$.

Since the general line $d$ could not then be either identical with $a$ or be a separation line normal to $a$, it follows that $d$ could not be normal to $P$, contrary to hypothesis.

Thus $X$ must lie in $W$ and therefore $d$ must lie in $W$.

But now $c$ and $d$ determine a general plane $Q'$ which has three elements in common with $W$ which are not all in one general line and therefore $Q'$ must lie in $W$.

Further, since $P$ and $Q'$ have the element $O$ in common, therefore by Theorem 150, they have a general line in common, which we may call $a'$.

But now $b$ is normal to both $c$ and $d$ and, since these intersect in $O$, it follows that $b$ is normal to $Q'$ and therefore normal to $a'$.

But $b$ and $a'$ lie in the optical plane $P$ and, since $b$ is a separation line, $a'$ must be an optical line.

Thus, since $a'$ passes through $O$, it must be identical with $a$ and so $Q'$ must be identical with $Q$.

It follows that $d$ lies in $Q$ and accordingly every general line which passes through $O$ and is normal to $P$ must lie in $Q$.

Thus any general plane which passes through $O$ and is completely normal to $P$ must be identical with $Q$, or there is only one general plane passing through $O$ and completely normal to $P$.

## THEOREM 154

*If $P$ be a separation plane and $O$ be any element in it, there is only one general plane passing through $O$ and completely normal to $P$.*

We already know that there is at least one inertia plane, say $Q$, passing through $O$ and completely normal to $P$.

Suppose, if possible, that there is a general line, say $a$, passing through $O$ and normal to $P$ but not lying in $Q$.

Then $Q$ and $a$ will determine an inertia threefold, say $W$.

If $b$ and $c$ be any two distinct general lines in $P$ which both pass through $O$, then $b$ and $c$ will each be normal to three distinct general lines passing through $O$ and lying in $W$ but not all lying in one general plane.

Thus, by Theorem 152, $b$ and $c$ must each be normal to every general line in $W$.

But now we have seen that any inertia threefold contains optical planes and so there would always be at least one optical plane, say $R$, passing through $O$ and lying in $W$.

But then both $b$ and $c$ would be normal to every general line in $R$ and, since $b$ and $c$ are intersecting general lines in $P$, we should have every general line in $R$ normal to every general line in $P$.

Thus $P$ would be completely normal to $R$ and would pass through the element $O$ in it.

But $P$ is a separation plane and we already know by Theorem 153 that there could be only one general plane passing through $O$ and completely normal to $R$, and that one must itself be an optical plane and could not be a separation plane.

Thus the assumption that there is a general line $a$ passing through $O$ and normal to $P$ but not lying in $Q$, leads to a contradiction and therefore is not true.

It follows that every general line passing through $O$ and normal to $P$ must lie in $Q$.

Thus $Q$ is the only general plane which passes through $O$ and is completely normal to $P$.

Thus the theorem is proved.

## THEOREM 155

*If $P$ be an inertia plane and $O$ be any element in it, there is only one general plane passing through $O$ and completely normal to $P$.*

We already know that there is at least one separation plane, say $Q$, passing through $O$ and completely normal to $P$.

Let $b$ be any separation line in $Q$ which passes through $O$ and let $c$ be the one separation line lying in $Q$ and passing through $O$ which is normal to $b$.

Suppose now, if possible, that there is a general line $d$ passing through $O$ and normal to $P$ but not lying in $Q$.

Then, since any inertia line in $P$ would be normal to $d$, it would follow that $d$ must be a separation line and, since then any inertia line in $P$ which passed through $O$ would be conjugate to the two intersecting separation lines $b$ and $d$, it would follow, as a consequence of Theorem 99, that $b$ and $d$ must lie in a separation plane, say $Q'$.

Now $Q'$ would require to be distinct from $Q$, since $d$ is supposed not to lie in $Q$.

Since however we should then have two intersecting separation lines in $Q'$, namely $b$ and $d$, normal to $P$, it would follow that $Q'$ was completely normal to $P$.

Now suppose $c'$ to be the one separation line in $Q'$ and passing through $O$ which would be normal to $b$.

Then $c$ and $c'$ would be distinct separation lines, since $b$ is the only general line common to $Q$ and $Q'$.

Further, since any inertia line in $P$ which passes through $O$ would be conjugate to both $c$ and $c'$, it follows that $c$ and $c'$ would lie in a separation plane, say $S$.

But now $P$ and $b$ would determine an inertia threefold, say $W$, and since both $c$ and $c'$ would be normal to $P$ and to the separation line $b$ (which does not lie in $P$), it follows, by Theorem 152, that both $c$ and $c'$ would be normal to every general line in $W$.

But, as we have seen, there is at least one optical plane passing through $O$ and lying in $W$, and if $T$ be such an optical plane we should have both $c$ and $c'$ normal to $T$.

Thus the separation plane $S$ would be completely normal to $T$ and this we know by Theorem 153 is impossible, since only an optical plane can have an element in common with an optical plane and be completely normal to it.

It follows that no such general line as $d$ can exist and so every general line which passes through $O$ and is normal to $P$ must lie in $Q$.

Thus $Q$ is the only general plane which passes through $O$ and is completely normal to $P$ and so the theorem is proved.

## REMARKS

Combining these last three theorems we get the general result:

*If $P$ be any general plane and $O$ be any element in it, there is one and only one general plane $Q$ passing through $O$ and completely normal to $P$.*

Further:

If $P$ be an optical plane, $Q$ is an optical plane.

If $P$ be a separation plane, $Q$ is an inertia plane.

If $P$ be an inertia plane, $Q$ is a separation plane.

Again we know that if $O'$ be any element outside $P$ there is at least one general plane through $O'$ which is completely normal to $P$.

If we call this general plane $Q'$, then $Q'$ is either identical with $Q$ or parallel to $Q$ according as $O'$ does or does not lie in $Q$.

Now there cannot be any other general plane than $Q'$ which passes through $O'$ and is completely normal to $P$.

For if $Q''$ were such another general plane it would either pass through $O$ or else there would be a general plane parallel to $Q''$ and passing through $O$, which would also be completely normal to $P$.

Thus there would be two distinct general planes passing through $O$ and completely normal to $P$; which is impossible.

Thus we can say:

*If $P$ be any general plane and $O$ be any element of the set, there is one and only one general plane passing through $O$ and completely normal to $P$.*

## THEOREM 156

(1) *If $P$ be an inertia or separation plane and $O$ be any element outside it, then the general plane through $O$ and completely normal to $P$ has one single element in common with $P$.*

(2) *If $P$ be an optical plane and $O$ be any element outside it, then the optical plane through $O$ and completely normal to $P$ has an optical line in common with $P$ if $O$ be neither* before *nor* after *any element of $P$ and has no element in common with $P$ if $O$ be either* before *or* after *any element of $P$.*

Let $P$ be an inertia or separation plane and $O$ any element outside it.

Then, by Theorem 142, there is one single general line passing through $O$ and normal to $P$ which has an element in common with $P$.

Let $O'$ be this element.

Then, by Theorem 155 or 154, there is one single separation or inertia plane, say $Q$, which passes through $O'$ and is completely normal to $P$; and $Q$ has only one element in common with $P$.

Thus $Q$ must contain the general line $O'O$ and therefore it must be identical with the one single general plane which passes through $O$ and is completely normal to $P$.

Thus the general plane through $O$ and completely normal to $P$ has one single element in common with $P$, and so the first part of the theorem is proved.

Next let $P$ be an optical plane and $O$ any element outside it.

Then, by Theorem 143, if $O$ be neither *before* nor *after* any element of $P$ there is one single generator of $P$ such that every general line which passes through $O$ and intersects this generator is normal to $P$.

Thus if $a$ be this generator and $O'$ be any element in $a$, the general lines $a$ and $OO'$ determine an optical plane, say $Q$, which passes through $O$, is completely normal to $P$ and has the optical line $a$ in common with $P$.

Since there is only one optical plane through $O$ and completely normal to $P$, this must be identical with $Q$ and it has the optical line $a$ in common with $P$ if $O$ be neither *before* nor *after* any element of $P$.

Next consider the case where $O$ is either *before* or *after* some element of $P$.

Here, by Theorem 143, there is no general line passing through $O$ and having an element in common with $P$ which is normal to $P$.

Thus the optical plane through $O$ and completely normal to $P$ can, in this case, have no element in common with $P$.

Thus all parts of the theorem are proved.

## Theorem 157

*If a general line $a$ have an element $O$ in common with a general threefold $W$, then there is at least one general plane lying in $W$ and passing through $O$ to which $a$ is normal.*

Let $Q$ be any general plane in $W$ and passing through $O$.

Then, by Theorem 144 there is at least one general line, say $b$, passing through $O$ and lying in $Q$ which is normal to $a$.

Let $c$ be any other general line distinct from $b$, lying in $Q$ and passing through $O$, and let $A$ be any element lying in $W$ but not in $Q$.

Then $c$ and $A$ determine a general plane, say $R$, which must lie in $W$, since it contains three elements in common with $W$ which do not all lie in one general line.

Further, $R$ must be distinct from $Q$, since $R$ contains the element $A$ which does not lie in $Q$, and moreover $R$ does not contain $b$.

But again, by Theorem 144, there is at least one general line, say $d$, passing through $O$ and lying in $R$ which is normal to $a$.

Then $d$ must be distinct from $b$ which it intersects in the element $O$ and so $d$ and $b$ determine a general plane, say $P$, which must lie in $W$, since it contains three elements in common with $W$ which do not all lie in one general line.

But, since $a$ is normal to the two intersecting general lines $d$ and $b$, therefore $a$ is normal to $P$, and thus there is at least one general plane $P$ lying in $W$ and passing through $O$ to which $a$ is normal.

It is to be observed in connexion with the above theorem that if $a$ were normal to any other general line passing through $O$ and lying in $W$ but not in $P$, then, by Theorem 152, $a$ would be normal to every general line in $W$.

It is also to be observed that the above theorem holds both when the general line $a$ lies in $W$ and when it has only one element in common with $W$.

### Theorem 158

(1) *If $W$ be a general threefold and $P$ be a general plane lying in $W$, while $O$ is any element in $P$, then there is at least one general line passing through $O$ and lying in $W$ which is normal to $P$.*

(2) *There is only one such general line except in the case where $W$ is an optical threefold and $P$ an optical plane, in which case there are an infinite number.*

To prove the first part of the theorem consider first the case where $P$ is an optical plane.

In this case the generator of $P$ which passes through $O$ is normal to $P$ and lies in $W$.

Next let $P$ be an inertia or separation plane and let $A$ be any element lying in $W$ but not in $P$.

Then by Theorem 142 there is one single general line passing through $A$ and normal to $P$ which has an element in common with $P$.

Let $B$ be this element.

Then the general line $AB$ has two distinct elements in common with $W$ and therefore lies in $W$, but does not lie in $P$.

If $B$ should be identical with $O$, then $AB$ passes through $O$, lies in $W$ and is normal to $P$.

If $B$ be not identical with $O$, then there is a general line passing through $O$ and parallel to $AB$ which must also be normal to $P$.

But, by Theorem 148, this general line must also lie in $W$.

Thus in all cases there is at least one general line passing through $O$ and lying in $W$ which is normal to $P$.

Proceeding now to the second part of the theorem, let us consider first the case where $P$ is either an inertia or separation plane.

Suppose, if possible, that $a$ and $b$ are two distinct general lines both of which pass through $O$, lie in $W$ and are normal to $P$.

Then $a$ and $b$ would determine a general plane, say $Q$, which would have three elements in common with $W$ not all lying in one general line, and so $Q$ would lie in $W$.

Thus, by Theorem 150, since $Q$ and $P$ have the element $O$ in common, they would have a general line in common.

But, since $Q$ is supposed to contain the two intersecting general lines $a$ and $b$ each of which is normal to $P$, it would follow that $Q$ must be completely normal to $P$, and since $P$ is by hypothesis either an inertia or separation plane, it would follow that $Q$ must be either a separation or inertia plane.

But we already know that if an inertia plane and a separation plane be completely normal to one another, they cannot have more than one element in common.

Thus $P$ and $Q$ could not have a general line in common, and so the supposition that more than one general line can pass through $O$, lie in $W$, and be normal to $P$ leads in this case to a contradiction and therefore is not true.

Thus if $P$ be an inertia or separation plane there cannot be more than one such general line.

Suppose next that $P$ is an optical plane and let $a$ be the generator of $P$ which passes through $O$ and let $b$ be any other general line lying in $W$ but not in $P$ and which passes through $O$.

Let $A$ be any element in $b$ distinct from $O$.

Then if $A$ be either *before* or *after* any element of $P$ the general threefold $W$ must be an inertia threefold and $a$ and $A$ must lie in an inertia plane.

Thus, since $a$ is an optical line and, since $b$ intersects $a$ and lies in an inertia plane with it, it follows that $b$ cannot be normal to $a$ and therefore cannot be normal to $P$.

Further, since $a$ is the only general line in $P$ which passes through $O$ and is normal to $P$, it follows that in this case there is only one general line in $W$ which passes through $O$ and is normal to $P$.

Consider now the case where the element $A$ is neither *before* nor *after* any element of $P$.

In this case the general threefold $W$ must be an optical threefold and the general line $b$ must be a separation line.

Let $c$ be any general line in $P$ and passing through $O$ but distinct from $a$.

Then $c$ is a separation line and $b$ and $c$ determine a separation plane, say $S$, which must lie in $W$.

Now $a$ must, in this case, be normal to both $b$ and $c$ and therefore normal to $S$.

Let $d$ be the one single separation line in $S$ which passes through $O$ and is normal to $c$.

Then $d$ is normal to both $a$ and $c$ and therefore is normal to $P$.

If then $Q$ be the general plane containing $a$ and $d$, it contains two intersecting general lines each of which is normal to $P$ and therefore it follows that $Q$ is completely normal to $P$.

Thus every general line which passes through $O$ and lies in $Q$ must be normal to $P$.

But, since $a$ and $d$ are two intersecting general lines which both lie in $W$, it follows that $Q$ contains three distinct elements in common with $W$ which do not all lie in one general line and therefore $Q$ must lie in $W$.

Thus in this case there are an infinite number of general lines which pass through $O$, lie in $W$, and are normal to $P$.

This exhausts all the different cases and so the second part of the theorem is proved.

### Theorem 159

*If $W$ be a general threefold and $O$ be any element which does not lie in it, then:*

*(1) If $W$ be an inertia or separation threefold there is one single general line passing through $O$ and normal to $W$ which has an element in common with $W$.*

*(2) If $W$ be an optical threefold there is no general line passing through $O$ and normal to $W$ which has an element in common with $W$.*

If $W$ be an inertia threefold it contains inertia lines.

Let $f$ be any inertia line in $W$.

Then $f$ and $O$ lie in an inertia plane, say $R$, and if $a$ be any inertia line in $R$ and passing through $O$ but not parallel to $f$, then $a$ and $f$ will intersect in some element, say $A$, which is an element of $W$.

If on the other hand $W$ be a separation threefold, let $A$ be *any* element in $W$ and let $a$ be the general line $OA$.

Now whether $W$ be an inertia or separation threefold, it follows, by Theorem 157, that there is at least one general plane, say $P$, lying in $W$ and passing through $A$ to which $a$ is normal.

Now if $W$ be an inertia threefold, $a$ has been selected so as to be an inertia line and, since only separation lines can be normal to an inertia line, it follows that $P$ is a separation plane.

If on the other hand $W$ be a separation threefold it can contain no other type of general plane, and so in this case also $P$ must be a separation plane.

Now, by Theorem 158, whether $W$ be an inertia or a separation threefold, there is one general line, say $b$, passing through $A$ and lying in $W$ which is normal to $P$, and, since $P$ is a separation plane, $b$ must intersect it.

Now $a$ and $b$ must be distinct, since $b$ lies in $W$ while $a$ can only have the one element $A$ in common with $W$.

Thus $a$ and $b$ lie in a general plane, say $Q$, and, since $Q$ contains two intersecting general lines each of which is normal to $P$, it follows that $Q$ must be completely normal to $P$.

Further, since $P$ is a separation plane, it follows that $Q$ is an inertia plane.

Now, since $b$ is normal to $P$ and lies in $W$, the general threefold $W$ might be determined by $P$ and $b$ and we know that if $b$ be a separation line, $W$ must be a separation threefold, while if $b$ be an optical line, $W$ must be an optical threefold, and if $b$ be an inertia line, $W$ must be an inertia threefold.

It follows that if $W$ be an inertia threefold then $b$ must be an inertia line, while if $W$ be a separation threefold, $b$ must be a separation line.

But now in either of these cases there is a general line, say $c$, which passes through $O$, lies in $Q$ and is normal to $b$, and in both cases $c$ intersects $b$ in some element, say $O'$, which is an element of $W$.

Further, $c$ will be a separation line if $b$ be an inertia line: that is, if $W$ be an inertia threefold; while $c$ will be an inertia line if $b$ be a separation line: that is, if $W$ be a separation threefold.

Now, since $c$ lies in $Q$, and since $Q$ is completely normal to $P$, it follows that $c$ is normal to $P$.

If then $P'$ be a general plane passing through $O'$ and parallel to $P$ or identical with it, it follows, by Theorem 148, that $P'$ must also lie in $W$.

Thus $c$ will be normal to $P'$ and to the general line $b$ which intersects $P'$ in $O'$.

It is thus evident that $c$ is normal to three distinct general lines in $W$ which have the element $O'$ in common and which do not all lie in one general plane and therefore, by Theorem 152, $c$ is normal to $W$.

Also $c$ passes through $O$ and has the element $O'$ in common with $W$.

Now there can be no other general line passing through $O$ and normal to $W$; for suppose, if possible, that $c'$ is such another general line.

Then $c$ and $c'$ would determine a general plane, say $T$, which would contain two intersecting general lines each of which would be normal to every general line in $W$ and therefore normal to every general plane in $W$.

Thus $T$ would be completely normal to every general plane in $W$.

But through any element of $W$ there passes more than one general plane which lies in $W$ and so we should have more than one general plane passing through any element of $W$ and completely normal to $T$, which, as we have seen, is impossible.

Thus the supposition that more than one general line can pass through $O$ and be normal to $W$ leads to a contradiction and therefore is not true.

Thus there is one and only one general line which passes through $O$ and is normal to $W$ when $W$ is an inertia or separation threefold, and this general line has an element in common with $W$.

Suppose next that $W$ is an optical threefold.

Then, by Post. XX, $O$ must be either *before* or *after* some element of $W$ and, as we have seen, if $O$ be *before* any element of $W$ it must be *before* elements of every generator of $W$, while if $O$ be *after* any element of $W$ it must be *after* elements of every generator of $W$.

If then $a$ be any general line which passes through $O$ and has an element $A$ in common with $W$, then $A$ must lie in some generator of $W$, say $f$, and $f$ and $a$ will lie in an inertia plane.

But, since $f$ is an optical line and $a$ is a general line intersecting $f$ and lying in an inertia plane with it, it follows that $a$ cannot be normal to $f$ and therefore cannot be normal to $W$.

Thus in this case there is no general line passing through $O$ and normal to $W$ which has an element in common with $W$.

Thus both parts of the theorem are proved.

REMARKS

If $W$ be an inertia or separation threefold and $O$ be any element *in* $W$, it is easy to see that there is one and only one general line passing through $O$ and normal to $W$.

For if $A$ be any element outside $W$ and $a$ be the one general line passing through $A$ and normal to $W$, then $a$ will have an element $B$ in common with $W$.

If $B$ should coincide with $O$, then $a$ is a general line passing through $O$ and normal to $W$.

If $B$ does not coincide with $O$, then a general line $a'$ passing through $O$ and parallel to $a$ must be normal to every general line in $W$ and must therefore be normal to $W$.

Thus we have shown that there is at least one general line passing through $O$ and normal to $W$, and the same considerations employed in the last theorem show that there is only one such general line.

Further, the general line through $O$ normal to $W$ cannot have more than the one element $O$ in common with $W$; for if it had a second element in common with $W$ it would lie entirely in $W$, and, by Theorem 148, it would follow that $a$ must lie in $W$, contrary to the hypothesis that the element $A$ of $a$ lies outside $W$.

In this respect an optical threefold is quite different.

Through any element $O$ in an optical threefold $W$ there passes one single generator of $W$, say $a$.

Now $a$ is normal to any separation line in $W$ and is also normal to itself.

Thus $a$ is normal to $W$ and passes through $O$, but lies entirely in $W$.

If $O'$ be any element outside $W$ and $a'$ be an optical line parallel to $a$, then $a'$ is also normal to $W$ but can have no element in common with $W$.

We may also show, by similar considerations to those employed in the case of an inertia or separation threefold, that there cannot be more than one general line passing through any element and normal to a given optical threefold.

Thus for all three types of general threefold we have the result:

*If $W$ be any general threefold and $O$ be any element of the set, there is one and only one general line passing through $O$ and normal to $W$.*

## THEOREM 160

*If a be a general line and O be any element in it, there is one and only one general threefold passing through O and normal to a.*

Let $P$ be any inertia plane containing $a$ and let $Q$ be the separation plane passing through $O$ and completely normal to $P$.

Then $P$ and $Q$ have only the one element $O$ in common.

Now through $O$ and lying in $P$ there is one single general line, say $b$, which is normal to $a$.

But $b$ and $Q$ can have only one element in common and therefore they determine a general threefold, say $W$.

Since, however, $a$ is normal to every general line in $Q$ and is also normal to the general line $b$ which passes through $O$ and does not lie in $Q$, it follows, by Theorem 152, that $a$ is normal to $W$.

Thus there is at least one general threefold passing through $O$ and normal to $a$.

We shall next show that every general line which passes through $O$ and is normal to $a$ must lie in $W$.

Since every such general line which lies in $Q$ must lie in $W$, it will be sufficient to consider any general line $c$ passing through $O$ normal to $a$ and not lying in $Q$.

Then $c$ and $Q$ determine a general threefold, say $W'$, and by Theorem 158 there is at least one general line, say $d$, passing through $O$ and lying in $W'$ which is normal to $Q$.

Further, since $Q$ is a separation plane, $d$ must lie in the inertia plane through $O$ which is completely normal to $Q$, and since there is only one such inertia plane, it follows that $d$ must lie in $P$.

But, since $a$ is normal to $c$ and $Q$, it follows that $a$ is normal to $W'$ and therefore is normal to $d$.

But there is only one general line passing through $O$ and lying in $P$ which is normal to $a$, and by hypothesis $b$ is this general line.

It follows that $d$ must be identical with $b$ and so, by Theorem 149, since $d$ and $Q$ must determine the same general threefold as do $c$ and $Q$, it follows that $W'$ must be identical with $W$.

Thus $c$ must lie in $W$.

But if there were any other general threefold distinct from $W$ which passed through $O$ and was normal to $a$, such general threefold would require to contain a general line which passed through $O$ and was normal to $a$ but which did not lie in $W$, and this we have shown to be impossible.

Thus there is one and only one general threefold which passes through $O$ and is normal to $a$.

## REMARKS

In the above theorem it is to be observed that: if $a$ be an inertia line, $b$ must be a separation line; if $a$ be a separation line, $b$ must be an inertia line; while if $a$ be an optical line, $b$ must be the same optical line.

Thus it follows that: if $a$ be a general line and $O$ be any element in it, while $W$ is a general threefold passing through $O$ and normal to $a$, then:

(1) If $a$ be an inertia line, $W$ is a separation threefold.

(2) If $a$ be a separation line, $W$ is an inertia threefold.

(3) If $a$ be an optical line, $W$ is an optical threefold containing $a$.

On the other hand we have already seen that if $W$ be a general threefold and $O$ be any element in it, there is one and only one general line $a$ passing through $O$ and normal to $W$.

Thus it follows that:

(1) If $W$ be a separation threefold, $a$ is an inertia line.

(2) If $W$ be an inertia threefold, $a$ is a separation line.

(3) If $W$ be an optical threefold, $a$ is an optical line lying in $W$.

Again if $a$ be a general line and $O$ be any element which does not lie in $a$, then, through $O$ there is one single general line, say $a'$, which is parallel to $a$ and is accordingly a general line of the same type.

Thus through $O$ there is a general threefold which is normal to $a'$ and therefore also normal to $a$.

Further, there cannot be a second general threefold passing through $O$ and normal to $a$, for such general threefold would also be normal to $a'$ and so we should have two general threefolds passing through $O$ and normal to $a'$ contrary to Theorem 160.

Thus we can extend Theorem 160 and say:

*If $a$ be a general line and $O$ be any element of the set, there is one and only one general threefold passing through $O$ and normal to $a$.*

## THEOREM 161

*If $W$ be an optical threefold and $A$ be any element outside it, then every optical line through $A$, except the one parallel to the generators of $W$, has one single element in common with $W$.*

Let $a$ be the optical line through $A$ parallel to the generators of $W$ and let $b$ be any such generator.

Then by Post. XX $A$ must be either *before* or *after* some element of $W$ and we have already seen that if $A$ be *before* an element of $W$ it must be *before* elements of every generator of $W$; while if $A$ be *after* an element of $W$ it must be *after* elements of every generator of $W$.

Thus $a$ must be either a before- or after-parallel of $b$.

It will be sufficient to consider the case where $a$ is a before-parallel of $b$ since the proof in the other case is quite analogous.

Then $a$ and $b$ lie in an inertia plane and so there is one single optical line passing through $A$ and intersecting $b$ in some element, say $B$.

If we call this optical line $c$, then $c$ has the element $B$ in common with $W$.

If then $d$ be any optical line passing through $A$ but distinct from $c$ and $a$, it follows, by Post. XII, that there is one single element in $d$, say $D$, which is neither *before* nor *after* any element of $b$.

Now if $D$ were outside $W$ it would be either *before* or *after* elements of every generator of $W$, as we have already seen.

Thus, since $D$ is neither *before* nor *after* any element of the generator $b$, it follows that $D$ must lie in $W$.

It follows that every optical line through $A$ with the exception of $a$ has at least one element in common with $W$.

But if any optical line has more than one element in common with $W$ it must lie entirely in $W$, which is not possible for any optical line which passes through the element $A$.

It follows that every optical line through $A$ with the exception of $a$ has one single element in common with $W$, as was to be proved.

## Theorem 162

*If $W$ be a general threefold and $A$ be any element outside it, then any general line through $A$ is either parallel to a general line in $W$ or else has one single element in common with $W$.*

It will be observed that the last theorem is a special case of this one.

Let $a$ be any general line which passes through $A$.

Now $a$ cannot have more than one element in common with $W$, for then it would require to lie entirely in $W$ and therefore could not pass through $A$.

Let $B$ be a second element in $a$ distinct from $A$.

In case $W$ be an inertia or separation threefold, let the general line through $A$ normal to $W$ meet $W$ in the element $A'$, as we have seen in Theorem 159 that it must.

Now in case the general line $a$ should coincide with $AA'$ it would have an element in common with $W$, and so we shall suppose it is distinct from it.

Again let the general line through $B$ normal to $W$ meet $W$ in the element $B'$.

Then since $B$ does not lie in $AA'$ we must have $BB'$ parallel to $AA'$.

In case $W$ be an optical threefold, then by Theorem 161 any optical line through $A$ except the one parallel to the generators of $W$ must have an element in common with $W$.

Let any optical line through $A$ which is not parallel to the generators of $W$ meet $W$ in the element $A'$.

In case the general line $a$ should coincide with $AA'$ it would have an element in common with $W$, and so we shall suppose it is distinct from it.

Let the optical line through $B$ parallel to $AA'$ meet $W$ in the element $B'$.

Now both in the cases where $W$ is an inertia or separation threefold and where $W$ is an optical threefold, since $BB'$ is parallel to $AA'$, it follows that $BB'$ and $AA'$ lie in a general plane which we may call $Q$.

But $A'B'$ and $a$ must also lie in $Q$, and therefore $a$ is either parallel to $A'B'$ or intersects $A'B'$ in some element, say $C$.

But $A'B'$ has two distinct elements $A'$ and $B'$ in common with $W$, and therefore $A'B'$ must lie in $W$, and if the element $C$ exists it must lie in $W$.

Thus the general line $a$ is either parallel to a general line in $W$ or else $a$ has one single element in common with $W$.

*Definition.* If a general line and a general threefold have *one single element* in common, they will be said to *intersect* in that element.

## REMARKS

Since a separation threefold contains neither an inertia nor an optical line it is evident that it can contain no general line which is parallel to either of these.

Thus it follows from the last theorem that: *every inertia and every optical line intersects every separation threefold.*

Again, an optical threefold does not contain any inertia line, and all the optical lines which it contains are parallel to one another.

Thus: *every inertia line and every optical line which is not parallel to a generator of an optical threefold intersects the optical threefold.*

Analogous results to these may be deduced from Theorem 151, with regard to the intersection of certain types of general lines with certain types of general planes.

Thus, since a separation plane contains neither an inertia nor an optical line, it follows from Theorem 151 that: *if W be an inertia three-fold, every inertia and every optical line in W intersects every separation plane in W.*

Similarly: *if W be an inertia threefold, every inertia line in W and every optical line in W which is not parallel to a generator of an optical plane in W intersects the optical plane.*

Again: *if W be an optical threefold, every optical line in W intersects every separation plane in W.*

## THEOREM 163

*If W be a general threefold and P be a general plane which does not lie in W, then if P has one element in common with W, it has a general line in common with W.*

Let $P$ and $W$ have the element $A$ in common and let $B$ be any element in $P$ which does not lie in $W$.

Let $b$ be any general line in $P$ which passes through $B$ but is distinct from $BA$.

Then by Theorem 162 $b$ must either intersect $W$ in some element, say $C$, or else $b$ must be parallel to some general line, say $b'$, which lies in $W$.

In the first case $P$ and $W$ have the two distinct elements $A$ and $C$ in common and therefore have the general line $AC$ in common.

In the second case a general line $b''$ passing through $A$ and parallel to $b'$ or identical with it must lie in $W$.

But $b''$ must be parallel to $b$ and since it passes through the element $A$ of $P$ it must lie in $P$.

Thus in this case $P$ and $W$ have the general line $b''$ in common and so the theorem holds in general.

## THEOREM 164

*If $W_1$ and $W_2$ be two distinct general threefolds having an element $A$ in common, then they have a general plane in common.*

Let $B$ be any element which lies in $W_1$ but not in $W_2$.

Then the general line $AB$ lies in $W_1$.

Let $Q$ and $R$ be any two distinct general planes which contain the general line $AB$ and which lie in $W_1$.

Then $Q$ does not lie in $W_2$ but has the element $A$ in common with $W_2$ and therefore, by Theorem 163, $Q$ has a general line, say $a$, in common with $W_2$.

Similarly $R$ has a general line, say $b$, in common with $W_2$.

Now both $a$ and $b$ must be distinct from the general line $AB$ since the latter does not lie in $W_2$ and, since $Q$ and $R$ have only the general line $AB$ in common, it follows that $b$ is distinct from $a$.

Thus $a$ and $b$ are two general lines intersecting in $A$ and each of them lying both in $W_1$ and $W_2$ and so they determine a general plane, say $P$.

But $P$ contains three elements in common both with $W_1$ and $W_2$ and which do not all lie in one general line and so $P$ lies both in $W_1$ and $W_2$.

Thus $W_1$ and $W_2$ have a general plane in common.

## THEOREM 165

*If $P_1$ and $P_2$ be two general planes having no element in common, then through any element of either of them there is at least one general line lying in that general plane which is parallel to a general line in the other general plane.*

Let $O_1$ be any element in $P_1$ and let $O_2$ be any element in $P_2$ and let the general line $O_1O_2$ be denoted by $a$.

Then $P_1$ and $a$ determine a general threefold, say $W_1$, while $P_2$ and $a$ determine a general threefold, say $W_2$.

If $W_2$ should be identical with $W_1$, then $P_1$ and $P_2$ lie in one general threefold and, since they have no element in common, it follows, by Theorem 151, that any general line in $P_1$ is parallel to a general line in $P_2$, and so $P_1$ and $P_2$ are parallel to one another.

If $W_2$ be not identical with $W_1$, then, since $W_1$ and $W_2$ have all the elements of $a$ in common, it follows, by Theorem 164, that they have a general plane, say $Q$, in common which must contain $a$.

But now $Q$ must be distinct from both $P_1$ and $P_2$, for otherwise $P_1$ or $P_2$ would contain $a$ and so $P_1$ and $P_2$ would have an element in common, contrary to hypothesis.

But now $P_1$ and $Q$ both lie in $W_1$ and they have the element $O_1$ in common, and therefore, by Theorem 150, they have a general line, say $b_1$, in common, which passes through $O_1$.

Similarly $P_2$ and $Q$ have a general line, say $b_2$, in common, which passes through $O_2$.

But, since $b_1$ and $b_2$ lie in $P_1$ and $P_2$ respectively, they can have no element in common and, since they both lie in the general plane $Q$, they must be parallel to one another.

Thus the theorem is proved.

## REMARKS

It is easy to see that if two general planes $P_1$ and $P_2$ have one single element $O$ in common, then no general line in $P_1$ can be parallel to any general line in $P_2$.

For let $a_1$ and $a_2$ be two general lines in $P_1$ and $P_2$ respectively, then $a_1$ cannot be parallel to $a_2$ if both pass through $O$.

Further, they cannot be parallel if one passes through $O$ and the other does not, for then they could not lie in one general plane.

Finally they cannot be parallel if neither of them passes through $O$, for then a general line $a_1'$ passing through $O$ and parallel to $a_1$ would lie in $P_1$ and so could not be parallel to $a_2$ as it would require to be if $a_2$ were parallel to $a_1$.

## THEOREM 166

*If $W$ be a general threefold and $O$ be any element outside it, and, if further, $a$ and $b$ be two distinct general lines intersecting in $O$ and each of them parallel to a general line in $W$, then:*

(1) *The general plane containing $a$ and $b$ has no element in common with $W$.*

(2) *The general plane containing $a$ and $b$ is parallel to a general plane in $W$.*

Neither $a$ nor $b$ can have any element in common with $W$, since, by Theorem 148, if either of them had an element in common with $W$, it would require to lie entirely in $W$ and so could not contain the element $O$.

But, if $P$ be the general plane containing $a$ and $b$, any element in $P$ must lie either in $a$ or in a general line parallel to $a$ and intersecting $b$.

But every general line of this character must be parallel to the general line in $W$ to which $a$ is parallel, and therefore can have no element in common with $W$.

Thus $P$ can have no element in common with $W$.

In order to prove the second part of the theorem, let $a'$ and $b'$ be general lines in $W$ to which $a$ and $b$ are respectively parallel.

Then $a'$ and $b'$ either intersect, in which case they lie in a general plane which lies in $W$ and is parallel to $P$, or else a general line, say $b''$, parallel to $b'$, may be taken through any element of $a'$ and then $b''$ must lie in $W$, by Theorem 148.

Thus in this case $a'$ and $b''$ will lie in a general plane which will lie in $W$ and be parallel to $P$.

Thus in all cases $P$ will be parallel to a general plane in $W$.

*Definition.* If $W$ be a general threefold and if through any element $A$ outside $W$ a general line $a$ be taken parallel to any general line in $W$, then the general line $a$ will be said to be *parallel* to the general three-fold $W$.

*Definition.* If $W$ be a general threefold and if through any element $A$ outside $W$ a general plane $P$ be taken parallel to any general plane in $W$, then the general plane $P$ will be said to be *parallel* to the general threefold $W$.

### Theorem 167

*If $W$ be a general threefold and $O$ be any element outside it, and if through $O$ there pass three general lines $a$, $b$, and $c$, which do not all lie in one general plane and which are respectively parallel to three general lines in $W$, then $a$, $b$ and $c$ determine a general threefold $W'$, such that every general line in $W'$ is parallel to a general line in $W$.*

Let $P$ be the general plane containing $b$ and $c$.

Then, since $a$, $b$ and $c$ do not lie in one general plane, it follows that $a$ can only have the one element $O$ in common with $P$.

Now the general line $a$ can have no element in common with $W$, for then, since it is parallel to a general line in $W$, it would, by Theorem 148, require to lie in $W$ and so could not contain the element $O$.

Again, by Theorem 166, the general plane $P$ can contain no element in common with $W$, nor can any general plane which is parallel to $P$ and which intersects $a$.

But now any element in $W'$ must either lie in $P$ or in a general plane parallel to $P$ and intersecting $a$.

Thus no element in $W'$ can lie in $W$, and so no general line in $W'$ can have an element in common with $W$.

Thus, by Theorem 162, any such general line must be parallel to a general line in $W$.

Similarly any general line in $W$ must be parallel to a general line in $W'$.

*Definition.* If $W$ be a general threefold and if through any element $A$ outside $W$ three general lines be taken not all lying in one general plane but respectively parallel to three general lines in $W$, then the three general lines through $A$ determine a general threefold which will be said to be *parallel* to $W$.

## REMARKS

Since a general line can only be parallel to a general line of the same kind, and since if one general threefold be parallel to another, any general line in either of them is parallel to a general line in the other, it follows that a general threefold can only be parallel to a general threefold of the same kind.

Again, if $W$ be a general threefold and $A$ be any element outside it, while $W'$ is a general threefold through $A$ parallel to $W$, then since $W'$ contains the general line through $A$ parallel to any general line in $W$, the general threefold $W'$ must be uniquely determined when we know $W$ and $A$.

Also, since two distinct general lines which are parallel to a third general line are parallel to one another, it follows that: *two distinct general threefolds which are parallel to a third general threefold are parallel to one another*.

Again, from Theorem 162, it is evident that: *if $W$ be a general threefold and $A$ be any element outside it, then any general line through $A$ must either lie in the general threefold passing through $A$ and parallel to $W$, or else must intersect $W$*.

If $W$ and $W'$ be two distinct general threefolds and if $A$ be any element in $W'$ but not in $W$, then, if $W'$ be not parallel to $W$, there must be at least one general line passing through $A$ and lying in $W'$ which is not parallel to any general line in $W$ and which therefore, by Theorem 162, must intersect $W$.

Thus $W'$ will have an element in common with $W$ and so, by Theorem 164, $W$ and $W'$ must have a general plane in common. Thus *any two distinct general threefolds must either be parallel or else must have a general plane in common*.

It is also to be noted that if a general threefold $W$ be normal to a general line $a$, then any general threefold $W'$ parallel to $W$ must also be normal to $a$.

## OTHER CASES OF NORMALITY

We have already considered the normality of a general line to a general line, a general plane, or a general threefold.

We have also considered the complete normality of a general plane to a general plane.

These are the only cases in our geometry in which the normality of $n$-folds is complete.

Thus it is not possible to have every general line in a general plane $P$ normal to every general line in a general threefold $W$, for then we should have more than one general plane passing through any element of $W$ and completely normal to $P$, which, as we have seen, is impossible.

For a similar reason we cannot have every general line in a general threefold $W_1$ normal to every general line in a general threefold $W_2$.

The most possible in these directions is to have a general plane $P$ through any element of which there is one single general line lying in $P$ which is normal to a general threefold $W$; or to have a general threefold $W_1$ through any element of which there is one single general line lying in $W_1$ which is normal to a general threefold $W_2$.

Again, we may have a general plane $P_1$ through any element of which there is one single general line lying in $P_1$ which is normal to a general plane $P_2$.

In these cases we have what may be described as partial normality.

In ordinary three dimensional geometry the normality of two planes is of this partial character.

Since it is desirable, so far as is possible, to have our nomenclature in conformity with that employed in ordinary geometry, we shall find it convenient to describe the general planes and general threefolds in the above cases as *normal* to one another.

Thus we may have general planes *normal* to one another or *completely normal* to one another: the expression 'normal' by itself being taken to mean partially normal.

In the case of a general plane or a general threefold which is partially normal to a general threefold the word *normal* may be used by itself without any ambiguity.

Thus we have the following definitions:

*Definition.* A general plane $P_1$ will be said to be *normal* to a general plane $P_2$ if through any element of $P_1$ there is one single general line lying in $P_1$ which is normal to $P_2$.

*Definition.* A general plane $P$ will be said to be *normal* to a general threefold $W$ if through any element of $P$ there is one single general line lying in $P$ which is normal to $W$.

*Definition.* A general threefold $W_1$ will be said to be *normal* to a general threefold $W_2$ if through any element of $W_1$ there is one single general line lying in $W_1$ which is normal to $W_2$.

It is evident in the above three definitions we might substitute the word *every* for the word *any*.

It is easy to see that if a general plane $P_1$ be normal to a general plane $P_2$, then $P_2$ will be normal to $P_1$.

It will be sufficient to consider the case where $P_1$ and $P_2$ have an element $A$ in common.

Let $a$ be the one single general line lying in $P_1$ and passing through $A$ which is normal to $P_2$ and let $b$ be any other general line in $P_1$ which passes through $A$.

Then by Theorem 144 there is at least one general line, say $c$, passing through $A$ and lying in $P_2$ which is normal to $b$.

But $c$ must also be normal to $a$ and therefore $c$ must be normal to $P_1$.

Further, there cannot be more than one general line passing through $A$ and lying in $P_2$ which is normal to $b$, unless $P_1$ be completely normal and not merely normal to $P_2$.

Again, a separation line $a$ may be normal to all three types of general plane and also may lie in all three types of general plane.

If then $a$ be normal to any general plane $P_1$ and if $P_2$ be any general plane containing $a$ but not completely normal to $P_1$, then $P_2$ will be normal to $P_1$.

Thus any type of general plane may be normal to any type of general plane.

In particular, since an optical plane contains a series of optical lines which are normal to it, it follows that an optical plane is normal to itself.

It is evident from the definitions that, if a general plane $P$ be normal to a general threefold $W$, then $P$ will be either simply normal or completely normal to any general plane in $W$.

### Theorem 168

*If a general plane $P$ be normal to a general threefold $W$, then through any element of $W$ there is one single general plane lying in $W$ and completely normal to $P$.*

By definition, since $P$ is normal to $W$, it follows that there is a general line passing through any element of $P$ and lying in $P$ which is normal to $W$.

Let $O$ be any element of $W$ and let $a$ be the one single general line passing through $O$ which is normal to $W$.

Now all general lines which are normal to $W$ are co-directional, so that $a$ is co-directional with a set of general lines in $P$.

Let $b$ be any general line other than $a$ which passes through $O$ and is co-directional with some general line in $P$.

Then $a$ and $b$ lie in a general plane, say $P'$, which is either parallel to $P$ or identical with it.

But now, by Theorem 157, there is a general plane, say $Q$, passing through $O$ and lying in $W$ to which $b$ is normal.

Since, however, $a$ is normal to $W$, it follows that $a$ is also normal to $Q$.

Thus we have the two intersecting general lines $a$ and $b$ both lying in $P'$ and both normal to every general line in $Q$.

It follows that $Q$ is completely normal to $P'$ and, since $P'$ is either parallel to $P$ or identical with it, it follows that $Q$ is completely normal to $P$.

Thus, since $Q$ is taken through any arbitrary element of $W$ and lies in $W$, the theorem is proved.

### THEOREM 169

*If a general threefold $W_1$ contain a general line which is normal to a general threefold $W_2$, then $W_2$ contains a general line which is normal to $W_1$.*

Let $a$ be a general line lying in $W_1$ and which is normal to $W_2$.

Let $P_1$ and $P_2$ be any two distinct general planes both of which contain $a$ and lie in $W_1$.

Then $P_1$ and $P_2$ are both normal to $W_2$ and accordingly, by Theorem 168, if $O$ be any element in $W_2$, there is a general plane, say $Q_1$, passing through $O$ and lying in $W_2$ which is completely normal to $P_1$ and similarly there is a general plane, say $Q_2$, passing through $O$ and lying in $W_2$ which is completely normal to $P_2$.

Now $Q_1$ and $Q_2$ cannot be identical, for then we should have the two distinct general planes $P_1$ and $P_2$ both containing $a$ and both completely normal to the same general plane, which we know to be impossible.

Thus, since $Q_1$ and $Q_2$ both lie in $W_2$ and have the element $O$ in common, it follows, by Theorem 150, that $Q_1$ and $Q_2$ have a general line in common which we shall call $b$.

Then $b$ must be normal to both $P_1$ and $P_2$ and, since these are distinct intersecting general planes in $W_1$, it follows that $b$ is normal to $W_1$ and lies in $W_2$.

The above theorem might also be stated in the form:

*If a general threefold $W_1$ be normal to a general threefold $W_2$, then $W_2$ is normal to $W_1$.*

## SOME ANALOGIES

Before proceeding with the next part of our subject we shall point out a few analogies which exist between an inertia plane, an inertia threefold and the whole set of elements.

We have seen that: if $P$ be an inertia plane and $A$ be any element in it there are two and only two optical lines passing through $A$ and lying in $P$.

We have an analogue to this in the case of an inertia threefold.

We shall show that if $W$ be an inertia threefold and $a$ be any separation line in it there are two and only two optical planes containing $a$ and lying in $W$.

In order to prove this: let $O$ be any element in $a$.

Then, by Theorem 157, there is at least one general plane lying in $W$ and passing through $O$ to which $a$ is normal.

Further, there cannot be more than one such general plane, for otherwise $a$ would require to be normal to the inertia threefold $W$ and would therefore intersect $W$ contrary to the hypothesis that $a$ lies in $W$.

Let $P$ be this one general plane.

Then $P$ cannot be a separation plane, for, since $a$ is a separation line, this would require $W$ to be a separation threefold, contrary to hypothesis.

Again, $P$ cannot be an optical plane for this would require $W$ to be an optical threefold, contrary to hypothesis.

Thus $P$ must be an inertia plane and so there are two and only two optical lines, say $c_1$ and $c_2$, which pass through $O$ and lie in $P$.

Thus $a$ must be normal to both $c_1$ and $c_2$ and it cannot be normal to any other optical line passing through $O$ and lying in $W$; for such an optical line could not lie in $P$, and if $a$ were normal to such an optical line in addition to $c_1$ and $c_2$, it would require to be normal to $W$, which we know to be impossible.

But now $c_1$ and $a$ lie in one optical plane, say $R_1$, while $c_2$ and $a$ lie in another optical plane, say $R_2$.

Now $R_1$ and $R_2$ are the only optical planes in $W$ which contain $a$; for the existence of a third would require the existence of a third optical line passing through $O$, lying in $W$ and normal to $a$, which, as we have seen, is impossible.

This proves the required result.

Again we have a corresponding result for the whole set of elements.

We shall show that if $S$ be any separation plane there are two and only two optical threefolds containing $S$.

For let $O$ be any element in $S$ and let $P$ be the one single inertia plane which passes through $O$ and is completely normal to $S$.

Further let $c_1$ and $c_2$ be the two generators of $P$ which pass through $O$.

Then $c_1$ and $c_2$ are each normal to $S$, and accordingly $c_1$ and $S$ determine one optical threefold, say $W_1$, while $c_2$ and $S$ determine another optical threefold, say $W_2$.

Now $W_1$ and $W_2$ are the only optical threefolds which contain $S$, for the existence of a third would require the existence of a third optical line passing through $O$ and normal to $S$.

But if there were three optical lines passing through $O$ and normal to $S$, there would be more than one inertia plane passing through $O$ and completely normal to $S$, which we have seen is impossible.

Thus there are two and only two optical threefolds containing $S$, and so we see that we have here a certain analogy between an inertia plane, an inertia threefold, and the whole set of elements.

It was pointed out in another part of this work, that if $W$ be an inertia threefold and $A$ be any element in it, then there are an infinite number of optical lines which pass through $A$ and lie in $W$.

It is easy to show that if $a$ be any separation line, there are an infinite number of optical planes which contain $a$, although, as we have seen, there are only two in any one inertia threefold containing $a$.

Thus let $O$ be any element in $a$ and let $W$ be the one single inertia threefold which passes through $O$ and is normal to $a$.

Then there are an infinite number of optical lines passing through $O$ and lying in $W$, and each of these must be normal to $a$.

Thus each of these optical lines along with $a$ determines an optical plane and all these latter must be distinct.

It follows that there are an infinite number of optical planes containing any separation line.

It is easy to show that if $W$ be an inertia threefold and $a$ be any optical line in it, then there is one and only one optical plane containing $a$ and lying in $W$.

For let $O$ be any element in $a$; then, since there are an infinite number of optical lines passing through $O$ and lying in $W$, there are an infinite number of inertia planes lying in $W$ and containing $a$.

Let $P$ be any such inertia plane.

Then, by Theorem 158, there is one and only one general line, say $b$, passing through $O$ and lying in $W$ which is normal to $P$.

Then $b$ must be normal to $a$, and so $a$ and $b$ determine an optical plane, say $R$, which lies in $W$.

Now $R$ is the only optical plane which contains $a$ and lies in $W$; for let $R'$ be any other optical plane containing $a$.

Then any element $X$ lying in $R'$ but not in $a$ would be neither *before* nor *after* any element of $R$, and so $X$ and $R$ would lie in an optical threefold and could not lie in $W$.

This proves that $R$ is the only optical plane containing $a$ and lying in $W$.

If now we consider the whole set of elements we can easily show that if $a$ be an optical line there is one and only one optical threefold containing $a$.

In order to prove this we have only to remember that $a$ is normal to any optical threefold containing it and, by Theorem 160, if $O$ be any element in $a$, there is one and only one optical threefold passing through $O$ and normal to $a$.

Again, if $P$ be an optical plane, there is one and only one optical threefold containing $P$; for if $A$ be any element which is neither *before* nor *after* any element of $P$, then $P$ and $A$ determine an optical threefold, say $W$.

Also $W$ is the only optical threefold containing $P$, for otherwise we should have more than one optical threefold containing any optical line in $P$.

### THEOREM 170

*If $A$, $B$, $C$, $D$ be the corners of an optical parallelogram ($AC$ being the inertia diagonal line) and if $A$, $B'$, $C$, $D'$ be the corners of a second optical parallelogram, while $A'$, $B'$, $C'$, $D'$ are the corners of a third optical parallelogram whose diagonal line $A'C'$ is conjugate to $BD$, then $A'$, $B$, $C'$, $D$ will be the corners of a fourth optical parallelogram.*

In order to prove this important theorem, we shall first prove the following lemma.

If $O$, $C$ and $C'$ be three distinct elements in an inertia plane $P$ such that $OC$ and $OC'$ are inertia lines while $CC'$ is a separation line, and if further, $CC''$ be another separation line intersecting $OC'$ in $C''$, and if $M$ be the mean of $C$ and $C'$ while $N$ is the mean of $C$ and $C''$, then if $MO$ be conjugate to $CC'$ we cannot have $NO$ conjugate to $CC''$.

It will be sufficient to consider the case where $O$ is *before* $C$, since the case where $O$ is *after* $C$ is quite analogous.

Since $CC'$ is a separation line, while $OC'$ is an inertia line, and since $O$ is *before* $C$, it follows that $O$ must also be *before* $C'$.

Let $E, C, F, C'$ be the corners of an optical parallelogram in the inertia plane $P$ and let $F$ be *after* $E$.

Then $FE$ is conjugate to $CC'$ and intersects it in $M$ and must therefore by hypothesis be identical with $MO$.

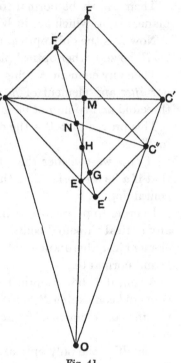

Fig. 41.

Now $E$ must be *after* $O$, for in the first place $E$ cannot be identical with $O$ since $EC'$ is an optical line while $OC'$ is an inertia line.

Again, $O$ cannot be *after* $E$, for then we should have $O$ *after* one element of the optical line $EC'$ and *before* another element of it without lying in the optical line, contrary to Theorem 12.

Thus since $OE$ is an inertia line we must have $E$ after $O$.

Now the element $C''$ is distinct from $C'$, and since $C''$ and $C'$ lie in an inertia line we must have one *after* the other.

Suppose first that $C'$ is *after* $C''$.

Let the optical line through $C''$ parallel to $C'E$ intersect $CE$ in $E'$ and let the optical line through $C''$ parallel to $E'C$ intersect $CF$ in $F'$.

Then $E', C, F', C''$ are the corners of an optical parallelogram, and, since $CC''$ is a separation line, $E'F'$ must be an inertia line conjugate to it and intersecting it in the element $N$.

But now $C''E'$ is a before-parallel of $C'E$ while $C''F'$ is a before-parallel of $C'F$.

Thus we must have $E'$ *before* $E$ and $F'$ *before* $F$.

Now let the inertia line $E'F'$ intersect the optical line $EC'$ in the element $G$.

Then $E'$ is *before* $E$ and is therefore in the $\beta$ sub-set of $E$ and so $G$ must be in the $\alpha$ sub-set of $E$.

Thus $G$ must be *after* $E$.

But, since we have also $F$ *after* $F'$, it follows that $EF$ and $GF'$ intersect in an element, say $H$, which is between $EC'$ and $CF$.

Thus $H$ is linearly between $E$ and $F$ and is therefore *after* $E$.

But $E$ is *after* $O$ and therefore $H$ is *after* $O$.

Thus the conjugate to $CC''$ through $N$ in the inertia plane $P$ intersects $MO$ in an element which is *after* $O$ and so $NO$ cannot be conjugate to $CC''$.

This proves the lemma provided that $C'$ is *after* $C''$.

Next consider the case where $C''$ is *after* $C'$.

Suppose, if possible, that $NO$ is conjugate to $CC''$.

Then, by the case already proved, $MO$ could not be conjugate to $CC'$, contrary to hypothesis, and so the lemma is proved in general.

We shall now make use of this lemma in order to prove the theorem.

We shall suppose that $C$ is *after* $A$ and $C'$ *after* $A'$.

Now, since the first and second optical parallelograms have the pair of opposite corners $A$ and $C$ in common, it follows, by Theorem 60, that they have a common centre, say O.

Further, since the second and third optical parallelograms have the pair of opposite corners $B'$ and $D'$ in common, they have also the same centre $O$.

Thus $AC$ and $A'C'$ intersect in the element $O$, and, since they are both inertia lines, they must lie in one inertia plane, say $P$.

But $C$ and $C'$ are distinct elements lying in the $\alpha$ sub-sets of the distinct elements $B'$ and $D'$, of which the one is neither *before* nor *after* the other, and therefore $C'$ is neither *before* nor *after* $C$, and, in an analogous way, $A'$ is neither *before* nor *after* $A$.

Thus $CC'$ and $AA'$ are both separation lines.

Let $M$ be the mean of $C$ and $C'$.

Then $B'$, $C$ and $C'$ are three corners of an optical parallelogram having $M$ as centre, while $D'$, $C$ and $C'$ are three corners of another optical parallelogram of which $M$ is the centre.

Further, $MB'$ and $MD'$ must both be inertia lines and are each conjugate to $CC'$.

Thus, by Theorem 103, $CC'$ is conjugate to every inertia line which passes through $M$ and lies in the inertia plane containing $MB'$ and $MD'$.

But $O$ is linearly between $B'$ and $D'$ while $M$ is *after* both $B'$ and $D'$ but is not in the general line $B'D'$ and so, by Theorem 73($b$), $MO$ is an inertia line.

Thus, since $MO$ is in the inertia plane containing $MB'$ and $MD'$, it follows that $CC'$ is conjugate to $MO$.

R                                                                 18

If now we consider the optical parallelogram having $B$ and $D$ as opposite corners and lying in the inertia plane containing $BD$ and $A'C'$, it follows, since $O$ is the mean of $B$ and $D$, that $O$ must be the centre of this optical parallelogram.

Further, since by hypothesis $A'C'$ is conjugate to $BD$, it follows that the remaining two corners of this optical parallelogram must lie in $A'C'$.

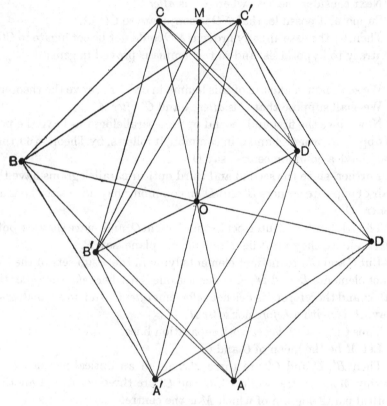

Fig. 42.

Let $A''$ and $C''$ be these remaining corners and let $C''$ be *after* $A''$.

Then just as $CC'$ was shown to be a separation line, we may show that $CC''$ is a separation line, and if $N$ be the mean of $C$ and $C''$ we may show that $CC''$ is conjugate to $NO$, which may be proved to be an inertia line as was $MO$.

But if $CC''$ were distinct from $CC'$, our lemma shows that this would not be possible, and so $CC''$ must be identical with $CC'$.

Thus, since $C''$ lies in $A'C'$, it follows that $C''$ is identical with $C'$.

Similarly $A''$ is identical with $A'$ and therefore $A'$, $B$, $C'$, $D$ are the corners of an optical parallelogram as was to be proved.

## THEORY OF CONGRUENCE

We are now in a position to consider the problems of *congruence* and *measurement* in our system of geometry.

The first point to be examined is the *congruence* of pairs of elements, and we shall find that there are several cases which have to be considered separately.

Two distinct elements $A$ and $B$ will be spoken of briefly as *a pair* and will be denoted by the symbols $(A, B)$ or $(B, A)$.

The order in which the letters are written will be taken advantage of in order to symbolize a certain correspondence between the elements of pairs, as we shall shortly explain.

Since any two distinct elements determine a general line, there will always be one general line associated with any given pair, but different pairs will be associated with the same general line.

If we set up a correspondence between the elements of a pair $(A, B)$ and a pair $(C, D)$ we might either take $C$ to correspond to $A$ and $D$ to $B$, or else take $D$ to correspond to $A$ and $C$ to $B$.

The first of these might be symbolized briefly by:

$$(A, B) \text{ corresponds to } (C, D),$$
or $\qquad\qquad (B, A) \text{ corresponds to } (D, C).$

The second might be symbolized by:

$$(A, B) \text{ corresponds to } (D, C),$$
or $\qquad\qquad (B, A) \text{ corresponds to } (C, D).$

If we consider the case of pairs which have a common element, say $(A, B)$ and $(A, C)$, and if

$$(A, B) \text{ corresponds to } (A, C),$$

then the element $A$ corresponds to itself.

Now the *congruence* of pairs is a correspondence which can be set up in a certain way between certain pairs lying in general lines of the same type.

In dealing with this subject it will be found convenient to have a systematic notation for optical parallelograms, so that we may be able to distinguish how the different corners are related.

If $A$, $B$, $C$, $D$ be the corners of an optical parallelogram we shall use the notation $A\overline{BC}D$ when we wish to signify that the corners $A$ and $D$ lie in the inertia diagonal line and that $A$ is *before* $D$, while $B$ and $C$

lie in the separation diagonal line so that the one is neither *before* nor *after* the other.

If $O$ be the centre of the optical parallelogram $A\overline{BC}D$, then it is obvious that $O$ will be *after A* and *before D*.

*Definition.* A pair $(A, B)$ will be spoken of as an *optical pair*, an *inertia pair*, or a *separation pair* according as $AB$ is an optical, an inertia or a separation line.

We shall first give a definition of the congruence of inertia pairs having a self-corresponding element.

*Definition.* If $A_1\overline{BC}D_1$ and $A_2\overline{BC}D_2$ be optical parallelograms having the common pair of opposite corners $B$ and $C$ and the common centre $O$, then the inertia pair $(O, D_1)$ will be said to be *congruent* to the inertia pair $(O, D_2)$.

This will be written:

$$(O, D_1) \ (\equiv) \ (O, D_2).$$

Similarly the inertia pair $(O, A_1)$ will be said to be *congruent* to the inertia pair $(O, A_2)$.

If $(O, D_1)$ be any inertia pair and $a$ be any inertia line intersecting $OD_1$ in $O$, then the above definition enables us to show that there is one and only one element, say $X$, in $a$ which is distinct from $O$ and such that:

$$(O, D_1) \ (\equiv) \ (O, X).$$

For, by Theorem 106, there is at least one separation line, say $c$, which passes through $O$ and is conjugate to both $OD_1$ and $a$.

Thus $OD_1$ and $c$ determine an inertia plane, say $P_1$, while $a$ and $c$ determine an inertia plane, say $P_2$.

Now if $D_1$ be *after O* there is one single optical parallelogram in $P$ having $O$ as centre and $D_1$ as one of its corners.

If $A_1$ be the corner opposite $D_1$ and if $B$ and $C$ be the remaining corners, this optical parallelogram will be $A_1\overline{BC}D_1$, where $B$ and $C$ will lie in $c$.

Again in the inertia plane $P_2$ there will be one single optical parallelogram having $B$ and $C$ as a pair of opposite corners and $O$ as centre.

If $A_2$ and $D_2$ be the remaining corners they will lie in $a$, and if $D_2$ be *after A_2*, this optical parallelogram will be $A_2\overline{BC}D_2$.

Thus we may identify $D_2$ with $X$ and can say that there is *at least one* element $X$ lying in $a$ and distinct from $O$ and such that:

$$(O, D_1) \ (\equiv) \ (O, X).$$

We have now to show that the element $X$ is unique in this respect in the inertia line $a$.

Let $c'$ be any other separation line distinct from $c$ which passes through $O$ and is conjugate to both $OD_1$ and $a$.

Then $OD_1$ and $c'$ determine an inertia plane, say $P_1'$, while $a$ and $c'$ determine an inertia plane, say $P_2'$.

There is one single optical parallelogram in $P_1'$, having $A_1$ and $D_1$ as a pair of opposite corners, and this optical parallelogram has also $O$ as its centre.

If $B'$ and $C'$ be the remaining corners this optical parallelogram will be $A_1\overline{B'C'}D_1$.

But now we have the optical parallelograms $A_1\overline{B'C'}D_1$, $A_1\overline{BC}D_1$, $A_2\overline{BC}D_2$ and the diagonal line $A_2D_2$ of the last of these is conjugate to $B'C'$ and so it follows, by Theorem 170, that the elements $A_2$, $B'$, $D_2$, $C'$ form the corners of a fourth optical parallelogram $A_2\overline{B'C'}D_2$.

Now $A_2\overline{B'C'}D_2$ will lie in the inertia plane $P_2'$ and will have $O$ as its centre, and further $A_2\overline{B'C'}D_2$ is the only optical parallelogram which lies in $P_2'$ and has $B'$ and $C'$ as a pair of opposite corners.

Thus the element $D_2$ or $X$ is independent of the particular separation line passing through $O$ and conjugate to both $OD_1$ and $a$, which we may select as the separation diagonal line of our optical parallelograms.

It follows that there is one and only one element $X$ in $a$ which is such that:

$$(O, D_1)\,(\equiv)\,(O, X).$$

The same result follows if $D_1$ be *before* $O$ instead of *after* it.

Again, if $(O, D_1)$, $(O, D_2)$ and $(O, D_3)$ be inertia pairs such that:

$$(O, D_1)\,(\equiv)\,(O, D_2)$$

and $$(O, D_2)\,(\equiv)\,(O, D_3),$$

we may easily show that:

$$(O, D_1)\,(\equiv)\,(O, D_3).$$

In order to see this we have only to remember that whether the inertia lines $OD_1$, $OD_2$, $OD_3$ all lie in one inertia plane or in one inertia threefold, there must be at least one general line passing through $O$ and normal to all three.

Since only a separation line can be normal to an inertia line, this separation line will be conjugate to $OD_1$, $OD_2$ and $OD_3$, and if we call it $c$, then $OD_1$ and $c$ will determine an inertia plane, say $P_1$, $OD_2$ and $c$ will determine an inertia plane, say $P_2$, and $OD_3$ and $c$ will determine an inertia plane, say $P_3$.

Now in $P_1$ there will be one single optical parallelogram having $O$ as centre and $D_1$ as one of its corners, while in $P_2$ there will be one single optical parallelogram having $O$ as centre and $D_2$ as one of its corners, and finally in $P_3$ there will be one single optical parallelogram having $O$ as centre and $D_3$ as one of its corners.

Since $(O, D_1) (\equiv) (O, D_2)$ and $(O, D_2) (\equiv) (O, D_3)$ these three optical parallelograms will have a common pair of opposite corners, and so it follows from the definition that:

$$(O, D_1) (\equiv) (O, D_3).$$

Thus *for inertia pairs having a self-corresponding element, the relation of congruence is a transitive relation.*

It is to be observed that if $(O, D_1)$ be an inertia pair we may write:

$$(O, D_1) (\equiv) (O, D_1),$$

or an inertia pair is to be regarded as congruent to itself.

We shall next consider the congruence of separation pairs having a self-corresponding element.

This case differs somewhat from the one we have considered.

While two intersecting inertia lines always lie in an inertia plane, two intersecting separation lines may lie either in a separation plane, an optical plane, or an inertia plane.

An inertia line can only be conjugate to two intersecting separation lines if these lie in a separation plane, as follows from Theorem 99.

Thus if we were to give a definition of the congruence of separation pairs having a self-corresponding element which was strictly analogous to that given for inertia pairs, such a definition would be incomplete.

It is however possible, by a slight modification, to give a definition which will hold for all cases.

In order to avoid complication we shall first explain what we mean by an inertia pair being "conjugate" to a separation pair or a separation pair being "conjugate" to an inertia pair.

*Definition.* If $A\overline{B}CD$ be an optical parallelogram and $O$ be its centre, then the inertia pairs $(O, D)$ and $(O, A)$ will be spoken of as *conjugates* to the separation pairs $(O, B)$ and $(O, C)$ and also conversely.

The pair $(O, D)$ will be called an *after-conjugate* to the pairs $(O, B)$, $(O, C)$, while $(O, A)$ will be called a *before-conjugate* to the pairs $(O, B)$, $(O, C)$.

Further, either of the separation pairs $(O, B)$, $(O, C)$ will be called an *after-conjugate* to $(O, A)$ and a *before-conjugate* to $(O, D)$.

Now we know that there are an infinite number of inertia planes which contain any given separation line, and so there are always inertia pairs which are conjugate to any given separation pair.

Knowing this we can give the following definition of the "congruence" of separation pairs having a self-corresponding element.

*Definition.* If $(O, B_1)$ and $(O, B_2)$ be separation pairs and if $(O, D_1)$ and $(O, D_2)$ be inertia pairs which are after-conjugates to $(O, B_1)$ and $(O, B_2)$ respectively, then if $(O, D_1) (\equiv) (O, D_2)$ we shall say that $(O, B_1)$ is *congruent* to $(O, B_2)$ and shall write this:

$$(O, B_1)\{\equiv\}(O, B_2).$$

If $(O, D_1')$ be any inertia pair which is an after-conjugate to $(O, B_1)$, but is distinct from $(O, D_1)$, then it is obvious by definition that:

$$(O, D_1) (\equiv) (O, D_1').$$

But since $$(O, D_1) (\equiv) (O, D_2),$$

and, since these are inertia pairs, it follows that:

$$(O, D_1') (\equiv) (O, D_2).$$

Thus the congruence of $(O, B_1)$ to $(O, B_2)$ is independent of the particular after-conjugate to $(O, B_1)$ which we may select, and similarly, it is independent of the particular after-conjugate to $(O, B_2)$ which we may select.

Again, if $(O, B_1)$, $(O, B_2)$ and $(O, B_3)$ be separation pairs such that:

$$(O, B_1)\{\equiv\}(O, B_2),$$

and $$(O, B_2)\{\equiv\}(O, B_3),$$

we may easily show that:

$$(O, B_1)\{\equiv\}(O, B_3).$$

In order to prove this, let $(O, D_1)$, $(O, D_2)$ and $(O, D_3)$ be inertia pairs which are after-conjugates to $(O, B_1)$, $(O, B_2)$ and $(O, B_3)$ respectively.

Then we must have:

$$(O, D_1) (\equiv) (O, D_2),$$

and $$(O, D_2) (\equiv) (O, D_3),$$

and, since these are inertia pairs, it follows, as previously shown, that:

$$(O, D_1) (\equiv) (O, D_3).$$

Thus, by the definition:

$$(O, B_1)\{\equiv\}(O, B_3),$$

and so, *for separation pairs having a self-corresponding element, the relation of congruence is a transitive relation.*

Again, if $(O, B)$ be any separation pair and $a$ be any separation line passing through $O$, there are two and only two elements, say $X_1$ and $Y_1$, in $a$ which are distinct from $O$ and such that:

$$(O, B)\{\equiv\}(O, X_1),$$

and
$$(O, B)\{\equiv\}(O, Y_1).$$

This may be easily shown as follows.

Let $(O, D)$ be any inertia pair which is an after-conjugate to $(O, B)$ and let $b$ be any inertia line which passes through $O$ and is conjugate to $a$.

Then, as we have already seen, there is one and only one element, say $D_1$, lying in $b$ and distinct from $O$ and such that:

$$(O, D)(\equiv)(O, D_1).$$

But now $a$ and $b$ determine an inertia plane and in this inertia plane there is one and only one optical parallelogram having $O$ as centre and $D_1$ as one of its corners.

If this optical parallelogram be $A_1\overline{B_1 C_1}D_1$, then the elements $B_1$ and $C_1$ will lie in $a$ and the inertia pair $(O, D_1)$ will be an after-conjugate to each of the separation pairs $(O, B_1)$ and $(O, C_1)$.

Thus since $(O, D)(\equiv)(O, D_1)$ it follows that:

$$(O, B)\{\equiv\}(O, B_1)$$

and
$$(O, B)\{\equiv\}(O, C_1).$$

Again, if there were any other element, say $B_2$, lying in $a$ and distinct from both $B_1$ and $C_1$ and such that we had

$$(O, B)\{\equiv\}(O, B_2),$$

then there would be an element, say $D_2$, lying in $b$ and such that $(O, D_2)$ was an after-conjugate to $(O, B_2)$.

Since $B_2$ is supposed distinct from both $B_1$ and $C_1$, therefore $D_2$ would require to be distinct from $D_1$.

But since we have supposed $(O, B)\{\equiv\}(O, B_2)$, therefore we should have $(O, D)(\equiv)(O, D_2)$ and so we should have the two distinct elements $D_1$ and $D_2$ lying in the inertia line $b$ and such that:

$$(O, D)(\equiv)(O, D_1) \quad \text{and} \quad (O, D)(\equiv)(O, D_2),$$

which we have already shown to be impossible.

Thus we may identify $B_1$ with $X_1$ and $C_1$ with $Y_1$ and say that there are two and only two elements $X_1$ and $Y_1$ lying in $a$ and distinct from $O$ and such that:

$$(O, B)\{\equiv\}(O, X_1) \quad \text{and} \quad (O, B)\{\equiv\}(O, Y_1).$$

If $A\overline{BC}D$ be an optical parallelogram and $O$ be its centre, we observe that according to our definitions we have

$$(O, B)\{\equiv\}(O, C),$$

but not $\qquad (O, A)(\equiv)(O, D).$

The reason why we make this distinction is that in the separation pairs we have $O$ neither *before* nor *after* $B$ and also $O$ neither *before* nor *after* $C$, while in the inertia pairs we have $O$ *after* $A$ and $O$ *before* $D$.

Thus in the first case the relations are alike in respect of *before* and *after*, while in the second case the relations are different.

The question now arises as to the "congruence" of optical pairs.

In this case constructions such as those by which we defined the congruence of inertia and separation pairs having a self-corresponding element, entirely fail and there is nothing at all analogous to them.

*We are thus led to regard optical pairs as not determinately comparable with one another in respect of congruence, except when they lie in the same, or in parallel, optical lines.*

As regards the "congruence" of pairs lying in the same general line, we have as yet given no definition, except for the very special case of inertia or separation pairs having a self-corresponding element; while no definition whatever has been given of the "congruence" of pairs lying in parallel general lines.

A definition covering all these omitted cases can be given, which applies to all three types of pair.

We must first however define what we mean when we say that one pair is opposite to another.

*Definition.* A pair $(A, B)$ will be said to be *opposite* to a pair $(C, D)$ if and only if the elements $A$, $B$, $C$, $D$ form the corners of a general parallelogram in such a way that $AB$ and $CD$ are one pair of opposite sides, while $AC$ and $BD$ are the other pair of opposite sides.

This will be denoted by the symbols

$$(A, B)\,\square\,(C, D).$$

It will be observed that the use of the symbol $\square$ implies that the pairs $(A, B)$ and $(C, D)$ lie in distinct general lines which are parallel to one another.

If however we have

$$(A, B)\,\square\,(C, D),$$

and $\qquad (E, F)\,\square\,(C, D),$

then the pairs $(A, B)$ and $(E, F)$ may lie either in the same or in parallel general lines.

If $(A, B)$ and $(E, F)$ do not lie in the same general line, it follows from Theorem 126 that we may write

$$(A, B) \square (E, F).$$

We have now to prove the following theorem:

### THEOREM 171

*If $(A, B)$, $(A', B')$ and $(C, D)$ be pairs such that:*

$$(A, B) \square (C, D),$$

*and*        $$(A', B') \square (C, D),$$

*and if $(C', D')$ be any other pair such that:*

$$(A, B) \square (C', D'),$$

*and which does not lie in the general line $A'B'$, then we shall also have*

$$(A', B') \square (C', D').$$

We shall first consider the case where $(A, B)$ and $(A', B')$ do not lie in one general line.

In this case since

$$(A, B) \square (C, D),$$
and        $$(A', B') \square (C, D),$$

it follows, by Theorem 126, that:

$$(A', B') \square (A, B).$$

But        $$(C', D') \square (A, B)$$

by hypothesis, and so, since $(C', D')$ and $(A', B')$ do not lie in one general line, it follows that:

$$(A', B') \square (C', D').$$

Next consider the case where $(A, B)$ and $(A', B')$ lie in one general line.

There are two sub-cases of this:

(1) $(C, D)$ and $(C', D')$ do not lie in one general line.

(2) $(C, D)$ and $(C', D')$ do lie in one general line.

Consider first sub-case (1).

Here since        $$(C, D) \square (A, B),$$
and        $$(C', D') \square (A, B),$$

and since $(C, D)$ and $(C', D')$ do not lie in one general line, it follows that:

$$(C', D') \square (C, D).$$

But        $$(A', B') \square (C, D),$$

and so, since $(C', D')$ and $(A', B')$ do not lie in one general line, it follows that:

$$(A', B') \square (C', D').$$

Next consider sub-case (2).

Let $E$ be any element in the general line $AC'$ distinct from both $A$ and $C'$ and let a general line through $E$ parallel to $AB$ intersect $D'B$ in the element $F$.

Then we shall have

$$(E, F) \square (A, B),$$

and also $\qquad (E, F) \square (C', D').$

But now, since $E$ is distinct from $A$ and also from $C'$, it follows that the general line $EF$ must be distinct from the general line containing $(A, B)$ and $(A', B')$ and must also be distinct from the general line containing $(C', D')$ and $(C, D)$.

Thus since $\qquad (E, F) \square (A, B),$

and $\qquad (C, D) \square (A, B),$

and, since $(E, F)$ and $(C, D)$ do not lie in one general line, it follows that:

$$(E, F) \square (C, D).$$

Also since $\qquad (A', B') \square (C, D),$

and, since $(E, F)$ and $(A', B')$ do not lie in one general line, it follows that:

$$(A', B') \square (E, F).$$

But $\qquad (C', D') \square (E, F),$

and since $(A', B')$ and $(C', D')$ lie respectively in the distinct general lines $AB$ and $CD$, it follows that:

$$(A', B') \square (C', D').$$

Thus the theorem holds in all cases.

We are now in a position to introduce the following definition:

*Definition.* A pair $(A, B)$ will be said to be *co-directionally congruent* to a pair $(A', B')$ provided a pair $(C, D)$ exists such that:

$$(A, B) \square (C, D),$$

and $\qquad (A', B') \square (C, D).$

The theorem just proved shows that we are at liberty to replace the pair $(C, D)$ by any other pair $(C', D')$ such that:

$$(A, B) \square (C', D'),$$

provided $(C', D')$ does not lie in the general line $A'B'$.

It is evident that $(A, B) \square (C, D)$ implies that $(A, B)$ is co-direction-ally congruent to $(C, D)$, but $(A, B)$ being co-directionally congruent to $(C, D)$ does not imply that $(A, B) \square (C, D)$, since $(A, B)$ and $(C, D)$ might lie in the same general line.

It is also obvious that $(A, B)$ is co-directionally congruent to $(A, B)$.

We shall ultimately represent co-directional congruence by the same symbol $\equiv$ as we shall use for the other cases of congruence, but when we wish to make it clear that the congruence is co-directional we shall use the symbol $|\equiv|$.

Thus we see that: $(A, B) \square (C, D)$ implies $(A, B) |\equiv| (C, D)$, but $(A, B) |\equiv| (C, D)$ does not imply $(A, B) \square (C, D)$, except when $AB$ and $CD$ are distinct general lines.

We have next to show that if
$$(A, B) |\equiv| (C, D),$$
and
$$(C, D) |\equiv| (E, F),$$
then must
$$(A, B) |\equiv| (E, F).$$

This is easily proved; for if $a$ be any general line parallel to $AB$ but distinct from $CD$ and $EF$ and therefore also parallel to them, we may select any pair $(G, H)$ in $a$, such that:
$$(A, B) \square (G, H) \quad \dots\dots\dots\dots\dots\dots\dots(1).$$
Then, since $(A, B) |\equiv| (C, D)$,
it follows, by Theorem 171, that:
$$(C, D) \square (G, H).$$
Similarly since $(C, D) |\equiv| (E, F)$,
it follows that: $(E, F) \square (G, H) \quad \dots\dots\dots\dots\dots\dots\dots(2).$

Thus from (1) and (2) it follows that:
$$(A, B) |\equiv| (E, F),$$
and so we see that: *the relation of co-directional congruence of pairs is a transitive relation.*

If $(A, B) \square (C, D)$ and if $B$ be *after* $A$ then it is easy to see that $D$ must be *after* $C$.

In the first place $AB$ must be either an optical or inertia line and, since $CD$ is parallel to $AB$, it follows that $CD$ must be the same type of general line as $AB$.

Suppose first that $AB$ is an optical line.

Then $C$ could not be *after* $D$, for then, by Theorem 58 or Theorem 92, $AC$ and $BD$ would intersect, contrary to the hypothesis that they are parallel.

Thus, since $C$ and $D$ are distinct, and since $CD$ is an optical line, it follows that $D$ must be *after* $C$.

Next suppose that $AB$ is an inertia line.

Then $AB$ and $CD$ must lie in an inertia plane, say $P$.

If $AC$ and $BD$ should happen to be optical lines then, since $B$ is *after* $A$ it follows that $BD$ would be an after-parallel of $AC$ and so, since $CD$ is an inertia line, it would follow that $D$ must be *after* $C$.

Next suppose that $AC$ and $BD$ are not optical lines.

Let $AE$ and $BE$ be generators of $P$ of opposite sets passing through $A$ and $B$ respectively and intersecting in $E$.

Let $CF$ be an optical line through $C$ parallel to $AE$ and let it intersect the general line through $E$ parallel to $AC$ in $F$.

Then $EF$ must be parallel to $BD$ and so, by Theorem 126, $DF$ must be parallel to $BE$ and therefore must be an optical line.

But now, since $B$ is *after* $A$, we must have $E$ after $A$ and $B$ after $E$.

Thus, by the first case, we must have $F$ *after* $C$ and $D$ *after* $F$ and therefore $D$ *after* $C$ as was to be proved.

Thus in all cases if $B$ be *after* $A$ we must have $D$ *after* $C$ and similarly if $B$ be *before* $A$ we must have $D$ *before* $C$.

It follows directly from this that if

$$(A, B) \square (C, D),$$

and
$$(A', B') \square (C, D),$$

then if $B$ be *after* $A$ we must have $D$ *after* $C$ and therefore $B'$ after $A'$.

Thus *if* $(A, B) \,|\equiv|\, (A', B')$ *and if* $B$ *be* after $A$ *we must have* $B'$ *after* $A'$, *while if* $B$ *be* before $A$ *we must have* $B'$ *before* $A'$.

Again if three corners of a general parallelogram $A'$, $C$ and $D$ be given and if we know that two of the side lines are $A'C$ and $CD$, then the general parallelogram is uniquely determined.

If then any pair $(A', X)$ be co-directionally congruent to a pair $(A, B)$, where $A$, $B$ and $A'$ are given, it is easy to see that $X$ is uniquely determinate, provided we know that $A'$ corresponds to $A$.

For let $a$ be a general line parallel to $AB$, but which does not pass through $A'$, and let $(C, D)$ be any pair in $a$ such that:

$$(A, B) \square (C, D).$$

Then there is one single general parallelogram having $A'$, $C$ and $D$ as three of its corners and $A'C$ and $CD$ as two of its side lines.

If $B'$ be the remaining corner we shall have

$$(A', B') \square (C, D),$$

and so
$$(A', B') \,|\equiv|\, (A, B).$$

Thus $X$ must be identified with $B'$, which is a definite element.

## THEOREM 172

*If $(O_1, A_1)$ and $(O_2, A_2)$ be inertia pairs such that:*

$$(O_1, A_1) \,|\equiv|\, (O_2, A_2),$$

*and if $(O_1, B_1)$ be any separation pair which is conjugate to $(O_1, A_1)$, then there is a separation pair, say $(O_2, B_2)$, which is conjugate to $(O_2, A_2)$, and such that:*

$$(O_1, B_1) \,|\equiv|\, (O_2, B_2).$$

It is evident that $(O_1, A_1)$ and $(O_1, B_1)$ must lie in an inertia plane, say $P_1$.

Since the inertia line $O_2 A_2$ must be either parallel to the inertia line $O_1 A_1$, or else identical with it, it follows that $O_2 A_2$ must either lie in $P_1$ or in an inertia plane parallel to $P_1$.

We shall first consider the case where $O_2 A_2$ lies in an inertia plane $P_2$ parallel to $P_1$.

If now we take the one single optical parallelogram in $P_1$ having $O_1$ as centre and $A_1$ as one of its corners, then $B_1$ will be another corner.

If $(O_1, B_1)$ be an after-conjugate to $(O_1, A_1)$ we may take this optical parallelogram to be $A_1 \overline{B_1 C_1} D_1$, while if $(O_1, B_1)$ be a before-conjugate to $(O_1, A_1)$ we may take the optical parallelogram to be $D_1 \overline{B_1 C_1} A_1$.

Now the inertia plane $P_1$ and the general line $O_1 O_2$ determine a general threefold containing $P_2$ and so, as we have already seen, if through any element of $P_1$ distinct from $O_1$ a general line be taken parallel to $O_1 O_2$, then this general line will intersect $P_2$.

Now through the elements $A_1$, $B_1$, $C_1$ and $D_1$ let general lines be taken parallel to $O_1 O_2$ and let these intersect $P_2$ in the elements $A_2$, $B_2$, $C_2$ and $D_2$ respectively.

Then any two of the general lines $O_1 O_2$, $A_1 A_2$, $B_1 B_2$, $C_1 C_2$, $D_1 D_2$ are parallel to one another and therefore any two of them lie in a general plane.

Since however the elements $A_1$, $O_1$ and $D_1$ lie in one general line, the three general lines $A_1 A_2$, $O_1 O_2$ and $D_1 D_2$ lie in one general plane, and since the elements $B_1$, $O_1$ and $C_1$ lie in one general line, the three general lines $B_1 B_2$, $O_1 O_2$ and $C_1 C_2$ lie in one general plane.

Thus the elements $A_2$, $O_2$ and $D_2$ lie in one general line parallel to the general line containing $A_1$, $O_1$ and $D_1$, while the elements $B_2$, $O_2$ and $C_2$ lie in another general line parallel to that containing $B_1$, $O_1$ and $C_1$.

Further the general lines $A_2 B_2$, $A_2 C_2$, $B_2 D_2$, $C_2 D_2$ must be respectively parallel to $A_1 B_1$, $A_1 C_1$, $B_1 D_1$, $C_1 D_1$ and, since these latter are all optical lines, it follows that $A_2B_2$, $A_2C_2$, $B_2D_2$, $C_2D_2$ are all optical lines.

Thus $A_2$, $B_2$, $C_2$, $D_2$ form the corners of an optical parallelogram having $O_2$ as centre.

Further, the diagonal line $A_2D_2$ is an inertia line, while the diagonal line $B_2C_2$ must be a separation line.

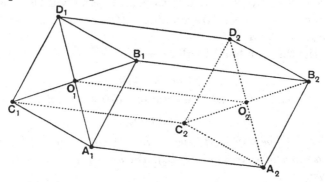

Fig. 43.

Thus the separation pair $(O_2, B_2)$ is conjugate to the inertia pair $(O_2, A_2)$, and, since $O_2$ is *after* or *before* $A_2$ according as $O_1$ is *after* or *before* $A_1$, it follows that $(O_2, B_2)$ is an after- or before-conjugate to $(O_2, A_2)$ according as $(O_1, B_1)$ is an after- or before-conjugate to $(O_1, A_1)$.

Also we have          $(O_1, B_1) \square (O_2, B_2)$

and so          $(O_1, B_1) \mid \equiv \mid (O_2, B_2)$.

This proves the theorem provided $O_2A_2$ does not lie in $P_1$.

Consider next the case where $O_2A_2$ does lie in $P_1$.

Let $P'$ be any inertia plane parallel to $P_1$ and let $(O', A')$ be any inertia pair in $P'$ such that:

$$(O_1, A_1) \square (O', A').$$

Then, by Theorem 171, since $(O_1, A_1) \mid \equiv \mid (O_2, A_2)$, we must have

$$(O_2, A_2) \square (O', A').$$

Thus, by the case already proved, it follows that there is an optical parallelogram lying in $P'$ which has $O'$ as centre and $A'$ as one of its corners and such that, if we denote it by $A'\overline{B'C'}D'$ or $D'\overline{B'C'}A'$ (according as the optical parallelogram in $P_1$ is $A_1\overline{B_1C_1}D_1$ or $D_1\overline{B_1C_1}A_1$), then:

$$(O_1, B_1) \square (O', B').$$

But in a similar manner we can show that there is an optical parallelogram lying in $P_1$ which has $O_2$ as centre and $A_2$ as one of its corners and such that if we denote it by $A_2 \overline{B_2 C_2} D_2$ or $D_2 \overline{B_2 C_2} A_2$ (according as the optical parallelogram in $P'$ is $A'B'C'D'$ or $D'B'C'A'$), then:

$$(O_2, B_2) \,\square\, (O', B').$$

Thus it follows from definition that:

$$(O_1, B_1) \,|\!\equiv\!|\, (O_2, B_2).$$

Further, $(O_2, B_2)$ is conjugate to $(O_2, A_2)$ and will be an after- or before-conjugate to $(O_2, A_2)$ according as $(O_1, B_1)$ is an after- or before-conjugate to $(O_1, A_1)$.

Thus the theorem holds in all cases.

### THEOREM 173

*If $(O_1, A_1)$ and $(O_2, A_2)$ be separation pairs such that:*

$$(O_1, A_1) \,|\!\equiv\!|\, (O_2, A_2),$$

*and if $(O_1, B_1)$ be any inertia pair which is conjugate to $(O_1, A_1)$, then there is an inertia pair, say $(O_2, B_2)$, which is conjugate to $(O_2, A_2)$ and such that:*

$$(O_1, B_1) \,|\!\equiv\!|\, (O_2, B_2).$$

The proof of this theorem is quite analogous to that of Theorem 172.

Also it will be seen that $(O_2, B_2)$ will be a before- or after-conjugate to $(O_2, A_2)$ according as $(O_1, B_1)$ is a before- or after-conjugate to $(O_1, A_1)$.

We have now to prove certain theorems involving both the co-directional congruence of pairs and the congruence of pairs having a self-corresponding element.

We shall make use of the symbols $(\equiv)$, $\{\equiv\}$ and $|\!\equiv\!|$ in the manner already explained in order to show clearly the types of congruence to which we refer.

### THEOREM 174

*If $(O_1, A_1)$, $(O_1, B_1)$ and $(O_2, A_2)$ be inertia pairs such that:*

$$(O_1, A_1) \,(\equiv)\, (O_1, B_1)$$

*and* $\qquad (O_1, A_1) \,|\!\equiv\!|\, (O_2, A_2),$

*then there is an inertia pair $(O_2, B_2)$ such that:*

$$(O_2, A_2) \,(\equiv)\, (O_2, B_2)$$

*and* $\qquad (O_1, B_1) \,|\!\equiv\!|\, (O_2, B_2).$

Let $c$ be any separation line passing through $O_1$ and normal to both the inertia lines $O_1 A_1$ and $O_1 B_1$, and let $C_1$ be an element in $c$ such that the separation pair $(O_1, C_1)$ is conjugate to $(O_1, A_1)$.

Then, since $(O_1, A_1)(\equiv)(O_1, B_1)$, it follows that $(O_1, C_1)$ must also be conjugate to $(O_1, B_1)$.

But, since $(O_1, A_1)\,|\equiv|\,(O_2, A_2)$, it follows, by Theorem 172, that there is a separation pair, say $(O_2, C_2)$, which is conjugate to $(O_2, A_2)$ and such that:

$$(O_1, C_1)\,|\equiv|\,(O_2, C_2).$$

But now, by Theorem 173, since $(O_1, B_1)$ is conjugate to $(O_1, C_1)$, it follows that there is an inertia pair, say $(O_2, B_2)$, which is conjugate to $(O_2, C_2)$ and such that:

$$(O_1, B_1)\,|\equiv|\,(O_2, B_2).$$

But now, since $(O_1, A_1)(\equiv)(O_1, B_1)$ and these are inertia pairs, we must have $A_1$ and $B_1$ either both *after* $O_1$ or both *before* $O_1$.

Further, $A_2$ must be *after* or *before* $O_2$ according as $A_1$ is *after* or *before* $O_1$, while $B_2$ must be *after* or *before* $O_2$ according as $B_1$ is *after* or *before* $O_1$.

Thus $A_2$ and $B_2$ are either both *after* $O_2$ or both *before* $O_2$.

Since therefore $(O_2, C_2)$ is conjugate to both $(O_2, A_2)$ and $(O_2, B_2)$, it follows that:

$$(O_2, A_2)(\equiv)(O_2, B_2).$$

Thus the theorem is proved.

### THEOREM 175

*If $(O_1, A_1)$, $(O_1, B_1)$ and $(O_2, A_2)$ be separation pairs such that:*

$$(O_1, A_1)\{\equiv\}(O_1, B_1)$$

*and*
$$(O_1, A_1)\,|\equiv|\,(O_2, A_2),$$

*then there is a separation pair $(O_2, B_2)$ such that:*

$$(O_2, A_2)\{\equiv\}(O_2, B_2)$$

*and*
$$(O_1, B_1)\,|\equiv|\,(O_2, B_2).$$

Let $(O_1, D_1)$ and $(O_1, E_1)$ be inertia pairs which are after-conjugates to $(O_1, A_1)$ and $(O_1, B_1)$ respectively.

Then since $\qquad (O_1, A_1)\{\equiv\}(O_1, B_1),$

we must have $\qquad (O_1, D_1)(\equiv)(O_1, E_1).$

But now since $\qquad (O_1, A_1)\,|\equiv|\,(O_2, A_2),$

and since $(O_1, D_1)$ is an inertia pair which is an after-conjugate to $(O_1, A_1)$, it follows, by Theorem 173, that there is an inertia pair, say $(O_2, D_2)$, which is an after-conjugate to $(O_2, A_2)$ and such that:

$$(O_1, D_1)\,|\equiv|\,(O_2, D_2).$$

But now $(O_1, D_1)$, $(O_1, E_1)$ and $(O_2, D_2)$ are inertia pairs such that:

$$(O_1, D_1)\,(\equiv)\,(O_1, E_1)$$

and $\qquad\qquad (O_1, D_1)\,|\equiv|\,(O_2, D_2),$

and so, by Theorem 174, there is an inertia pair $(O_2, E_2)$ such that:

$$(O_2, D_2)\,(\equiv)\,(O_2, E_2)$$

and $\qquad\qquad (O_1, E_1)\,|\equiv|\,(O_2, E_2).$

Since however $(O_1, B_1)$ is a separation pair which is a before-conjugate to the inertia pair $(O_1, E_1)$, it follows, by Theorem 172, that there is a separation pair, say $(O_2, B_2)$, which is a before-conjugate to $(O_2, E_2)$ and such that:

$$(O_1, B_1)\,|\equiv|\,(O_2, B_2).$$

But since $(O_2, D_2)$ and $(O_2, E_2)$ are after-conjugates to $(O_2, A_2)$ and $(O_2, B_2)$ respectively, and since

$$(O_2, D_2)\,(\equiv)\,(O_2, E_2),$$

it follows by definition that:

$$(O_2, A_2)\,\{\equiv\}\,(O_2, B_2).$$

Thus the theorem is proved.

### THEOREM 176

*If $(A, B)$ and $(A, C)$ be inertia pairs such that:*

$$(A, B)\,(\equiv)\,(A, C),$$

*then there is an inertia pair $(C, D)$ such that:*

$$(C, A)\,(\equiv)\,(C, D)$$

*and* $\qquad\qquad (B, A)\,|\equiv|\,(C, D).$

Let $a$ be any separation line which passes through $A$ and is normal to both $AB$ and $AC$.

Let $A_1$ be an element in $a$ such that the separation pair $(A, A_1)$ is conjugate to the inertia pair $(A, C)$.

Then since $\qquad\qquad (A, B)\,(\equiv)\,(A, C),$

it follows that $(A, A_1)$ is also conjugate to $(A, B)$.

Let the general line through $C$ parallel to $AA_1$ intersect the general line through $A_1$ parallel to $AC$ in the element $C_1$, and let the general line through $B$ parallel to $AA_1$ intersect the general line through $A_1$ parallel to $AB$ in the element $B_1$.

Thus $\qquad\qquad (C, C_1) \square (A, A_1)$

and $\qquad\qquad\quad (B, B_1) \square (A, A_1),$

and therefore $\qquad (B, B_1) \,|\equiv| \,(C, C_1).$

Now, since $(A, A_1)$ is conjugate to $(A, C)$, it follows that $A_1 C$ is an optical line and it is easy to show that $A C_1$ is also an optical line as follows:

Since $A_1 C$ and $A C_1$ are diagonal lines of the general parallelogram whose corners are $A$, $A_1$, $C_1$, $C$, it follows that they intersect in an element, say $E$, which is the mean of $A_1$ and $C$.

If $F$ be the mean of $A_1$ and $A$, then $EF$ is parallel to $CA$ and therefore $EF$ is conjugate to the separation line $AA_1$.

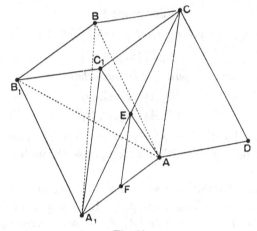

Fig. 44.

Thus $A$, $A_1$ and $E$ are three corners of an optical parallelogram whose centre $F$ lies in $AA_1$ and therefore $AE$ (that is $AC_1$) is an optical line.

Similarly, since $(A, A_1)$ is conjugate to $(A, B)$, it follows that $AB_1$ is an optical line.

But since $CC_1$ and $BB_1$ are parallel to $AA_1$ we have $CC_1$ conjugate to $CA$, and $BB_1$ conjugate to $BA$.

Thus the inertia pairs $(C, A)$ and $(B, A)$ are conjugate to the separation pairs $(C, C_1)$ and $(B, B_1)$ respectively.

But now, since $\qquad (B, B_1) \,|\equiv| \,(C, C_1),$

and since $(B, A)$ is an inertia pair which is conjugate to $(B, B_1)$, it follows, by Theorem 173, that there is an inertia pair, say $(C, D)$, which is conjugate to $(C, C_1)$ and such that:

$$(B, A) \,|\equiv| \,(C, D).$$

But from this it follows that $D$ must be *before* or *after* $C$ according as $A$ is *before* or *after* $B$.

Since however $\qquad (A, B)\,(\equiv)\,(A, C)$,

we have $A$ *before* or *after* $B$ according as $A$ is *before* or *after* $C$.

Thus we must have $D$ *before* or *after* $C$ according as $A$ is *before* or *after* $C$.

Since therefore the inertia pairs $(C, A)$ and $(C, D)$ are both conjugate to the separation pair $(C, C_1)$, it follows that:

$$(C, A)\,(\equiv)\,(C, D).$$

Thus the theorem is proved.

### THEOREM 177

*If $(A, B)$ and $(A, C)$ be separation pairs such that:*

$$(A, B)\,\{\equiv\}\,(A, C),$$

*then there is a separation pair $(C, D)$ such that:*

$$(C, A)\,\{\equiv\}\,(C, D)$$

*and* $\qquad\qquad (B, A)\,|\equiv|\,(C, D).$

Let $a$ be any inertia line which passes through $A$ and is normal to $AB$, and let $a'$ be any inertia line which passes through $A$ and is normal to $AC$.

Let $A_1$ be an element in $a$ such that the inertia pair $(A, A_1)$ is an after-conjugate to the separation pair $(A, B)$ and let $A'$ be an element in $a'$ such that the inertia pair $(A, A')$ is an after-conjugate to the separation pair $(A, C)$.

Then since $\qquad (A, B)\,\{\equiv\}\,(A, C)$

it follows that: $\qquad (A, A_1)\,(\equiv)\,(A, A').$

Let the general line through $C$ parallel to $AA'$ intersect the general line through $A'$ parallel to $AC$ in the element $C'$ and let the general line through $B$ parallel to $AA_1$ intersect the general line through $A_1$ parallel to $AB$ in the element $B_1$.

Then $\qquad\qquad (C, C')\,\square\,(A, A')$

and $\qquad\qquad\quad (B, B_1)\,\square\,(A, A_1),$

and so we may write: $\quad (C, C')\,|\equiv|\,(A, A'),$

and $\qquad\qquad\quad (B, B_1)\,|\equiv|\,(A, A_1).$

But, since $(A, A')$, $(A, A_1)$ and $(C, C')$ are inertia pairs such that

$$(A, A')\,(\equiv)\,(A, A_1)$$

and $\qquad\qquad (A, A')\,|\equiv|\,(C, C'),$

therefore, by Theorem 174, there is an inertia pair $(C, C_1)$ such that:

$$(C, C')\,(\equiv)\,(C, C_1)$$

and
$$(A, A_1)\,|\equiv|\,(C, C_1).$$

Thus since
$$(B, B_1)\,|\equiv|\,(A, A_1),$$

it follows that:
$$(B, B_1)\,|\equiv|\,(C, C_1).$$

But now we may show in the manner employed in the last theorem that, since $(A, A_1)$ is an after-conjugate to $(A, B)$, therefore $(B, B_1)$ is an after-conjugate to $(B, A)$, and, since $(A, A')$ is an after-conjugate to $(A, C)$, therefore $(C, C')$ is an after-conjugate to $(C, A)$.

Since, however, we have the inertia pairs $(B, B_1)$ and $(C, C_1)$ such that:

$$(B, B_1)\,|\equiv|\,(C, C_1),$$

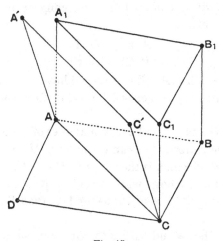

Fig. 45.

and, since $(B, A)$ is a separation pair which is conjugate to $(B, B_1)$, it follows, by Theorem 172, that there is a separation pair, say $(C, D)$, which is conjugate to $(C, C_1)$ and such that:

$$(B, A)\,|\equiv|\,(C, D).$$

But now $(A, A_1)$ is an after-conjugate to $(A, B)$ and so $A_1$ is *after* $A$.

Thus, since
$$(A, A_1)\,|\equiv|\,(C, C_1),$$

it follows that $C_1$ is *after* $C$, and so, since $(C, C_1)$ is conjugate to $(C, D)$, it must be an after-conjugate.

But $(C, C')$ is an after-conjugate to $(C, A)$ and so since

$$(C, C')\,(\equiv)\,(C, C_1),$$

it follows from the definition that:
$$(C, A)\{\equiv\}(C, D).$$
Thus the theorem is proved.

## THEOREM 178

(1) *If $A$, $B$ and $C$ be three distinct elements and the pairs $(A, B)$ and $(B, C)$ be such that:*
$$(A, B)\,|\equiv|\,(B, C),$$
*then $B$ is the mean of $A$ and $C$.*

(2) *If $A$, $B$ and $C$ be three distinct elements such that $B$ is the mean of $A$ and $C$, then the pairs $(A, B)$ and $(B, C)$ are such that:*
$$(A, B)\,|\equiv|\,(B, C).$$

First suppose that:     $(A, B)\,|\equiv|\,(B, C).$

Then, by the definition of co-directional congruence, there must be a pair, say $(D, E)$, such that:
$$(A, B)\,\square\,(D, E)$$
and$$\qquad\qquad (B, C)\,\square\,(D, E).$$

Now, since the pairs $(A, B)$ and $(B, C)$ have a common element $B$, they cannot lie in parallel general lines and so must lie in the same general line.

Then $BE$ and $CD$ must be the diagonal lines of the general parallelogram whose corners are $B$, $C$, $D$ and $E$ and so $BE$ and $CD$ must intersect in an element $F$ which is the mean of $D$ and $C$.

But $D$ does not lie in the general line $AC$, and so, since $BF$ is parallel to $AD$, it follows, by Theorem 78, 94, or 118, that $B$ is the mean of $A$ and $C$.

Next, to prove the second part of the theorem, suppose that $B$ is the mean of $A$ and $C$.

Let $(D, E)$ be any pair such that:
$$(B, C)\,\square\,(D, E).$$

Then the diagonal lines $BE$ and $CD$ of the general parallelogram, whose corners are $B$, $C$, $D$ and $E$, must intersect in an element $F$ which is the mean of $D$ and $C$.

But, since $D$ does not lie in the general line $AC$, it follows, by corollary to Theorem 78, 94 or 118, that $BF$ (that is $BE$) is parallel to $AD$.

Thus, since also $AB$ is parallel to $DE$, it follows that:
$$(A, B)\,\square\,(D, E).$$

Thus, by the definition of co-directional congruence, we have:
$$(A, B)\,|\equiv|\,(B, C).$$

Thus both parts of the theorem are proved.

We are now in a position to introduce general definitions of the congruence of inertia and separation pairs.

This is done by combining co-directional congruence with congruence in which an element is self-corresponding, in the following manner.

*Definition.* An inertia pair $(A_1, B_1)$ will be said to be *congruent* to an inertia pair $(A_2, B_2)$ provided an inertia pair $(A_2, C_2)$ exists such that:

$$(A_1, B_1) \mid \equiv \mid (A_2, C_2)$$

and $\qquad (A_2, B_2) ( \equiv ) (A_2, C_2).$

*Definition.* A separation pair $(A_1, B_1)$ will be said to be *congruent* to a separation pair $(A_2, B_2)$ provided a separation pair $(A_2, C_2)$ exists such that:

$$(A_1, B_1) \mid \equiv \mid (A_2, C_2)$$

and $\qquad (A_2, B_2) \{ \equiv \} (A_2, C_2).$

We shall denote the generalized congruence of inertia or of separation pairs by the symbol $\equiv$, thus:

$$(A_1, B_1) \equiv (A_2, B_2).$$

We shall also use the same symbol to denote the congruence of optical pairs, but, in the latter case, it is to be regarded as simply equivalent to the symbol $\mid \equiv \mid$, since the only congruence of optical pairs is taken to be co-directional.

Let us consider now two inertia pairs $(A_1, B_1)$ and $(A_2, B_2)$ such that:

$$(A_1, B_1) \equiv (A_2, B_2).$$

Then there exists an inertia pair $(A_2, C_2)$ such that:

$$(A_1, B_1) \mid \equiv \mid (A_2, C_2)$$

and $\qquad (A_2, B_2) ( \equiv ) (A_2, C_2).$

But, by Theorem 174, there exists an inertia pair $(A_1, C_1)$ such that:

$$(A_2, B_2) \mid \equiv \mid (A_1, C_1)$$

and $\qquad (A_1, B_1) ( \equiv ) (A_1, C_1).$

Thus we may write $\qquad (A_2, B_2) \equiv (A_1, B_1).$

Again, by Theorem 176, there is an inertia pair $(B_2, D_2)$ such that:

$$(B_2, A_2) ( \equiv ) (B_2, D_2)$$

and $\qquad (C_2, A_2) \mid \equiv \mid (B_2, D_2).$

Since however $\qquad (C_2, A_2) \mid \equiv \mid (B_1, A_1),$

we have $\qquad (B_1, A_1) \mid \equiv \mid (B_2, D_2),$

which together with the relation
$$(B_2, A_2)(\equiv)(B_2, D_2)$$
gives us
$$(B_1, A_1) \equiv (B_2, A_2).$$

If now we take instead two separation pairs $(A_1, B_1)$ and $(A_2, B_2)$ such that:
$$(A_1, B_1) \equiv (A_2, B_2),$$
then by using Theorem 175 in place of Theorem 174, we may prove that:
$$(A_2, B_2) \equiv (A_1, B_1).$$

Also, by a similar method to that employed in the case of inertia pairs, but using Theorem 177 in place of Theorem 176, we may prove that:
$$(B_1, A_1) \equiv (B_2, A_2).$$

Again, if $(A, B)$ be an inertia pair, we have
$$(A, B) | \equiv | (A, B)$$
and
$$(A, B)(\equiv)(A, B).$$
Thus we have
$$(A, B) \equiv (A, B).$$

A similar result obviously holds if $(A, B)$ be a separation pair.

Again if $(A, B)$ and $(A, C)$ be inertia pairs such that:
$$(A, B)(\equiv)(A, C),$$
then since
$$(A, C) | \equiv | (A, C),$$
we may write
$$(A, B) \equiv (A, C).$$

A similar result holds if $(A, B)$ and $(A, C)$ be separation pairs such that:
$$(A, B)\{\equiv\}(A, C).$$

Further, it is also clear that:
$$(A_1, B_1) | \equiv | (A_2, B_2)$$
implies
$$(A_1, B_1) \equiv (A_2, B_2),$$
both when $(A_1, B_1)$ and $(A_2, B_2)$ are inertia pairs and when they are separation pairs.

Again if $(A_1, B_1)$, $(A_2, B_2)$ and $(A_3, B_3)$ be inertia pairs such that:
$$(A_1, B_1) \equiv (A_2, B_2)$$
and
$$(A_2, B_2) \equiv (A_3, B_3),$$
then, by the definition of congruence, there is an inertia pair $(A_2, C_2)$ such that:
$$(A_1, B_1) | \equiv | (A_2, C_2) \qquad \ldots\ldots(1)$$
and
$$(A_2, B_2)(\equiv)(A_2, C_2). \qquad \ldots\ldots(2).$$

Also there is an inertia pair $(A_3, C_3)$ such that:

$$(A_2, B_2) | \equiv | (A_3, C_3) \qquad \ldots\ldots(3)$$

and
$$(A_3, B_3) ( \equiv ) (A_3, C_3) \qquad \ldots\ldots(4).$$

Now from (2) and (3) it follows, by Theorem 174, that there is an inertia pair $(A_3, D_3)$ such that:

$$(A_3, C_3) ( \equiv ) (A_3, D_3) \qquad \ldots\ldots(5)$$

and
$$(A_2, C_2) | \equiv | (A_3, D_3) \qquad \ldots\ldots(6).$$

But from (1) and (6) it follows that:

$$(A_1, B_1) | \equiv | (A_3, D_3) \qquad \ldots\ldots(7),$$

while from (4) and (5) it follows that:

$$(A_3, B_3) ( \equiv ) (A_3, D_3) \qquad \ldots\ldots(8).$$

Thus from (7) and (8) it follows that:

$$(A_1, B_1) \equiv (A_3, B_3).$$

A similar result may be proved for the case of separation pairs; using Theorem 175 in place of Theorem 174.

*Thus for inertia or separation pairs the general relation of congruence is a transitive relation.*

Again, if $(A, B)$ be any separation pair and $P$ be any inertia plane containing the separation line $AB$, there is one single optical parallelogram in $P$ having $B$ as centre and $A$ as one of its corners.

If $C$ be the corner opposite to $A$ then, by definition, $B$ is the mean of $A$ and $C$.

Thus, by Theorem 178, we have:

$$(A, B) | \equiv | (B, C).$$

But also by definition we have:

$$(B, A) \{ \equiv \} (B, C).$$

And so:
$$(A, B) \equiv (B, A).$$

We have not however a corresponding result in the case either of inertia or optical pairs since the elements in such pairs are asymmetrically related.

### THEOREM 179

*If $(O_1, D_1)$ and $(O_2, D_2)$ be inertia pairs while $(O_1, B_1)$ and $(O_2, B_2)$ are separation pairs which are before-conjugates to $(O_1, D_1)$ and $(O_2, D_2)$ respectively or else after-conjugates to $(O_1, D_1)$ and $(O_2, D_2)$ respectively; then:*

(1) *If* $\qquad (O_1, D_1) \equiv (O_2, D_2)$

*we shall also have* $\qquad (O_1, B_1) \equiv (O_2, B_2).$

(2) *If* $\qquad (O_1, B_1) \equiv (O_2, B_2)$

*we shall also have* $\qquad (O_1, D_1) \equiv (O_2, D_2)$.

Let us consider the first part of the theorem.

Since $\qquad (O_1, D_1) \equiv (O_2, D_2)$,

it follows by definition that there is a pair $(O_2, D')$ such that:

$$(O_1, D_1) \,|\equiv| \, (O_2, D')$$

and $\qquad (O_2, D_2)\,(\equiv)\,(O_2, D')$.

Then $(O_2, D')$ is an inertia pair and so, since $(O_1, B_1)$ is a separation pair which is conjugate to $(O_1, D_1)$, it follows, by Theorem 172, that there is a separation pair, say $(O_2, B')$, which is conjugate to $(O_2, D')$ and such that:

$$(O_1, B_1) \,|\equiv| \, (O_2, B').$$

Now if $D_2$ be *after* $O_2$ we shall also have $D'$ *after* $O_2$ and so $(O_2, D_2)$ and $(O_2, D')$ will be after-conjugates to $(O_2, B_2)$ and $(O_2, B')$ respectively.

Thus we shall have

$$(O_2, B_2)\{\equiv\}(O_2, B').$$

If, on the other hand, $D_2$ be *before* $O_2$ we shall also have $D'$ *before* $O_2$, and so $(O_2, D_2)$ and $(O_2, D')$ will be before-conjugates to $(O_2, B_2)$ and $(O_2, B')$ respectively.

Now by completing the optical parallelograms implied in the relation:

$$(O_2, D_2)\,(\equiv)\,(O_2, D'),$$

we see that in this case there will be inertia pairs, say $(O_2, A_2)$ and $(O_2, A')$ which will be after-conjugates to $(O_2, B_2)$ and $(O_2, B')$ respectively, and such that:

$$(O_2, A_2)\,(\equiv)\,(O_2, A').$$

Thus we have also in this case

$$(O_2, B_2)\{\equiv\}(O_2, B').$$

Combining this with the relation:

$$(O_1, B_1) \,|\equiv| \, (O_2, B'),$$

it follows by definition that:

$$(O_1, B_1) \equiv (O_2, B_2).$$

Thus the first part of the theorem is proved.

Consider now the second part of the theorem.

Since $\qquad (O_1, B_1) \equiv (O_2, B_2)$,

it follows by definition that there is a pair $(O_2, B')$ such that:

$$(O_1, B_1) \,|\equiv| \, (O_2, B')$$

and $\qquad (O_2, B_2)\{\equiv\}(O_2, B')$.

Then $(O_2, B')$ is a separation pair and so, since $(O_1, D_1)$ is an inertia pair which is conjugate to $(O_1, B_1)$, it follows, by Theorem 173, that there is an inertia pair, say $(O_2, D')$, which is conjugate to $(O_2, B')$ and such that:

$$(O_1, D_1) \, | \equiv | \, (O_2, D').$$

Now if $(O_1, B_1)$ and $(O_2, B_2)$ be before-conjugates to $(O_1, D_1)$ and $(O_2, D_2)$ respectively we shall have $D_1$ *after* $O_1$ and therefore $D'$ *after* $O_2$, and also we shall have $D_2$ *after* $O_2$.

Thus $(O_2, D_2)$ and $(O_2, D')$ will be after-conjugates to $(O_2, B_2)$ and $(O_2, B')$ respectively and so, since

$$(O_2, B_2) \{ \equiv \} (O_2, B'),$$

it follows that: $\qquad (O_2, D_2) \, ( \equiv ) \, (O_2, D').$

If, on the other hand, $(O_1, B_1)$ and $(O_2, B_2)$ be after-conjugates to $(O_1, D_1)$ and $(O_2, D_2)$ respectively, we shall have $D_1$ *before* $O_1$ and therefore $D'$ *before* $O_2$ and also we shall have $D_2$ *before* $O_2$.

Thus $(O_2, D_2)$ and $(O_2, D')$ will be before-conjugates to $(O_2, B_2)$ and $(O_2, B')$ respectively.

Now by completing the optical parallelograms implied in these relations we see that there are inertia pairs, say $(O_2, A_2)$ and $(O_2, A')$, which are after-conjugates to $(O_2, B_2)$ and $(O_2, B')$ respectively and such that $D_2, O_2$ and $A_2$ lie in one inertia line and also $D', O_2$ and $A'$ lie in one inertia line.

Now, since $\qquad (O_2, B_2) \{ \equiv \} (O_2, B'),$

we must have $\qquad (O_2, A_2) \, ( \equiv ) \, (O_2, A'),$

and therefore also in this case

$$(O_2, D_2) \, ( \equiv ) \, (O_2, D').$$

Combining this with the relation

$$(O_1, D_1) \, | \equiv | \, (O_2, D')$$

it follows by definition that:

$$(O_1, D_1) \equiv (O_2, D_2).$$

Thus the second part of the theorem is proved.

From Theorems 178 and 179 it follows that: if $(O_1, D_1)$ and $(O_2, D_2)$ be inertia pairs while $(O_1, B_1)$ and $(O_2, B_2)$ are separation pairs such that $(O_1, B_1)$ is a before-conjugate to $(O_1, D_1)$ and $(O_2, B_2)$ is an after-conjugate to $(O_2, D_2)$, then:

(1) If $\qquad\qquad (O_1, D_1) \equiv (D_2, O_2),$

we shall also have $\qquad (O_1, B_1) \equiv (O_2, B_2).$

(2) If $\qquad\qquad (O_1, B_1) \equiv (O_2, B_2),$

we shall also have $\qquad (O_1, D_1) \equiv (D_2, O_2).$

## THEOREM 180

*If $(A_1, B_1)$, $(A_2, B_2)$, $(B_1, C_1)$, $(B_2, C_2)$ be pairs such that:*

$$(A_1, B_1) \mid \equiv \mid (A_2, B_2)$$

*and* $$(B_1, C_1) \mid \equiv \mid (B_2, C_2),$$

*and if $C_1$ be distinct from $A_1$, then we shall also have*

$$(A_1, C_1) \mid \equiv \mid (A_2, C_2).$$

The elements $A_1$, $B_1$ and $C_1$ must lie in at least one general plane, say $P_1$, and since $A_2 B_2$ must either be parallel to $A_1 B_1$ or identical with it, while $B_2 C_2$ must either be parallel to $B_1 C_1$ or identical with it, it follows that there is a general plane, say $P_2$, either parallel to $P_1$ or identical with it which contains the elements $A_2$, $B_2$ and $C_2$.

Let $P'$ be a general plane parallel to $P_1$ and $P_2$, and therefore distinct from both, and let $(A', B')$ and $(B', C')$ be pairs in $P'$ such that:

$$(A_1, B_1) \square (A', B')$$

and $$(B_1, C_1) \square (B', C').$$

Then, by Theorem 126, since $A_1 A'$ and $C_1 C'$ cannot lie in the same general line (owing to $C_1$ being distinct from $A_1$ and both of them lying in $P_1$), it follows that:

$$(A_1, C_1) \square (A', C').$$

Thus $C'$ must be distinct from $A'$.

But now, since $A_2$, $B_2$ and $C_2$ lie in $P_2$ while $A'$, $B'$ and $C'$ lie in the parallel general plane $P'$, it follows that $A_2 B_2$ cannot be identical with $A'B'$, and $B_2 C_2$ cannot be identical with $B'C'$.

Thus, since $$(A_1, B_1) \mid \equiv \mid (A_2, B_2)$$

and $$(B_1, C_1) \mid \equiv \mid (B_2, C_2),$$

it follows that we must have

$$(A_2, B_2) \square (A', B')$$

and $$(B_2, C_2) \square (B', C').$$

Thus, since $C'$ is distinct from $A'$, it follows that:

$$(A_2, C_2) \square (A', C').$$

But we have seen that:

$$(A_1, C_1) \square (A', C'),$$

and so $$(A_1, C_1) \mid \equiv \mid (A_2, C_2).$$

Thus the theorem is proved.

## Remarks

One special case of this theorem deserves attention.

If $B_1$ be linearly between $A_1$ and $C_1$, it follows, by Theorems 72, 91 and 117, that $B'$ must be linearly between $A'$ and $C'$ and similarly $B_2$ must be linearly between $A_2$ and $C_2$.

We shall require this result in proving the next theorem.

Again, since the only congruence of optical pairs is co-directional, we may state the following result:

If $(A_1, B_1)$, $(A_2, B_2)$, $(B_1, C_1)$, $(B_2, C_2)$ be optical pairs such that:

$$(A_1, B_1) \equiv (A_2, B_2)$$

and
$$(B_1, C_1) \equiv (B_2, C_2),$$

then if $B_1$ be linearly between $A_1$ and $C_1$ we shall have $B_2$ linearly between $A_2$ and $C_2$ and also have

$$(A_1, C_1) \equiv (A_2, C_2).$$

## Theorem 181

*If $(A_1, B_1)$, $(A_2, B_2)$, $(B_1, C_1)$, $(B_2, C_2)$ be inertia pairs such that:*

$$(A_1, B_1) \equiv (A_2, B_2)$$

*and*
$$(B_1, C_1) \equiv (B_2, C_2),$$

*then if $B_1$ be linearly between $A_1$ and $C_1$, while $B_2$ is linearly between $A_2$ and $C_2$, we shall also have*

$$(A_1, C_1) \equiv (A_2, C_2).$$

If $a$ be an inertia line passing through $A_2$ and co-directional with $A_1 B_1$ we may take a separation line $b$ which passes through $A_2$ and is normal to both $a$ and $A_2 B_2$.

Let $D_2$ be an element in $b$ such that $(A_2, D_2)$ is conjugate to $(A_2, B_2)$ and let $(A_1, D_1)$ be a separation pair such that:

$$(A_1, D_1) \,|\equiv|\, (A_2, D_2).$$

Now $A_1 D_1$ is co-directional with $A_2 D_2$ while $A_1 B_1$ is co-directional with $a$ and so $A_1 D_1$ must be normal to $A_1 B_1$.

Since $A_1 B_1$ and $A_2 B_2$ are inertia lines, it follows that $A_1 B_1$ and $A_1 D_1$ lie in an inertia plane and also $A_2 B_2$ and $A_2 D_2$ lie in an inertia plane.

Then, since $(A_2, D_2)$ is conjugate to $(A_2, B_2)$, it follows that $D_2 B_2$ is an optical line.

Let $B_1'$ be an element in $A_1 B_1$ such that $(A_1, B_1')$ is an after- or before-conjugate to $(A_1, D_1)$ according as $(A_2, B_2)$ is an after- or before-conjugate to $(A_2, D_2)$.

Then, by Theorem 179, since

$$(A_2, D_2) \equiv (A_1, D_1),$$

we must have          $(A_2, B_2) \equiv (A_1, B_1').$

But               $(A_1, B_1) \equiv (A_2, B_2)$

and so            $(A_1, B_1) \equiv (A_1, B_1').$

Thus, since $A_1 B_1$ is an inertia line we must have $B_1'$ identical with $B_1$ and so, since $(A_1, B_1)$ is conjugate to $(A_1, D_1)$, it follows that $D_1 B_1$ is an optical line.

Now let the optical line through $C_1$ parallel to $B_1 D_1$ intersect $A_1 D_1$ in $F_1$ and let the separation line through $B_1$ parallel to $A_1 F_1$ intersect $C_1 F_1$ in $E_1$.

Then, since $B_1$ is linearly between $A_1$ and $C_1$, it follows, by Theorem 72, that $D_1$ is linearly between $A_1$ and $F_1$.

Let $(D_2, F_2)$ be a pair such that:

$$(D_1, F_1)\,|\equiv|\,(D_2, F_2).$$

Then, by the remarks at the end of Theorem 180, $D_2$ will be linearly between $A_2$ and $F_2$ and we shall also have

$$(A_1, F_1)\,|\equiv|\,(A_2, F_2).$$

Now let the optical line through $F_2$ parallel to $D_2 B_2$ intersect $A_2 B_2$ in $C_2'$, and let the separation line through $B_2$ parallel to $A_2 F_2$ intersect $F_2 C_2'$ in $E_2$.

Then we have          $(B_1, E_1)\,|\equiv|\,(D_1, F_1)$

and               $(D_1, F_1)\,|\equiv|\,(D_2, F_2),$

and therefore          $(B_1, E_1)\,|\equiv|\,(D_2, F_2).$

But we have also          $(D_2, F_2)\,|\equiv|\,(B_2, E_2),$

and so          $(B_1, E_1)\,|\equiv|\,(B_2, E_2).$

But now, since $B_1 E_1$ is parallel to $A_1 D_1$, it must be normal to $B_1 C_1$, and since $E_1 C_1$ is an optical line, it follows that $(B_1, E_1)$ is conjugate to $(B_1, C_1)$.

Similarly, since $B_2 E_2$ is parallel to $A_2 D_2$, it must be normal to $B_2 C_2'$ and, since $E_2 C_2'$ is an optical line, it follows that $(B_2, E_2)$ is conjugate to $(B_2, C_2')$.

But now, since $D_2$ is linearly between $A_2$ and $F_2$, it follows that $B_2$ is linearly between $A_2$ and $C_2'$.

If then $B_2$ be *after* $A_2$ we must have $C_2'$ *after* $B_2$, while if $B_2$ be *before* $A_2$ we must have $C_2'$ *before* $B_2$.

But, since $B_2$ is linearly between $A_2$ and $C_2$, it follows that if $B_2$ be *after* $A_2$ we must have $C_2$ *after* $B_2$, while if $B_2$ be *before* $A_2$ we must have $C_2$ *before* $B_2$.

Thus $C_2'$ is *after* or *before* $B_2$ according as $C_2$ is *after* or *before* $B_2$.

But, since $(B_1, C_1)$ and $(B_2, C_2)$ are inertia pairs such that:

$$(B_1, C_1) \equiv (B_2, C_2),$$

it follows that $C_2$ is *after* or *before* $B_2$ according as $C_1$ is *after* or *before* $B_1$ and therefore $C_2'$ is *after* or *before* $B_2$ according as $C_1$ is *after* or *before* $B_1$.

Thus, since $\qquad (B_1, E_1) \equiv (B_2, E_2),$

it follows, by Theorem 179, that:

$$(B_1, C_1) \equiv (B_2, C_2'),$$

and since $\qquad\qquad (B_1, C_1) \equiv (B_2, C_2),$

it follows that: $\qquad\qquad (B_2, C_2) \equiv (B_2, C_2').$

Thus, since these pairs lie in the same inertia line, we must have $C_2'$ identical with $C_2$.

But now $C_2$ will be *after* or *before* $A_2$ according as $C_1$ is *after* or *before* $A_1$ and so $(A_2, C_2)$ will be an after- or before-conjugate to $(A_2, F_2)$ according as $(A_1, C_1)$ is an after- or before-conjugate to $(A_1, F_1)$.

Thus, since $\qquad\qquad (A_1, F_1) \equiv (A_2, F_2),$

it follows, by Theorem 179, that:

$$(A_1, C_1) \equiv (A_2, C_2),$$

and so the theorem is proved.

### THEOREM 182

*If $(A_1, B_1)$, $(A_2, B_2)$, $(B_1, C_1)$, $(B_2, C_2)$ be separation pairs such that:*

$$(A_1, B_1) \equiv (A_2, B_2)$$

*and* $\qquad\qquad (B_1, C_1) \equiv (B_2, C_2),$

*then if $B_1$ be linearly between $A_1$ and $C_1$ while $B_2$ is linearly between $A_2$ and $C_2$ we shall also have*

$$(A_1, C_1) \equiv (A_2, C_2).$$

Let $(A_1, D_1)$ and $(A_2, D_2)$ be inertia pairs which are after-conjugates to $(A_1, B_1)$ and $(A_2, B_2)$ respectively.

Then, since $A_1 D_1$ and $A_2 D_2$ are inertia lines, it follows that $A_1 D_1$ and $A_1 B_1$ lie in an inertia plane and $A_2 D_2$ and $A_2 B_2$ lie in an inertia plane.

Since $(A_1, D_1)$ is conjugate to $(A_1, B_1)$, it follows that $B_1 D_1$ is an optical line, and similarly $B_2 D_2$ is an optical line.

Now let the optical line through $C_1$ parallel to $B_1 D_1$ intersect $A_1 D_1$

in $F_1$, and let the optical line through $C_2$ parallel to $B_2 D_2$ intersect $A_2 D_2$ in $F_2$.

Then, since $B_1$ is linearly between $A_1$ and $C_1$, it follows, by Theorem 72, that $D_1$ is linearly between $A_1$ and $F_1$.

Similarly $D_2$ is linearly between $A_2$ and $F_2$.

Let the inertia line through $B_1$ parallel to $A_1 F_1$ intersect $C_1 F_1$ in $E_1$, and let the inertia line through $B_2$ parallel to $A_2 F_2$ intersect $C_2 F_2$ in $E_2$.

Then, since $(A_1, D_1)$ is an after-conjugate to $(A_1, B_1)$, we must have $D_1$ *after* $A_1$ and, since $D_1$ is linearly between $A_1$ and $F_1$, we must have $F_1$ *after* $D_1$.

Thus we must have $E_1$ *after* $B_1$ and in a similar manner we can show that $E_2$ must be *after* $B_2$.

But now, since $(A_1, D_1)$ is conjugate to $(A_1, B_1)$, it follows that $A_1 D_1$ is normal to $A_1 B_1$, and since $B_1 E_1$ is parallel to $A_1 D_1$ while $A_1$, $B_1$ and $C_1$ lie in one general line, it follows that $B_1 E_1$ is normal to $B_1 C_1$.

Thus, since $C_1 E_1$ (that is $C_1 F_1$) is an optical line, it follows that $(B_1, E_1)$ is conjugate to $(B_1, C_1)$ and $(A_1, F_1)$ is conjugate to $(A_1, C_1)$.

Further, $D_1$ is *after* $A_1$ and $F_1$ is *after* $D_1$ and so $F_1$ is *after* $A_1$.

Thus $(B_1, E_1)$ and $(A_1, F_1)$ are after-conjugates to $(B_1, C_1)$ and $(A_1, C_1)$ respectively.

Similarly $(B_2, E_2)$ and $(A_2, F_2)$ are after-conjugates to $(B_2, C_2)$ and $(A_2, C_2)$ respectively.

But now, since $$(A_1, B_1) \equiv (A_2, B_2),$$

while $(A_1, D_1)$ and $(A_2, D_2)$ are after-conjugates to $(A_1, B_1)$ and $(A_2, B_2)$ respectively, it follows, by Theorem 179, that:

$$(A_1, D_1) \equiv (A_2, D_2).$$

Similarly, since $$(B_1, C_1) \equiv (B_2, C_2),$$

while $(B_1, E_1)$ and $(B_2, E_2)$ are after-conjugates to $(B_1, C_1)$ and $(B_2, C_2)$ respectively, it follows that:

$$(B_1, E_1) \equiv (B_2, E_2).$$

But we clearly have $(B_1, E_1)\,|\equiv|\,(D_1, F_1)$

and $$(B_2, E_2)\,|\equiv|\,(D_2, F_2).$$

Thus we have $$(D_1, F_1) \equiv (D_2, F_2).$$

But, since $D_1$ is linearly between $A_1$ and $F_1$, while $D_2$ is linearly between $A_2$ and $F_2$, it follows, by Theorem 181, that:

$$(A_1, F_1) \equiv (A_2, F_2).$$

We have however seen that $(A_1, F_1)$ and $(A_2, F_2)$ are after-conjugates to $(A_1, C_1)$ and $(A_2, C_2)$ respectively, and so it follows, by Theorem 179, that
$$(A_1, C_1) \equiv (A_2, C_2).$$
Thus the theorem is proved.

### THEOREM 183

*If $A$ and $B$ be two distinct elements and $E$ be any element in $AB$ distinct from $A$ and $B$, while $F$ is an element in $AB$ such that:*
$$(A, E) \mid \equiv \mid (F, B),$$
*then we shall have* $\qquad (A, F) \mid \equiv \mid (E, B).$

Let $a$ be a general line parallel to $AB$ and let a general line through $A$ intersect $a$ in $A'$ while parallel general lines through $B$ and $E$ intersect $a$ in $B'$ and $E'$ respectively.

Finally let a general line through $F$ parallel to $AA'$ or $BB'$ intersect $a$ in $F'$.

Now we clearly have
$$(E, E') \square (B, B'),$$
and so $\qquad (E, E') \mid \equiv \mid (B, B').$

But, since $E'$ and $A$ lie in parallel general lines, they must be distinct, and so, by Theorem 180,
$$(A, E') \mid \equiv \mid (F, B').$$

Now $F$ cannot coincide with $A$, for then $E$ would require to coincide with $B$, contrary to hypothesis, and so $FB'$ must be parallel to $AE'$. Thus we must have
$$(A, F) \square (E', B').$$

But we obviously have
$$(E, B) \square (E', B')$$
and so $\qquad (A, F) \mid \equiv \mid (E, B).$

Thus the theorem is proved.

### THEOREM 184

*If $A_0$, $A_1$ and $C$ be three distinct elements such that $A_1$ is linearly between $A_0$ and $C$ and if $A_2$, $A_3$, $A_4$, ... be elements such that:*

$A_1$ *is linearly between $A_0$ and $A_2$,*
$A_2$ *is linearly between $A_0$ and $A_3$,*
............................................
............................................

*and such that:* $\qquad (A_0, A_1) \equiv (A_1, A_2) \equiv (A_2, A_3) ...,$

R 20

*then there are not more than a finite number of the elements* $A_1$, $A_2$, $A_3$, ... *linearly between* $A_0$ *and* $C$.

It is evident that all the series of elements $A_1$, $A_2$, $A_3$, ... lie in the general line $A_0 C$ which we shall call $a$.

We shall first prove the theorem for the case where $a$ is an inertia line and $C$ is *after* $A_0$.

We shall suppose that $a$ lies in an inertia plane $P$.

Now since $A_1$ is linearly between $A_0$ and $C$ we must have $A_1$ *after* $A_0$, and so if we take two generators of $P$ of opposite sets passing through $A_0$ and $A_1$ respectively, they will intersect in some element, say $B_0$, which will be *after* $A_0$ and *before* $A_1$ and must lie outside $a$.

Let $b$ be an inertia line passing through $B_0$ and parallel to $a$ and let optical lines parallel to $A_0 B_0$ and passing through $A_1$, $A_2$, $A_3$, ... intersect $b$ in the elements $B_1$, $B_2$, $B_3$, ... respectively.

Now, since $A_1$ is *after* $A_0$ and, since further:

$A_1$ is linearly between $A_0$ and $A_2$,

$A_2$ is linearly between $A_0$ and $A_3$,

..........................................

..........................................

it follows that:

$A_1$ is *after* $A_0$;     $A_2$ is *after* $A_1$;     $A_3$ is *after* $A_2$;   ....

Thus, since        $(A_0, A_1) \equiv (A_1, A_2) \equiv (A_2, A_3) ...,$

it follows that:        $A_1$ is the mean of $A_0$ and $A_2$,

$A_2$ is the mean of $A_1$ and $A_3$,

..........................................

..........................................

But now, by construction, we have

$$(A_0, A_2) \,\square\, (B_0, B_2),$$

and so, since $A_1$ is the mean of $A_0$ and $A_2$, and since $A_1 B_1$ is parallel to $A_0 B_0$ and $A_2 B_2$, it follows that $B_1$ is the mean of $B_0$ and $B_2$ and so, by Theorem 80, $A_2 B_1$ is parallel to $A_1 B_0$.

Similarly $A_3 B_2$ is parallel to $A_2 B_1$ and so on.

Thus, since $A_1 B_0$ is an optical line, it follows that $A_2 B_1$, $A_3 B_2$, ... are all optical lines and so $A_1$, $A_2$, $A_3$, ... mark steps taken along $a$ with respect to $b$.

But, since $a$ and $b$ do not intersect, it follows, by Post. XVII, that $C$ may be surpassed in a finite number of steps taken from $A_0$.

Thus there cannot be more than a finite number of the elements $A_1$, $A_2$, $A_3$, ... linearly between $A_0$ and $C$.

Similarly if $C$ be *before* $A_0$ the same result follows by using the $(b)$ form of Post. XVII, and so the theorem is proved for the case where $a$ is an inertia line.

Consider next the case where $a$ is a separation line and let $b$ be any inertia line which passes through $A_0$.

Then $a$ and $b$ determine an inertia plane which we shall call $P$.

Now one of the generators of $P$ which pass through $A_1$ intersects $b$ in an element which lies in the $\alpha$ sub-set of $A_1$, while the other generator intersects $b$ in an element which lies in the $\beta$ sub-set of $A_1$.

Let the former of these generators be called $f_1$, and let it intersect $b$ in the element $A_1'$.

Then, since $A_1$ does not lie in $b$, it follows that $A_1'$ is *after* $A_1$ and so, since $A_0 A_1$ is a separation line, we must also have $A_1'$ *after* $A_0$.

Let $f_0, f_2, f_3, f_4, \ldots$ and $f_c$ be generators of $P$ parallel to $f_1$ and passing through $A_0, A_2, A_3, A_4, \ldots$ and $C$ respectively.

Further, let $f_2, f_3, f_4, \ldots$ and $f_c$ intersect $b$ in $A_2', A_3' A_4', \ldots$ and $C'$ respectively.

Then, since $A_1'$ is *after* $A_0$, it follows that $f_1$ is an after-parallel of $f_0$, and since $A_1$ is linearly between $A_0$ and $C$, it follows that $f_c$ is an after-parallel of $f_1$ and so $C'$ is *after* $A_1'$.

Further,    $A_1$ is linearly between $A_0$ and $A_2$,

$A_2$ is linearly between $A_0$ and $A_3$,

............................................

............................................

Thus we have    $f_1$ is an after-parallel of $f_0$,

$f_2$ is an after-parallel of $f_1$,

$f_3$ is an after-parallel of $f_2$,

............................................

............................................

Thus we have

$A_1$ is linearly between $A_0$ and $A_2$,

$A_2$ is linearly between $A_1$ and $A_3$,

$A_3$ is linearly between $A_2$ and $A_4$,

............................................

............................................

But, since    $(A_0, A_1) \equiv (A_1, A_2) \equiv (A_2, A_3) \ldots$,

it follows that:    $A_1$ is the mean of $A_0$ and $A_2$,

$A_2$ is the mean of $A_1$ and $A_3$,

............................................

............................................

Thus, by Theorem 81,

$$A_1' \text{ is the mean of } A_0 \text{ and } A_2',$$
$$A_2' \text{ is the mean of } A_1' \text{ and } A_3',$$
$$\dots\dots\dots\dots\dots\dots\dots\dots\dots\dots\dots\dots\dots\dots\dots\dots$$
$$\dots\dots\dots\dots\dots\dots\dots\dots\dots\dots\dots\dots\dots\dots\dots\dots$$

and so　　　　$(A_0, A_1') \equiv (A_1', A_2') \equiv (A_2', A_3') \dots .$

Thus, by the first case of the theorem, there cannot be more than a finite number of the elements $A_1'$, $A_2'$, $A_3'$ linearly between $A_0$ and $C'$.

But each of these elements which is linearly between $A_0$ and $C'$ corresponds to one of the series $A_1, A_2, A_3, \dots$ which is linearly between $A_0$ and $C$, while any one which is not linearly between $A_0$ and $C'$ corresponds to one of the series $A_1, A_2, A_3, \dots$ which is not linearly between $A_0$ and $C$.

Thus there are not more than a finite number of the elements $A_1$, $A_2, A_3, \dots$ linearly between $A_0$ and $C$, and so the theorem holds when $a$ is a separation line.

As regards the case where $a$ is an optical line and $C$ is *after* $A_0$ we may proceed just as we have done for the case where $a$ is a separation line.

In this case $a$ is one of the generators of the inertia plane $P$, while $f_0, f_1, f_2, \dots f_c$ will be generators of the opposite set.

The result then follows in a similar manner.

In the case where $a$ is an optical line and $C$ is *before* $A_0$ we also make use of a similar method except that the element $C'$ in the inertia line $b$ will be *before* $A_0$ instead of *after* it.

Thus the theorem holds in all cases.

## REMARKS

It will be observed that the above theorem is equivalent to the *Axiom of Archimedes* and has been deduced by the help of Post. XVII.

In our remarks on the introduction of this postulate, its analogy to the Axiom of Archimedes was pointed out together with the fact that the postulate contains no reference to *congruence*.

Having defined congruence of pairs we are able to deduce the Axiom of Archimedes in the usual form as given above.

*Definitions.* If $A$ and $B$ be two distinct elements, then the set of all elements lying linearly between $A$ and $B$ will be called the *segment AB*.

The elements $A$ and $B$ will be called the *ends* of the segment, but are not included in it.

The set of elements obtained by including the ends will be called a *linear interval*.

Since an element which is linearly between two elements $A$ and $B$ is linearly between $B$ and $A$, it follows, from the definition, that the segment $BA$ is the same as the segment $AB$, even in those cases where one of the two elements: $A$, $B$ is *after* the other.

The same remark applies to linear intervals.

If $A$ and $B$ be two distinct elements, then the set of elements such as $X$ where $B$ is linearly between $A$ and $X$ may be called the *prolongation of the segment $AB$ beyond $B$.*

Such a set of elements will also be spoken of as a *general half-line.*

The element $B$ will be called the *end* of the general half-line.

We shall describe segments, linear intervals and general half-lines as *optical, inertia,* or *separation,* according as they lie in optical, inertia, or separation lines.

It is easy to see that any element $B$ in a general line $a$ divides the remaining elements of the general line into two sets such that $B$ is linearly between any two elements of opposite sets, but is not linearly between any two elements of the same set.

For let $P$ be any inertia plane which contains $a$ and let $b$ be a generator of $P$ which passes through $B$, and which, in case $a$ is an optical line, we shall suppose to be distinct from $a$.

Then through every element of $a$ which is distinct from $B$ there will pass an optical line which is parallel to $b$.

Those elements of $a$ which lie in optical lines which are after-parallels of $b$ constitute the one set and those which lie in optical lines which are before-parallels of $b$ constitute the other set.

It is then obvious, from the definition of linearly between, that $B$ is linearly between any two elements of $a$ belonging to opposite sets, but is not linearly between any two elements of $a$ belonging to the same set.

It is clear that these sets are general half-lines.

If elements $X$ and $Y$ lie in the same general half-line whose end is $B$, they will be said to lie *on the same side* of $B$.

If, on the other hand, $B$ be linearly between $X$ and $Y$, then these elements will be said to lie *on opposite sides* of $B$.

A general half-line whose end is $B$ and which contains an element $X$ may be denoted by the *general half-line $BX$.*

It is also easy to see that any general line $b$ in a general plane $P$ divides the remaining elements of $P$ into two sets such that if $A$ and $C$ be any two elements of opposite sets then $b$ will intersect $AC$ in an element linearly between $A$ and $C$; while if $A$ and $A'$ be two elements of the same set, then $b$ will not intersect $AA'$ in any element linearly between $A$ and $A'$.

For consider the elements of $P$ exclusive of those lying in $b$ and, when dealing with them, let us use the expression: "$A$ is opposite to $C$" as an abbreviation for: "$b$ intersects $AC$ in an element linearly between $A$ and $C$".

Let $B$ be such an element of intersection and let $A'$ be a second element which is opposite to $C$.

If $A'$ should happen to lie in $AC$ then, from what we have just shown, it follows that $A'$ is not opposite to $A$.

If next we take the case where $A'$ does not lie in $AC$ then, since $B$ is linearly between $A$ and $C$ and, since neither $A$, $C$ nor $A'$ lie in $b$, it follows, by Theorem 127 (2), that $A'$ is not opposite to $A$. Thus (i) *in all cases, if $A$ and $A'$ are both opposite to $C$, then $A'$ is not opposite to $A$.*

Next suppose that $A$ is opposite to $C$ while $A_1$ is another element of $P$ which is not opposite to $A$ and does not lie in $b$.

If $A_1$ should happen to lie in $AC$ then, from the linear analogue already considered it follows that $A_1$ is opposite to $C$.

If $A_1$ does not lie in $AC$ then, since $B$ is linearly between $A$ and $C$ and since neither $A$, $C$ nor $A_1$ lie in $b$, it follows, by Theorem 127 (1), that $A_1$ is opposite to $C$. Thus (ii) *in all cases, if $A$ is opposite to $C$ while $A_1$ is not opposite to $A$, we must have $A_1$ opposite to $C$.*

If $A_2$ be a second element which is not opposite to $A$, it follows by (ii) that $A_2$ is opposite to $C$.

But, since both $A_1$ and $A_2$ are opposite to $C$, it follows by (i) that $A_2$ is not opposite to $A_1$. Thus all the elements which are not opposite to $A$ are opposite to $C$ and no one of them is opposite to any other of them.

Similarly, all the elements which are not opposite to $C$ are opposite to $A$, $A_1$, $A_2$, etc. and no one of them is opposite to any other of them.

This shows that the general plane $P$ is divided by the general line $b$ in the manner above stated.

If elements $X$ and $Y$ lie in the general plane $P$, but not in the general line $b$, they will be said to lie *on the same side* of $b$ if they both lie in the same set and will be said to lie *on opposite sides* of $b$ if $X$ lies in one of the sets and $Y$ in the other set.

*Definition.* If a general line $b$ lies in a general plane $P$, then either of the sets of elements on one side of $b$ will be called a *general half-plane*.

The general line $b$ will be called the *boundary* of the general half-plane.

The following important result which may be conveniently expressed in the nomenclature of general half-lines can be easily proved.

If $(A_1, B_1)$, $(A_2, B_2)$, $(A_1, C_1)$, $(A_2, C_2)$ be inertia, optical or separation pairs such that:

$$(A_1, B_1) \equiv (A_2, B_2),$$
and $$(A_1, C_1) \equiv (A_2, C_2),$$

then if $B_1$ be linearly between $A_1$ and $C_1$ and if $C_2$ lies in the general half-line $A_2 B_2$, we shall also have $B_2$ linearly between $A_2$ and $C_2$.

In the case of optical pairs the above congruences imply that $A_1 B_1$ and $A_2 B_2$ are the same or parallel optical lines, but nothing of this sort is implied in the case of inertia or separation pairs.

In all cases there is an element, say $C_2'$, in $A_2 B_2$ and on the opposite side of $B_2$ to that on which $A_2$ lies and such that

$$(B_1, C_1) \equiv (B_2, C_2').$$

Then in all cases it follows that:

$$(A_1, C_1) \equiv (A_2, C_2'),$$
and so $$(A_2, C_2) \equiv (A_2, C_2').$$

But $C_2$ and $C_2'$ both lie in $A_2 B_2$ and on the same side of $A_2$, and must therefore be identical.

Thus $B_2$ is linearly between $A_2$ and $C_2$, and

$$(B_1, C_1) \equiv (B_2, C_2).$$

*Definitions.* If $(A, B)$ and $(C, D)$ be inertia or optical pairs in which $B$ is *after* $A$ and $D$ *after* $C$, or if $(A, B)$ and $(C, D)$ be separation pairs, then, in respect of magnitude:

(1) If $(A, B) \equiv (C, D)$ we shall say that the segment $AB$ is *equal to* the segment $CD$.

(2) If $(A, B) \equiv (C, E)$ where $E$ is any element linearly between $C$ and $D$, we shall say that the segment $AB$ is *less than* the segment $CD$.

(3) If $(A, B) \equiv (C, F)$ where $F$ is any element such that $D$ is linearly between $C$ and $F$, we shall say that the segment $AB$ is *greater than* the segment $CD$.

In the case of separation or inertia segments we must always have either:

$$AB \text{ is equal to } CD,$$
or $$AB \text{ is less than } CD,$$
or $$AB \text{ is greater than } CD.$$

In the case of optical segments, however, this is only true provided they lie in co-directional optical lines.

Again, if $(A, B)$ and $(C, D)$ be inertia or optical pairs in which $B$ is *after* $A$ and $D$ *after* $C$, or if they be separation pairs, and if $E$, $F$, $G$ be elements such that $F$ is linearly between $E$ and $G$ while

$$(A, B) \equiv (E, F)$$

and $$(C, D) \equiv (F, G),$$

we shall say that *the segment $EG$ is equal to the sum of the segments $AB$ and $CD$ in respect of magnitude.*

It is evident that two optical segments can only have a sum in this sense provided they lie in co-directional optical lines, whereas two inertia segments or two separation segments always have a sum.

In the above definitions the words *linear interval* may be substituted for the word *segment*.

If $A$ and $B$ be any two distinct elements and if $C$ be any element such that $B$ is linearly between $A$ and $C$ and if $a$ and $b$ be taken to denote the segments $AB$ and $BC$ respectively, then we may express the result of Theorem 183 in the form:

$$a + b = b + a.$$

If $D$ be any other element such that $C$ is linearly between $B$ and $D$, and if we denote the segment $CD$ by $c$; then, by application of Theorem 183 alternately to a pair of segments lying towards one end of the total interval $AD$ and then to a pair of segments lying towards the other end, we obtain successively:

$$a + b + c = b + a + c,$$
$$b + a + c = b + c + a,$$
$$b + c + a = c + b + a,$$
$$c + b + a = c + a + b,$$
$$c + a + b = a + c + b,$$
$$a + c + b = a + b + c.$$

It thus appears that we may regard both the Commutative Law and the Associative Law as holding for the addition of segments, or of linear intervals.

Having thus introduced the idea of a segment (or linear interval) being equal to the sum of two others we can obviously have any *multiple* and also (as follows from the remarks at the end of Theorem 81) any *sub-multiple* of a given segment (or linear interval): using the terms "multiple" and "sub-multiple" in the ordinary sense.

We may also clearly have a segment (or linear interval) equal to any proper or improper fractional part of a given one.

The criterion of *proportion* given by Euclid, and which is probably due to Eudoxos, is clearly applicable in our geometry.

This criterion is as follows:

If we have four magnitudes of which the first and second are *of one kind**\* and the third and fourth also *of one kind* and if, for all values of $m$ and $n$ (where $m$ and $n$ are integers), we have $m$ (first magnitude) is greater than, equal to, or less than $n$ (second magnitude) according as $m$ (third magnitude) is greater than, equal to, or less than $n$ (fourth magnitude), then the four magnitudes are *in proportion* in such a way that the first magnitude is to the second as the third is to the fourth.

We shall also express the proportion of the four magnitudes by saying that the *ratio* of the first to the second is equal to the *ratio* of the third to the fourth.

This might be regarded merely as another form of words expressing the same fact, without assigning any specific definition to the term *ratio* taken by itself; since, in all cases where the term is used, it is possible to get rid of it by means of a circumlocution. As, however, it is desirable not to use a technical term without definition, we may define *ratio*, when employed in the Euclidean sense, thus:

*Definition.* The *ratio* of two magnitudes of one kind is the mode of distribution of the multiples $m$ of the one magnitude among the multiples $n$ of the other in respect of the relations of *greater than, equal to,* and *less than* for all integral values of $m$ and $n$.

Later on, for purposes of manipulation, we shall find it convenient to represent ratios by what are called *real numbers* and to associate positive and negative signs with them, but, in the mean time this is unnecessary.

If, in a proportion, for some particular values of $m$ and $n$ (say $m_1$ and $n_1$) we should have $m_1$ (first magnitude) is equal to $n_1$ (second magnitude) and, along with that, $m_1$ (third magnitude) is equal to $n_1$ (fourth magnitude), then the first and second magnitudes are commensurable as are also the third and fourth, and their common ratio is that of $n_1$ to $m_1$: written $n_1 : m_1$.

This, however, is by no means always so and, when it is not the case,

---

\* On this point see M. J. M. Hill's *Theory of Proportion*, Article 3.

It is very important to observe the sense in which the words "of one kind" are employed in the above criterion.

Thus, *segments of optical lines can only be deemed magnitudes of one kind provided that the optical lines are co-directional.*

the magnitudes are said to stand in an incommensurable ratio to one another.

The criterion then reduces to the form that, for all integral values of $m$ and $n$ we have $m$ (first magnitude) is greater than or less than $n$ (second magnitude) according as $m$ (third magnitude) is greater than or less than $n$ (fourth magnitude).

It has been shown by Stolz that this is a sufficient criterion even when the magnitudes stand in a commensurable ratio.

The abstract theory of proportion, as treated by Euclid in his fifth book (or by the late Professor M. J. M. Hill in his work on the subject), will be assumed in what follows and we shall concern ourselves only with the application of it to our geometry.

Certain results regarding the proportion of segments may easily be shown to hold for all types of general line, by using methods such as are employed in the following theorem.

## THEOREM 185

*If $O$, $A_1$ and $B_1$ be three distinct elements not lying in one general line, while $C_1$ is any other element lying in the general half-line $OA_1$, and if a general line through $C_1$ parallel to $A_1B_1$ intersects $OB_1$ in an element $D_1$, then*

(1) *segment $OA_1$ : segment $OC_1$ = segment $OB_1$ : segment $OD_1$;*

(2) *segment $OA_1$ : segment $OC_1$ = segment $A_1B_1$ : segment $C_1D_1$.*

In the general half-line $OA_1$ let elements $A_2$, $A_3$, ... $A_m$ be taken such that:

$$A_1 \text{ is the mean of } O \text{ and } A_2,$$
$$A_2 \text{ is the mean of } A_1 \text{ and } A_3,$$
$$\cdots\cdots\cdots\cdots\cdots\cdots\cdots\cdots\cdots\cdots\cdots$$
$$\cdots\cdots\cdots\cdots\cdots\cdots\cdots\cdots\cdots\cdots\cdots$$
$$A_{m-1} \text{ is the mean of } A_{m-2} \text{ and } A_m,$$

and let general lines through $A_2$, $A_3$, ... $A_m$ be taken parallel to $A_1B_1$ and meeting $OB_1$ in the elements $B_2$, $B_3$, ... $B_m$ respectively. Then we know that:

$$B_1 \text{ is the mean of } O \text{ and } B_2,$$
$$B_2 \text{ is the mean of } B_1 \text{ and } B_3,$$
$$\cdots\cdots\cdots\cdots\cdots\cdots\cdots\cdots\cdots\cdots\cdots$$
$$\cdots\cdots\cdots\cdots\cdots\cdots\cdots\cdots\cdots\cdots\cdots$$
$$B_{m-1} \text{ is the mean of } B_{m-2} \text{ and } B_m.$$

Then          segment $OA_m = m$ (segment $OA_1$)

and            segment $OB_m = m$ (segment $OB_1$).

Similarly in the general half-line $OC_1$ (i.e. $OA_1$) let elements $C_2$, $C_3$, ... $C_n$ be taken such that:

$C_1$ is the mean of $O$ and $C_2$,

$C_2$ is the mean of $C_1$ and $C_3$,

.......................................

.......................................

$C_{n-1}$ is the mean of $C_{n-2}$ and $C_n$,

and let general lines through $C_2$, $C_3$, ... $C_n$ be taken parallel to $C_1 D_1$

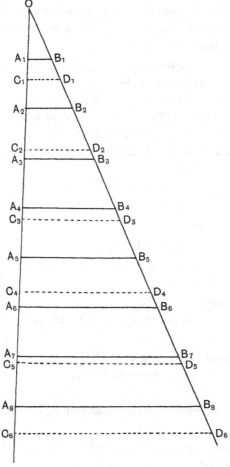

Fig. 46.

and meeting $OC_1$ in the elements $D_2$, $D_3$, ... $D_n$ respectively. Then, as before, we get

segment $OC_n = n$ (segment $OC_1$),

and        segment $OD_n = n$ (segment $OD_1$).

But now we shall have

        (i) $C_n$ linearly between $O$ and $A_m$,

or         (ii) $C_n$ identical with $A_m$,

or         (iii) $A_m$ linearly between $O$ and $C_n$,

according respectively as we have

        (i) $D_n$ linearly between $O$ and $B_m$,

or         (ii) $D_n$ identical with $B_m$,

or         (iii) $B_m$ linearly between $O$ and $D_n$.

Thus segment $OA_m$ is greater than, equal to, or less than segment $OC_n$ according as segment $OB_m$ is greater than, equal to, or less than segment $OD_n$; which is the criterion that

      segment $OA_1$ : segment $OC_1$ = segment $OB_1$ : segment $OD_1$.

This proves the first part of the theorem.

In order to prove the second part of the theorem let the general line through $D_1$ parallel to $OA_1$ intersect $A_1B_1$ in the element $E$ and let the general line through $E$ parallel to $OB_1$ intersect $OA_1$ in the element $F$.

Then clearly we have

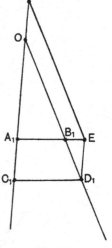

$$(C_1, D_1) \,|\equiv|\, (A_1, E)$$

and     $$(C_1, A_1) \,|\equiv|\, (O, F).$$

Thus, by Theorem 183,

$$(C_1, O) \,|\equiv|\, (A_1, F).$$

But, since $EF$ is parallel to $B_1O$, we have, by the first part of the theorem,

segment $A_1O$ : segment $A_1F =$

        segment $A_1B_1$ : segment $A_1E$.

Thus, since

      segment $A_1F =$ segment $C_1O$

and     segment $A_1E =$ segment $C_1D_1$,

we have

    segment $A_1O$ : segment $C_1O =$ segment $A_1B_1$ : segment $C_1D_1$.

That is

    segment $OA_1$ : segment $OC_1 =$ segment $A_1B_1$ : segment $C_1D_1$,

as was to be proved.

Fig. 47.

## REMARKS

It should be noted that certain transformations of the above results may be deduced by the abstract theory of proportion in all cases; while certain other transformations are only permissible in those cases where the four terms of the proportion are all segments of the same kind.

Thus if all four terms are either segments of separation lines, or else all segments of inertia lines, part (1) may be transformed into:

segment $OA_1$ : segment $OB_1$ = segment $OC_1$ : segment $OD_1$;

while part (2) may be transformed into:

segment $A_1B_1$ : segment $OA_1$ = segment $C_1D_1$ : segment $OC_1$;

but these would not be permissible in other cases.

Another important point to be observed is that if $OA_1$ should be a separation line and $OB_1$ an inertia line normal to $OA_1$ (or conversely), while $A_1B_1$ is an optical line, then $C_1D_1$ will also be an optical line and so it follows that: *separation segments are proportional to their conjugate inertia segments.*

## THEOREM 186

*If $B$ and $C$ be two distinct elements in a separation line and $O$ be their mean, and if $A$ be any element in a separation line $a$ which passes through $O$ and is normal to $BC$, then:*

$$(A, B) \equiv (A, C).$$

Since $a$ is normal to $BC$ and since they are both separation lines, it follows that $a$ and $BC$ lie in a separation plane, say $S$.

If the element $A$ should happen to coincide with $O$, then, since $BC$ is a separation line, the theorem obviously holds.

Suppose next that $A$ does not coincide with $O$ and let $d$ be an inertia line passing through $A$ and normal to $S$.

Let $P$ be the inertia plane containing $a$ and $d$.

Now, since $d$ is normal to $S$, it follows that $BC$ is normal to $d$, and since $BC$ is also normal to $a$, and since $a$ and $d$ intersect and lie in $P$, it follows that $BC$ is normal to $P$.

Let $D$ be the one single element common to $d$ and the $\alpha$ sub-set of $B$.

Then $BC$ and $BD$ determine a general plane, say $Q$, which must be either an optical plane or an inertia plane, since $BD$ is an optical line.

But now $P$ and $Q$ have the general line $OD$ in common, and since $BC$ is normal to $P$, it follows that $BC$ is normal to $OD$.

If $Q$ were an optical plane $OD$ would require to be an optical line, while if $Q$ were an inertia plane $OD$ would be an inertia line.

But, since $BD$ is an optical line in $Q$ and since $BD$ and $OD$ intersect, it follows that $Q$ cannot be an optical plane.

Thus $Q$ must be an inertia plane and $OD$ must be an inertia line normal to the separation line $BC$.

Thus, since $O$ is the mean of $B$ and $C$, it follows that $B$, $C$ and $D$ are three corners of an optical parallelogram of which $O$ is the centre.

Thus $CD$ is an optical line.

But, since $AD$ is normal to $S$, it must be normal to both $AB$ and $AC$.

Also, since $D$ is in the $\alpha$ sub-set of $B$ and is distinct from $B$, it follows that $D$ is *after* $B$.

Thus, since $AB$ is a separation line while $AD$ is an inertia line, it follows that $D$ is *after* $A$, and accordingly $(A, D)$ is an after-conjugate to both $(A, B)$ and $(A, C)$.

Thus we have $$(A, B) \equiv (A, C),$$

and so the theorem is proved.


## THEOREM 187

*If $A$, $B$ and $C$ be three distinct elements which lie in a separation plane $S$, but do not all lie in one general line, and if $O$ be the mean of $B$ and $C$ while*

$$(A, B) \equiv (A, C),$$

*then $AO$ is normal to $BC$.*

Let $d$ be an inertia line passing through $A$ and normal to $S$ and let $P$ be the inertia plane containing $d$ and $AO$.

Then $d$ is normal to both $AB$ and $AC$ and, since

$$(A, B) \equiv (A, C),$$

there is one definite element, say $D$, in $d$ such that $(A, D)$ is an after-conjugate to both $(A, B)$ and $(A, C)$.

Thus $BD$ and $CD$ are optical lines and, since they intersect, they must lie in an inertia plane, say $Q$.

But now, since $O$ is the mean of $B$ and $C$, it follows that $B$, $C$ and $D$ are three corners of an optical parallelogram whose centre is $O$, and therefore $BC$ is normal to $OD$.

But $OD$ is common to both $Q$ and $P$, while $BC$ (since it lies in $S$) is normal to $AD$, which also lies in $P$.

Thus $BC$ is normal to two intersecting general lines in $P$ and therefore $BC$ is normal to $P$.

But $AO$ lies in $P$ and therefore $AO$ is normal to $BC$.

Thus the theorem is proved.

## THEOREM 188

*If $A$, $B$ and $C$ be three distinct elements which lie in a separation plane S, but do not all lie in one general line, and if*

$$(A, B) \equiv (A, C),$$

*and if $O$ be an element in $BC$ such that $AO$ is normal to $BC$, then $O$ is the mean of $B$ and $C$.*

Let $O'$ be the mean of $B$ and $C$.

Then, by Theorem 187, $AO'$ is normal to $BC$, and, by hypothesis, $AO$ is normal to $BC$.

But both $AO'$ and $AO$ pass through the element $A$ and lie in the separation plane $S$ and we have already seen that there is only one general line in a given separation plane which passes through a given element and is normal to another general line in the separation plane.

Thus $AO$ must be identical with $AO'$ and therefore $O$ must be identical with $O'$.

It follows that $O$ is the mean of $B$ and $C$ and so the theorem is proved.

*Definition.* If $A$, $B$, $C$ be three distinct elements which do not all lie in one general line, then the three segments $AB$, $BC$, $CA$, together with the three elements $A$, $B$, $C$, will be called a *general triangle*, or briefly a *triangle* in an inertia optical, or separation plane, as the case may be.

The elements $A$, $B$, $C$ will be called the *corners* while the segments $AB$, $BC$, $CA$ will be called the *sides* of the general triangle.

## THEOREM 189

*If $A_1$, $B_1$, $C_1$ be the corners of a triangle in a separation plane $P_1$ and $A_2$, $B_2$, $C_2$ be the corners of a triangle in a separation plane $P_2$ and if further*

$$(C_1, A_1) \equiv (C_2, A_2),$$
$$(C_1, B_1) \equiv (C_2, B_2),$$

*while $B_1 C_1$ is normal to $A_1 C_1$, and $B_2 C_2$ is normal to $A_2 C_2$, then we shall also have*

$$(A_1, B_1) \equiv (A_2, B_2).$$

In order to prove this theorem we shall consider a number of special cases on which the general proof is made to depend.

CASE I. *$B_2$ identical with $B_1$ and $C_2$ identical with $C_1$, while $P_2$ is identical with $P_1$.*

In this case, since the separation lines $A_1 C_1$ and $A_2 C_1$ are both normal to $B_1 C_1$ and both lie in the separation plane $P_1$ and pass through the element $C_1$, they must be identical.

If further $A_2$ should coincide with $A_1$ the result is obvious, and so we shall suppose that $A_2$ does not coincide with $A_1$.

Now, since $(C_1, A_1)$, etc. lie in the separation plane $P_1$, they must all be separation pairs and, since in this case

$$(C_1, A_2) \equiv (A_2, C_1),$$

it follows that: $\qquad (A_2, C_1) \equiv (C_1, A_1).$

Thus $C_1$ must be the mean of $A_1$ and $A_2$ and therefore, by Theorem 186, we have

$$(B_1, A_1) \equiv (B_1, A_2),$$

or $\qquad (A_1, B_1) \equiv (A_2, B_2).$

CASE II.  *$B_2$ identical with $B_1$ and $C_2$ identical with $C_1$, while $P_1$ and $P_2$ lie in the same separation threefold $W$.*

If $P_2$ should be identical with $P_1$ this case reduces to Case I, and so we shall suppose them distinct.

Now, since $A_1$, $B_1$ and $A_2$ are three distinct elements in $W$ which do not lie in one general line, it follows that $A_1$, $B_1$ and $A_2$ lie in a separation plane, say $R$, which must be distinct from both $P_1$ and $P_2$, since these latter two separation planes are supposed distinct.

Similarly $A_1$, $C_1$ and $A_2$ lie in a separation plane, say $S$, which is also distinct from $P_1$ and $P_2$.

Now let $O$ be the mean of $A_1$ and $A_2$.

Then, by Theorem 187, since

$$(C_1, A_1) \equiv (C_1, A_2),$$

it follows that $C_1 O$ is normal to $A_1 A_2$.

But, since $B_1 C_1$ is normal to $A_1 C_1$ and to $A_2 C_1$ which are distinct intersecting separation lines, it follows that $B_1 C_1$ is normal to $S$ and therefore must be normal to $A_1 A_2$.

Thus $A_1 A_2$ is normal to the two intersecting separation lines $B_1 C_1$ and $C_1 O$ and must therefore be normal to every general line in the general plane containing them.

It follows that $A_1 A_2$ is normal to $B_1 O$.

But now, by Theorem 186, since $O$ is the mean of $A_1$ and $A_2$ and, since $B_1$, $A_1$, $A_2$ lie in a separation plane, it follows that:

$$(B_1, A_1) \equiv (B_1, A_2),$$

or $\qquad (A_1, B_1) \equiv (A_2, B_2).$

CASE III. *$C_2$ identical with $C_1$ and $P_2$ identical with $P_1$.*

Let $b$ be a separation line passing through $C_1$ and normal to $P_1$ and let $B'$ be an element in $b$ such that:

$$(C_1, B') \equiv (C_1, B_1).$$

Then we shall also have

$$(C_1, B') \equiv (C_1, B_2).$$

Now the separation plane $P$ and the separation line $b$ determine a separation threefold, say $W$, which contains $A_1$, $B_1$, $C_1$, $A_2$, $B_2$, $B'$.

Again, since $B'C_1$ is normal to $P$, it must be normal to $C_1 A_1$, $C_1 A_2$, $C_1 B_1$ and $C_1 B_2$.

Then since $\qquad (C_1, B_1) \equiv (C_1, B')$,

it follows, by Case II, that:

$$(A_1, B_1) \equiv (A_1, B') \qquad \qquad \ldots\ldots(1).$$

Again since $\qquad (C_1, A_1) \equiv (C_1, A_2)$,

it follows, by Case II, that:

$$(A_1, B') \equiv (A_2, B') \qquad \qquad \ldots\ldots(2).$$

Further, by Case II, since

$$(C_1, B') \equiv (C_1, B_2),$$

it follows that: $\qquad (A_2, B') \equiv (A_2, B_2) \qquad \qquad \ldots\ldots(3).$

Thus, from (1), (2) and (3), it follows that:

$$(A_1, B_1) \equiv (A_2, B_2).$$

CASE IV. *$P_2$ either identical with $P_1$ or parallel to $P_1$.*

There is, as we have already seen, one single element, say $A'$, such that:

$$(C_1, A_1) \,|\equiv| \,(C_2, A').$$

Similarly there is one single element, say $B'$, such that:

$$(C_1, B_1) \,|\equiv| \,(C_2, B').$$

Now, since $P_2$ is either identical with $P_1$ or parallel to $P_1$, it follows that $P_2$ must contain $C_2 A'$ and $C_2 B'$.

Also, since $C_2 A'$ must be either parallel to $C_1 A_1$ or identical with it, and since $C_2 B'$ must be either parallel to $C_1 B_1$ or identical with it, then since $B_1 C_1$ is normal to $A_1 C_1$, it follows that $B'C_2$ is normal to $A'C_2$.

But now, by Theorem 180, we must have

$$(A_1, B_1) \,|\equiv| \,(A', B').$$

Also since $\qquad (C_1, A_1) \equiv (C_2, A_2)$,

it follows that: $\qquad (C_2, A_2) \equiv (C_2, A')$,

and, since $\qquad (C_1, B_1) \equiv (C_2, B_2),$

it follows that: $\qquad (C_2, B_2) \equiv (C_2, B').$

Thus, by Case III, it follows that:

$$(A', B') \equiv (A_2, B_2).$$

Since however we have

$$(A_1, B_1) \equiv (A', B'),$$

it follows that: $\qquad (A_1, B_1) \equiv (A_2, B_2).$

CASE V. *$P_1$ and $P_2$ lie in the same separation threefold $W$.*

If $P_1$ and $P_2$ have no element in common, then since they both lie in $W$ they must be parallel to one another and the result follows from Case IV.

We shall therefore suppose that $P_1$ and $P_2$ have an element in common, but are distinct

Then, by Theorem 150, they have a second element in common, and therefore have a general line in common which we shall call $b$.

Let $C$ be any element in $b$ and let $a_1$ and $a_2$ be separation lines passing through $C$ and normal to $b$ and lying in $P_1$ and $P_2$ respectively.

Let $B$ be an element in $b$ such that:

$$(C_1, B_1) \equiv (C, B).$$

Then we shall also have

$$(C_2, B_2) \equiv (C, B).$$

Let $A_1'$ and $A_2'$ be elements in $a_1$ and $a_2$ respectively such that:

$$(C_1, A_1) \equiv (C, A_1')$$

and $\qquad (C_2, A_2) \equiv (C, A_2').$

Then since $\qquad (C_1, A_1) \equiv (C_2, A_2),$

we have $\qquad (C, A_1') \equiv (C, A_2').$

Thus, by Case II, it follows that:

$$(A_1', B) \equiv (A_2', B) \qquad \ldots\ldots(1).$$

But, by Case IV, it follows that:

$$(A_1, B_1) \equiv (A_1', B) \qquad \ldots\ldots(2),$$

and similarly it follows that:

$$(A_2, B_2) \equiv (A_2', B) \qquad \ldots\ldots(3).$$

Thus from (1), (2) and (3) it follows that:

$$(A_1, B_1) \equiv (A_2, B_2).$$

Thus whether $P_1$ and $P_2$ are identical, or parallel, or whether they have a general line in common, the theorem holds provided $P_1$ and $P_2$ lie in the same separation threefold $W$.

CASE VI. *$P_1$ and $P_2$ do not lie in one separation threefold.*

In this case we may take one separation threefold, say $W_1$, which contains $P_1$ and another separation threefold, say $W_2$, which contains $P_2$.

Now $W_2$ may be either parallel to $W_1$, or else not parallel to it and, if not parallel, we know that $W_1$ and $W_2$ must have a general plane in common, which must obviously be a separation plane.

Suppose then first that $W_2$ is parallel to $W_1$ and let $C$ be any element in $W_2$.

Then there is a general line, say $a$, passing through $C$ and lying in $W_2$ which is parallel to $C_1 A_1$.

Similarly there is a general line, say $b$, passing through $C$ and lying in $W_2$ which is parallel to $C_1 B_1$.

Thus, since $B_1 C_1$ is normal to $A_1 C_1$, it follows that $b$ is normal to $a$.

Now let $A$ and $B$ be elements in $a$ and $b$ respectively such that:
$$(C_1, A_1) \,|\equiv|\, (C, A)$$
and
$$(C_1, B_1) \,|\equiv|\, (C, B).$$

Then, by Theorem 180, we must have
$$(A_1, B_1) \,|\equiv|\, (A, B).$$
But now, since
$$(C_1, A_1) \equiv (C_2, A_2),$$
it follows that:
$$(C_2, A_2) \equiv (C, A),$$
and since
$$(C_1, B_1) \equiv (C_2, B_2),$$
it follows that:
$$(C_2, B_2) \equiv (C, B).$$

Thus since $A$, $B$ and $C$ lie in $W_2$ which also contains $P_2$, it follows by Case V that:
$$(A, B) \equiv (A_2, B_2).$$
Thus, since
$$(A_1, B_1) \equiv (A, B),$$
it follows that:
$$(A_1, B_1) \equiv (A_2, B_2).$$

Suppose next that $W_2$ is not parallel to $W_1$ and let $S$ be the separation plane which they have in common.

Let $a$ be any separation line in $S$, and $C$ be any element in it.

Let $b$ be the separation line which passes through $C$ and lies in $S$ and which is normal to $a$.

Let $A$ and $B$ be elements in $a$ and $b$ respectively such that:
$$(C_1, A_1) \equiv (C, A)$$
and
$$(C_1, B_1) \equiv (C, B).$$

Then we shall also have
$$(C_2, A_2) \equiv (C, A)$$
and
$$(C_2, B_2) \equiv (C, B).$$

But, since $P_1$ and $S$ lie in $W_1$, it follows by Case V that:

$$(A_1, B_1) \equiv (A, B).$$

Also, since $P_2$ and $S$ lie in $W_2$, it follows by Case V that:

$$(A_2, B_2) \equiv (A, B).$$

Thus we get finally:    $(A_1, B_1) \equiv (A_2, B_2)$.

Combining now Cases V and VI we see that whether $P_1$ and $P_2$ lie in one separation threefold or not, the theorem still holds.

Thus the theorem holds in all cases.

## Theorem 190

*If $A_1$, $B_1$, $C_1$ be the corners of a triangle in a separation plane $P_1$ and $A_2$, $B_2$, $C_2$ be the corners of a triangle in a separation plane $P_2$, and if further*

$$(C_1, A_1) \equiv (C_2, A_2),$$
$$(A_1, B_1) \equiv (A_2, B_2),$$

*while $B_1 C_1$ is normal to $A_1 C_1$, and $B_2 C_2$ is normal to $A_2 C_2$, then we shall also have*

$$(C_1, B_1) \equiv (C_2, B_2).$$

Let $B_2'$ be an element in $B_2 C_2$ and on the same side of $C_2$ as is $B_2$ and such that:

$$(C_1, B_1) \equiv (C_2, B_2').$$

Then by Theorem 189 we must have

$$(A_1, B_1) \equiv (A_2, B_2'),$$

and so we must have     $(A_2, B_2) \equiv (A_2, B_2')$.

Suppose now, if possible, that $B_2'$ is distinct from $B_2$ and let $O$ be the mean of $B_2$ and $B_2'$.

Then, by Theorem 187, $A_2 O$ must be normal to $B_2 C_2$.

But $A_2 C_2$ is normal to $B_2 C_2$ and so, since $P_2$ is a separation plane, we should have $A_2 O$ identical with $A_2 C_2$ and therefore $O$ would be identical with $C_2$.

Since, however, $O$ is supposed to be the mean of $B_2$ and $B_2'$, it would require to be linearly between them and so $B_2$ and $B_2'$ would be on opposite sides of $C_2$, contrary to hypothesis.

Thus the supposition that $B_2$ and $B_2'$ are distinct leads to a contradiction and so $B_2'$ must be identical with $B_2$.

But          $(C_1, B_1) \equiv (C_2, B_2'),$

and therefore     $(C_1, B_1) \equiv (C_2, B_2)$

as was to be proved.

## Theorem 191

*If $A_1$, $B_1$, $C_1$ be the corners of a triangle in a separation plane $P_1$ and $A_2$, $B_2$, $C_2$ be the corners of a triangle in a separation plane $P_2$, and if further*

$$(A_1, B_1) \equiv (A_2, B_2),$$
$$(A_1, C_1) \equiv (A_2, C_2),$$
$$(B_1, C_1) \equiv (B_2, C_2),$$

*while $A_1 C_1$ is normal to $B_1 C_1$, then we shall also have $A_2 C_2$ normal to $B_2 C_2$.*

Let $a$ be a separation line passing through $C_2$ and lying in $P_2$ and which is normal to $B_2 C_2$, and let $A_2{}'$ be an element in $a$ on the same side of $B_2 C_2$ as is $A_2$ and such that:

$$(A_2{}', C_2) \equiv (A_1, C_1).$$

Then, by Theorem 189, we shall have

$$(A_2{}', B_2) \equiv (A_1, B_1).$$

It follows that we must have

$$(A_2, B_2) \equiv (A_2{}', B_2),$$

and further

$$(A_2, C_2) \equiv (A_2{}', C_2).$$

Now if $A_2{}'$ lies in $A_2 B_2$ it must be identical with $A_2$ for there is only one element, say $A$, distinct from $A_2$ and lying in $A_2 B_2$ and such that:

$$(A_2, B_2) \equiv (A, B_2),$$

and this element $A$ lies on the opposite side of $B_2$ to that on which $A_2$ lies.

Thus, since $A_2$ and $A_2{}'$ lie on the same side of $B_2 C_2$ and therefore on the same side of $B_2$, it follows that $A_2{}'$ must be identical with $A_2$.

Similarly if $A_2{}'$ lies in $A_2 C_2$ it must be identical with $A_2$.

Suppose now, if possible, that $A_2{}'$ is distinct from $A_2$ and lies neither in $A_2 B_2$ nor in $A_2 C_2$, and let $O$ be the mean of $A_2$ and $A_2{}'$.

Then, by Theorem 187, $B_2 O$ must be normal to $A_2 A_2{}'$ and similarly $C_2 O$ must be normal to $A_2 A_2{}'$.

Thus, since $B_2 O$ and $C_2 O$ lie in the same separation plane as $A_2 A_2{}'$, it follows that $B_2 O$ and $C_2 O$ would be the same general line which accordingly would be identical with $B_2 C_2$, and so $O$ would require to lie in $B_2 C_2$.

But, since $O$ is supposed to be the mean of $A_2$ and $A_2{}'$, it would have to be linearly between them and so, $A_2 A_2{}'$ being distinct from $B_2 C_2$, we should have $A_2$ and $A_2{}'$ on opposite sides of $B_2 C_2$, contrary to hypothesis.

Thus the assumption that $A_2'$ is distinct from $A_2$ leads to a contradiction and so $A_2'$ is identical with $A_2$.

But $A_2'C_2$ is normal to $B_2C_2$ by hypothesis and so $A_2C_2$ is normal to $B_2C_2$ as was to be proved.

### Theorem 192

*If $A_1$, $B_1$, $C_1$ be the corners of a triangle in a separation plane $P_1$ and $A_2$, $B_2$, $C_2$ be the corners of a triangle in a separation plane $P_2$, and if further*

$$(A_1, B_1) \equiv (A_2, B_2),$$
$$(A_1, C_1) \equiv (A_2, C_2),$$
$$(B_1, C_1) \equiv (B_2, C_2),$$

*while $N_1$ is an element in $B_1C_1$ such that $A_1N_1$ is normal to $B_1C_1$, and $N_2$ is an element in $B_2C_2$ such that $A_2N_2$ is normal to $B_2C_2$; and if $N_1$ be distinct from both $B_1$ and $C_1$, then $N_2$ will be distinct from both $B_2$ and $C_2$, and we shall also have*

$$(A_1, N_1) \equiv (A_2, N_2),$$
$$(B_1, N_1) \equiv (B_2, N_2),$$
$$(C_1, N_1) \equiv (C_2, N_2).$$

If $N_1$ be linearly between $B_1$ and $C_1$, let $N_2'$ be an element in $B_2C_2$ and on the same side of $B_2$ as is $C_2$ and such that:

$$(B_1, N_1) \equiv (B_2, N_2').$$

Let $C_2'$ be an element in $B_2C_2$ and on the opposite side of $N_2'$ to that on which $B_2$ lies and such that:

$$(N_1, C_1) \equiv (N_2', C_2').$$

Then, by Theorem 182, we shall have

$$(B_1, C_1) \equiv (B_2, C_2'),$$

and so $\qquad (B_2, C_2) \equiv (B_2, C_2').$

But $C_2$ and $C_2'$ both lie on the same side of $B_2$ and so they must be identical.

Thus we must have

$$(C_1, N_1) \equiv (C_2, N_2').$$

Again, if $C_1$ be linearly between $B_1$ and $N_1$ let $N_2'$ be an element in $B_2C_2$ and on the opposite side of $C_2$ to that on which $B_2$ lies and such that:

$$(C_1, N_1) \equiv (C_2, N_2').$$

Then, by Theorem 182, we shall have

$$(B_1, N_1) \equiv (B_2, N_2').$$

Similarly if $B_1$ be linearly between $C_1$ and $N_1$, let $N_2'$ be an element in $C_2 B_2$ and on the opposite side of $B_2$ to that on which $C_2$ lies and such that:

$$(B_1, N_1) \equiv (B_2, N_2').$$

Then, by Theorem 182, we shall have

$$(C_1, N_1) \equiv (C_2, N_2').$$

Thus in all three cases $N_2'$ has been taken in $B_2 C_2$ in such a manner that:

$$(B_1, N_1) \equiv (B_2, N_2')$$
and $$(C_1, N_1) \equiv (C_2, N_2').$$

Now let $a$ be a separation line lying in $P_2$ and passing through $N_2'$ and normal to $B_2 C_2$.

Let an element $A_2'$ be selected in $a$ and on the same side of $B_2 C_2$ as $A_2$ lies and such that:

$$(N_1, A_1) \equiv (N_2', A_2').$$

Then, by Theorem 189, it follows that:

$$(A_1, B_1) \equiv (A_2', B_2)$$
and $$(A_1, C_1) \equiv (A_2', C_2).$$

Thus we must have

$$(A_2, B_2) \equiv (A_2', B_2)$$
and $$(A_2, C_2) \equiv (A_2', C_2).$$

Then, as in the last theorem, we may prove that the elements $A_2$ and $A_2'$ must be identical and, since $A_2 N_2'$ is normal to $B_2 C_2$ and intersects it in the element $N_2'$ and, since $P_2$ is a separation plane, it follows that $N_2'$ is identical with $N_2$, and therefore $N_2$ is distinct from both $B_2$ and $C_2$.

Thus we must have

$$(A_1, N_1) \equiv (A_2, N_2),$$
$$(B_1, N_1) \equiv (B_2, N_2),$$
$$(C_1, N_1) \equiv (C_2, N_2),$$

and so the theorem is proved.

It is also evident from the manner in which $N_2'$ was determined that we must have

|  | |
|---|---|
| | $N_2$ linearly between $B_2$ and $C_2$, |
| or | $C_2$ linearly between $B_2$ and $N_2$, |
| or | $B_2$ linearly between $C_2$ and $N_2$, |
| according as | $N_1$ is linearly between $B_1$ and $C_1$, |
| or | $C_1$ is linearly between $B_1$ and $N_1$, |
| or | $B_1$ is linearly between $C_1$ and $N_1$. |

## REMARKS

If $B$, $C_1$ and $C_2$ be three distinct elements in a separation plane which do not all lie in one separation line, and if

$$(B, C_1) \equiv (B, C_2)$$

while $N_1$ and $N_2$ are elements in $BC_1$ and $BC_2$ respectively, such that $C_2 N_1$ is normal to $BC_1$ and $C_1 N_2$ is normal to $BC_2$, then, if we make the restriction that $BC_1$ is not normal to $BC_2$, we may take the triangle whose corners are $B$, $C_1$ and $C_2$ and apply the results of the last theorem to the one triangle taken in two aspects.

Since $BC_1$ is not normal to $BC_2$, it follows that neither $N_1$ nor $N_2$ can be identical with $B$; so that we can speak of the pairs $(B, N_1)$ and $(B, N_2)$.

Further, we could not have both $N_1$ identical with $C_1$ and $N_2$ identical with $C_2$; for then we should have two intersecting separation lines: $BC_1$ and $BC_2$ both normal to one separation line $C_1 C_2$ lying in the same separation plane with them and this, we know, is impossible.

Thus either $N_1$ is distinct from $C_1$, or $N_2$ is distinct from $C_2$ and, without essential loss of generality, we may suppose that $N_1$ is distinct from $C_1$.

If then, in the last theorem, we take

$$B_2 \text{ identical with } B_1,$$
$$A_1 \text{ identical with } C_2,$$
$$A_2 \text{ identical with } C_1,$$

and write $B$ for $B_1$ or $B_2$, we get $N_2$ distinct from $C_2$ and

$$(C_2, N_1) \equiv (C_1, N_2),$$
$$(B, N_1) \equiv (B, N_2),$$
$$(C_1, N_1) \equiv (C_2, N_2).$$

Also the *linearly between* relations of $B$, $N_2$ and $C_2$ will be similar to those of $B$, $N_1$ and $C_1$ respectively.

*Definitions.* If $O$ and $X_0$ be two distinct elements in a separation plane $S$, then the set of all elements in $S$ such as $X$, where

$$(O, X) \equiv (O, X_0),$$

will be called a *separation circle*.

The element $O$ will be called the *centre* of the separation circle.

Any one of the linear intervals such as $OX$ will be called a *radius* of the separation circle.

If $X_1$ and $X_2$ be two elements of the separation circle such that

$X_1 X_2$ passes through $O$, then the linear interval $X_1 X_2$ will be called a *diameter* of the separation circle.

Any element which lies in a radius but which is not an element of the separation circle itself will be said to lie *inside* or in the *interior* of the separation circle.

Any element which lies in $S$ but not in a radius will be said to lie *outside* or *exterior to* the separation circle.

### THEOREM 193

(1) *If a separation line $a$ and a separation circle both lie in the same separation plane $S$, and if any element $A$ of $a$ lies within the separation circle, then the latter has two elements in common with $a$ and the element $A$ lies linearly between them.*

(2) *If a separation line $a$ has two elements $D$ and $E$ in common with a separation circle lying in a separation plane $S$, and if an element $A$ of $a$ lies linearly between $D$ and $E$, then $A$ lies within the separation circle.*

Consider the first part of the theorem.

Let $O$ be the centre of the separation circle and let a separation line in $S$ be taken through $O$ and $A$. Then there are two elements of this separation line, say $X_1$ and $X_2$, which lie on the circle and are therefore such that:

$$(O, X_1) \equiv (O, X_2),$$

and $A$ must lie linearly between $X_1$ and $X_2$.

Let $c$ be an inertia line through $O$ normal to $S$ and let $C$ be the element common to $c$ and the $\alpha$ sub-set of $X_1$.

Then $CX_1$ is an optical line and, since $X_2$ is also an element of the circle, we must also have $CX_2$ an optical line.

Further, $C$ is *after* $X_1$ and since $X_1 X_2$ is a separation line, we must also have $C$ *after* $X_2$.

Now, by Theorem 73, since $A$ is linearly between $X_1$ and $X_2$, it follows that $CA$ is an inertia line and $A$ is *before* $C$.

But, since $CA$ is an inertia line which intersects the separation line $a$, it follows that $CA$ and $a$ lie in an inertia plane, say $P$.

Then there are two optical lines passing through $C$ and lying in $P$ and these will intersect $a$ in two distinct elements, say $D$ and $E$.

Then, since $c$ is normal to $S$, it follows that:

$$(O, D) \equiv (O, E) \equiv (O, X_1) \equiv (O, X_2),$$

and so $D$ and $E$ lie on the separation circle and accordingly, since they also lie in $a$, the existence of the two elements is proved.

It also follows that $A$ must lie linearly between $D$ and $E$. For, in the first place, $A$ could not coincide with either $D$ or $E$, since $CD$ and $CE$ are optical lines while $CA$ is an inertia line. Again, $D$ could not lie linearly between $A$ and $E$ for, since $C$ is *after* both $A$ and $E$, it would follow, by Theorem 73, that $CD$ must be an inertia line; which is impossible. Similarly $E$ could not lie linearly between $A$ and $D$. It remains that $A$ must lie linearly between $D$ and $E$; which proves the first part of the theorem.

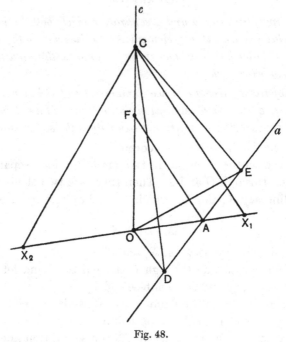

Fig. 48.

Consider now the second part of the theorem.

As before, let $O$ be the centre of the separation circle and let $c$ be an inertia line through $O$ normal to $S$.

Let $C$ be the element common to $c$ and the $\alpha$ sub-set of $D$. Then $CD$ is an optical line and $C$ is *after* $D$.

Also, since $D$ and $E$ are both elements of the circle centre $O$, it follows that $CE$ is also an optical line and, since $DE$ is a separation line, we must also have $C$ *after* $E$.

But now, since by hypothesis $A$ is linearly between $D$ and $E$, it follows, by Theorem 73, that $CA$ is an inertia line and $A$ is *before* $C$.

Now, in the very special case where $A$ is identical with $O$, it is obvious that $A$ lies within the separation circle; so we shall suppose that $A$ is not identical with $O$.

Let $F$ be the element common to $c$ and the $\alpha$ sub-set of $A$. Then $FA$ is an optical line and $F$ is *after* $A$.

But since $OA$ is a separation line and $FO$ is an inertia line we must also have $F$ *after* $O$.

Now $F$ could not be identical with $C$ since $FA$ is an optical line while $CA$ is an inertia line.

Also $F$ could not be *after* $C$ for then we should have $C$ *after* $A$ and *before* $F$: two distinct elements of the optical line $AF$ in which $C$ does not lie, and we know that this is impossible.

It remains that $C$ is *after* $F$ so that $F$ is linearly between $O$ and $C$.

Now, since $c$ is an inertia line, therefore $c$ and $OA$ lie in an inertia plane, and, if we take an optical line through $C$ parallel to $FA$, it will intersect the separation line $OA$ in some element $X_1$.

Then $X_1$ will be an element of the circle and, since $F$ is linearly between $O$ and $C$, it follows that $A$ is linearly between $O$ and $X_1$ and so $A$ lies within the separation circle.

Thus the second part of the theorem is proved.

THEOREM 194

*If $A$, $B$ and $C$ be the corners of a triangle in a separation plane, and if $BC$ be normal to $AC$, then the side $AB$ is greater than either of the other two sides of the triangle.*

It will be sufficient to prove that $AB$ is greater than $AC$.

Let $D$ be an element in $BC$ such that $C$ is the mean of $B$ and $D$.

Then, by Theorem 186, we must have

$$(A, D) \equiv (A, B);$$

and so $B$ and $D$ are two elements of a separation circle of centre $A$; while the separation line $BD$ has the two elements $B$ and $D$ in common with it.

Further, the element $C$ lies linearly between $B$ and $D$ and so, by Theorem 193, $C$ lies within the circle.

Thus $AB$ is greater than $AC$; and similarly we may prove that $AB$ is greater than $BC$.

REMARKS

If a separation line $a$ and a separation circle both lie in a separation plane $S$, they can either have no element in common, or one element in common, or two elements in common, but cannot have more than two.

Taking $O$ as centre, let a separation line passing through $O$, lying in $S$ and normal to $a$ intersect $a$ in an element $N$ and let $B$ be any element of $a$ distinct from $N$.

Then, by Theorem 194, $OB$ is greater than $ON$, so that if $ON$ is greater than a radius of the circle, then $OB$ is always greater than a radius and $a$ can have no element in common with the circle.

If $ON$ is equal to a radius, then the element $N$ lies on the circle but $B$ cannot do so and, in this case, $a$ has one element in common with the circle.

If $ON$ is less than a radius, then $N$ must lie within the circle, so that, by the first part of Theorem 193, $a$ must have two elements in common with it, say $D$ and $E$.

If $O$, $D$ and $E$ should happen to lie in one separation line, $N$ would coincide with $O$ and would therefore be the mean of $D$ and $E$; while if $O$, $D$ and $E$ do not lie in one separation line the same result follows by Theorem 188.

Now suppose, if possible, that there is an element $E'$ distinct from $D$ and $E$ and common to the circle and the separation line $a$. Then $N$ would require to be the mean of $D$ and $E'$ as well as of $D$ and $E$, which is impossible by Theorem 62.

Thus no such element as $E'$ can exist and so the separation circle cannot have more than two elements in common with the separation line $a$.

It follows very simply from Theorem 194 that: *If a triangle lies in a separation plane, then the sum of the lengths of any two sides is greater than that of the third side.*

This may be shown as follows:

Let $A$, $B$, $C$ be the corners of any triangle in a separation plane $S$ and let a separation line through $A$ normal to $BC$ intersect $BC$ in an element $N$.

Then (1) if $N$ is linearly between $B$ and $C$ we have $BA$ is greater than $BN$ and $AC$ is greater than $NC$ so that

$$BA + AC \text{ is greater than } BC.$$

(2) If $N$ coincides with $C$ we have $BA$ is greater than $BC$ and so

$$BA + AC \text{ is greater than } BC.$$

Similarly, if $N$ coincides with $B$, we have $AC$ greater than $BC$ and so

$$BA + AC \text{ is greater than } BC.$$

(3) If $C$ is linearly between $B$ and $N$ we have $BA$ is greater than $BN$ while $BN$ is greater than $BC$ and so $BA$ is greater than $BC$; from which it follows that:

$$BA + AC \text{ is greater than } BC.$$

Similarly, if $B$ is linearly between $C$ and $N$, we may prove the same result.

These cover all the possibilities which are open; so that in all cases we have

$$BA + AC \text{ is greater than } BC.$$

By a similar method we may prove that the sum of the lengths of any other two sides of the triangle is greater than that of the third side.

Another important result which can readily be obtained, using the notation of Theorem 194, is as follows:

*If the separation line through $C$ normal to $AB$ intersects $AB$ in $M$, then $M$ is linearly between $A$ and $B$.*

For we have $AB$ is greater than $BC$ and $BC$ is greater than $BM$, so that $AB$ is greater than $BM$. Similarly $AB$ is greater than $AM$. Thus, since $M$ lies in $AB$, it must be linearly between $A$ and $B$.

*Definition.* If $A$, $B$ and $C$ be the corners of a triangle in a separation or inertia plane and if $BC$ be normal to $AC$, then the side $AB$ will be called the *hypotenuse* of the triangle.

In case the triangle lies in an optical plane and $BC$ be normal to $AC$, then either $BC$ or $AC$ must be an optical line and, whichever it be, that one must also be normal to $AB$.

Thus, when the triangle lies in an optical plane, two of its sides would equally well be entitled to the name *hypotenuse*. This could never be the case either in a separation or in an inertia plane.

## ANGLE BOUNDARIES IN SEPARATION PLANES

We are now going to make three successive applications of the result proved in the remarks at the end of Theorem 192 to the case of a construction obtained from two intersecting separation lines lying in a separation plane and which are not normal to one another.

Let the separation lines be called $\bar{x}_1$ and $\bar{x}_2$ and let $O$ be their element of intersection.

Then $O$ will divide the separation line $\bar{x}_1$ into two half-lines, which we shall denote by $x_1$ and $x_1'$; while it will divide the separation line $\bar{x}_2$ into two half-lines which we shall denote by $x_2$ and $x_2'$.

Let $C_1$, $C_1'$, $C_2$, $C_2'$ be any elements in $x_1$, $x_1'$, $x_2$, $x_2'$ respectively such that

$$(O, C_1) \equiv (O, C_1') \equiv (O, C_2) \equiv (O, C_2'),$$

and let separation lines through $C_2$ and $C_2'$ normal to the separation line $\bar{x}_1$ intersect it in $N_{21}$ and $N'_{21}$ respectively; while separation lines through $C_1$ and $C_1'$ normal to the separation line $\bar{x}_2$ intersect it in $N_{12}$ and $N'_{12}$ respectively.

Then, since by hypothesis, $\bar{x}_1$ and $\bar{x}_2$ are supposed not to be normal to one another, it follows that none of the elements $N_{21}$, $N'_{21}$, $N_{12}$, $N'_{12}$ can coincide with $O$.

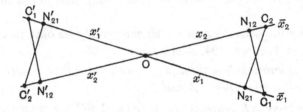

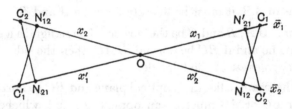

Fig. 49.

If we consider first the element $N_{21}$, then since, by Theorem 194, the segment $ON_{21}$ is less than the segment $OC_2$, it follows that $N_{21}$ must either lie linearly between $O$ and $C_1$, or else linearly between $O$ and $C_1'$.

Without any essential loss of generality we may suppose that $N_{21}$ is linearly between $O$ and $C_1$, so that $N_{21}$ lies in $x_1$.

Then it follows that $N_{12}$ must be linearly between $O$ and $C_2$, so that $N_{12}$ lies in $x_2$ and we must also have

$$(C_2, N_{21}) \equiv (C_1, N_{12})$$

and

$$(O, N_{21}) \equiv (O, N_{12}).$$

Now, since $N_{12}$ lies in $x_2$ while $C_2'$ lies in $x_2'$, it follows that $O$ is linearly between $C_2'$ and $N_{12}$.

Thus, by a second application of the theorem, it follows that $O$ is linearly between $C_1$ and $N'_{21}$, so that $N'_{21}$ lies in $x_1'$, while

$$(C_1, N_{12}) \equiv (C_2', N'_{21})$$
$$(O, N_{12}) \equiv (O, N'_{21}).$$

Now, since $N'_{21}$ lies in $x_1'$ and since the segment $ON'_{21}$ must be less than the segment $OC_2'$, it follows that $N'_{21}$ must be linearly between $O$ and $C_1'$.

Thus, by a third application of the theorem, it follows that $N'_{12}$ must be linearly between $O$ and $C_2'$, so that $N'_{12}$ must lie in $x_2'$, while

$$(C_2', N'_{21}) \equiv (C_1', N'_{12})$$

and

$$(O, N'_{21}) \equiv (O, N'_{12}).$$

Thus we must have

$$(C_2, N_{21}) \equiv (C_1, N_{12}) \equiv (C_2', N'_{21}) \equiv (C_1', N'_{12}),$$

and

$$(O, N_{21}) \equiv (O, N_{12}) \equiv (O, N'_{21}) \equiv (O, N'_{12});$$

while

$$N_{21} \text{ lies in } x_1,$$
$$N_{12} \text{ lies in } x_2,$$
$$N'_{21} \text{ lies in } x_1',$$
$$N'_{12} \text{ lies in } x_2'.$$

The congruences hold whether $N_{21}$ lies in $x_1$ or $x_1'$, but, if $N_{21}$ lies in $x_1'$, we shall have instead of the above

$$N_{21} \text{ lies in } x_1',$$
$$N'_{12} \text{ lies in } x_2,$$
$$N'_{21} \text{ lies in } x_1,$$
$$N_{12} \text{ lies in } x_2'.$$

Now any two separation lines in the separation plane which are normal to $\bar{x}_1$ must be parallel to one another, and similarly any two which are normal to $\bar{x}_2$ must be parallel to one another.

Accordingly if we take different positions for $C_1$, $C_2$, $C_1'$, $C_2'$ in the corresponding half-lines $x_1$, $x_2$, $x_1'$, $x_2'$ we may apply the results given in the remarks at the end of Theorem 185 and show that the ratios of the segments,

$$ON_{21} : OC_2,$$
$$C_2 N_{21} : OC_2,$$
$$C_2 N_{21} : ON_{21},$$

are independent of the position of $C_2$ in the half-line $x_2$, and we get corresponding constancy of ratios for all positions of $C_1$ in $x_1$, of $C_1'$ in $x_1'$ and of $C_2'$ in $x_2'$; so that these ratios are definite for any definite pair of such separation lines.

Again let $\bar{O}$, $\bar{C}$ and $\bar{N}$ be the corners of a triangle lying in any separation plane and let $\bar{C}\bar{N}$ ·be normal to $\bar{O}\bar{N}$.

(i) Suppose, in the first place that

$$\bar{O}\bar{C} = OC_2$$

and that $\qquad \bar{O}\bar{N} : \bar{O}\bar{C} = ON_{21} : OC_2.$

Then $\qquad \bar{O}\bar{N} : ON_{21} = \bar{O}\bar{C} : OC_2.$

Thus, since $\bar{O}\bar{C} = OC_2$, it follows from the criterion of proportion that $\bar{O}\bar{N} = ON_{21}$ and, by Theorem 190, we must have

$$\bar{C}\bar{N} = C_2 N_{21}.$$

It follows that

$$\bar{C}\bar{N} : \bar{O}\bar{C} = C_2 N_{21} : OC_2$$

and $\qquad \bar{C}\bar{N} : \bar{O}\bar{N} = C_2 N_{21} : ON_{21}.$

(ii) Next let us suppose that

$$\bar{O}\bar{C} = OC_2$$

and that $\qquad \bar{C}\bar{N} : \bar{O}\bar{C} = C_2 N_{21} : OC_2.$

Then again, since $\bar{O}\bar{C} = OC_2$, it follows that $\bar{C}\bar{N} = C_2 N_{21}$ and accordingly, by Theorem 190, it follows that

$$\bar{O}\bar{N} = ON_{21}.$$

Thus we see that

$$\bar{O}\bar{N} : \bar{O}\bar{C} = ON_{21} : OC_2$$

and $\qquad \bar{C}\bar{N} : \bar{O}\bar{N} = C_2 N_{21} : ON_{21}.$

(iii) Suppose finally that

$$\bar{O}\bar{N} = ON_{21}$$

and that $\qquad \bar{C}\bar{N} : \bar{O}\bar{N} = C_2 N_{21} : ON_{21}.$

Then we shall have $\bar{C}\bar{N} = C_2 N_{21}$ and since also $\bar{O}\bar{N} = ON_{21}$ it follows, by Theorem 189, that

$$\bar{O}\bar{C} = OC_2.$$

Thus we see that

$$\bar{O}\bar{N} : \bar{O}\bar{C} = ON_{21} : OC_2$$

and $\qquad \bar{C}\bar{N} : \bar{O}\bar{C} = C_2 N_{21} : OC_2.$

Thus we see that if any one of the three proportionalities:

$$\bar{O}\bar{N} : \bar{O}\bar{C} = ON_{21} : OC_2,$$
$$\bar{C}\bar{N} : \bar{O}\bar{C} = C_2 N_{21} : OC_2,$$
$$\bar{C}\bar{N} : \bar{O}\bar{N} = C_2 N_{21} : ON_{21},$$

holds, then the remaining two will also hold, and the pair of separation lines $\bar{O}\bar{N}$ and $\bar{O}\bar{C}$ will be characterised by the same triplet of ratios as were the original pair of separation lines: $\bar{x}_1$ and $\bar{x}_2$.

We shall find it convenient to denote these ratios by special names as follows:

The ratio $ON_{21}:OC_2$

we shall call the c-ratio of the separation lines $\bar{x}_1$ and $\bar{x}_2$; the ratio

$$C_2N_{21}:OC_2$$

we shall call the s-ratio; while the ratio

$$C_2N_{21}:ON_{21}$$

we shall call the t-ratio of the separation lines. The letters c, s and t are the initial letters of the words cosine, sine and tangent respectively, but, since they are ratios of absolute magnitudes, and the word "ratio" is used in the Euclidean sense, there is no question of sign involved.

Employing the above notation, but removing the restriction that the separation lines $\bar{x}_1$ and $\bar{x}_2$ are not normal and also any implied restriction that they are necessarily distinct, we may introduce the following definitions:

*Definition.* If $x_1$ and $x_2$ be separation half-lines lying in a separation plane S and having a common end O, then, together with the element O, they will be said to form an *angle-boundary* and the element O will be called its *vertex* while $x_1$ and $x_2$ will be called its *sides*.

For the sake of brevity we shall frequently speak of the sides as forming the angle-boundary, without explicit mention of the vertex.

If a separation line in S taken through any element of $x_2$ normal to the separation line $\bar{x}_1$ intersects the latter in an element of $x_1$, then the angle-boundary which $x_1$ and $x_2$ form will be said to be *acute*.

If such a separation line intersects the separation line $\bar{x}_1$ in an element of $x_1'$, then the angle-boundary which $x_1$ and $x_2$ form will be said to be *obtuse*.

If such a separation line intersects the separation line $\bar{x}_1$ in the element O, then the angle-boundary which $x_1$ and $x_2$ form will be said to be *right*.

This is equivalent to saying that $x_1$ and $x_2$ will form a right angle-boundary provided that the separation lines $\bar{x}_1$ and $\bar{x}_2$ are normal to one another.

If $x_1$ and $x_2$ form two distinct portions of the same separation line they will be said to form a *straight* or *flat* angle-boundary; while if $x_1$ and $x_2$ are identical they will be said to form a *null* angle-boundary.

It will be observed from the results obtained above that if $x_1$ and $x_2$ form an acute angle-boundary, then $x_1'$ and $x_2'$ will also form an acute angle-boundary, while $x_1'$ and $x_2$ will form an obtuse angle-boundary; as will also $x_1$ and $x_2'$.

It will be observed that we may interchange the rôles of $x_1$ and $x_2$ in our definitions without affecting the acute, right or obtuse character of the angle-boundary formed by them.

In case $x_1$ and $x_2$ form a right angle-boundary, then so also will $x_1'$ and $x_2'$, $x_1$ and $x_2'$, $x_2$ and $x_1'$.

*Definition.* The angle-boundary formed by $x_2$ and $x_1'$ will be called a *supplement* of the angle-boundary formed by $x_2$ and $x_1$ and conversely.

We shall speak of the *c*-ratio, *s*-ratio or *t*-ratio of an angle-boundary meaning thereby the *c*-ratio, *s*-ratio or *t*-ratio of the complete pair of separation lines of which the sides of the angle-boundary are parts.

*Definition.* Angle-boundaries which are not right angle-boundaries will be said to be *congruent* provided that their *c*-ratios are equal and their acute or obtuse characters are the same.

*Definition.* Any right angle-boundary will be said to be *congruent* to any right angle-boundary.

It is to be noted that the *congruency of angle-boundaries*, as here defined, is a similarity in the relationships of the pairs of half-lines which form the sides of the angle-boundaries which are said to be congruent.

It does not, in itself, imply more than this; and certain other things have to be taken into consideration before one can adequately treat such theorems as involve the "*addition of angles*" in separation planes.

The customary notation for an "angle", such as $\angle ABC$, is only properly applicable to the relationship which the pair of half-lines $BA$ and $BC$ stand in to one another, and, although the notation is continually employed in ordinary geometry to represent an angular magnitude, it cannot, strictly speaking, do so without ambiguity.

We shall accordingly make use of the notation $\angle ABC$ to denote what we have called an "angle-boundary" whose sides are the half-lines $BA$ and $BC$, and shall denote the congruence of angle-boundaries by the symbol $\equiv$ placed between the symbols for the latter.

According to the above definitions a null angle-boundary is to be regarded as acute, while a flat angle-boundary is to be regarded as obtuse.

We have employed the *c*-ratios in the definition of the congruence of acute or obtuse angle-boundaries, but we might also have used either the $s$-ratios or the $t$-ratios; were it not that we shall find the $c$-ratios more convenient when we come to introduce numerical measurement with $+$ and $-$ signs. It will then appear that the acute or obtuse character of the angle-boundary may be expressed by the sign of the cosine, but not by that of the sine or tangent.

It will be observed that, as a result of our definitions along with the congruence relations already proved, it follows that the angle-boundary made by $x_1$ and $x_2$ is congruent to the angle-boundary made by $x_1'$ and $x_2'$, while the angle-boundary made by $x_1$ and $x_2'$ is congruent to the angle-boundary made by $x_1'$ and $x_2$.

We are now in a position to prove various theorems involving angle-boundaries in separation planes.

It will be observed that the results of Theorem 192 enable us at once to write

$$B_1 N_1 : B_1 A_1 = B_2 N_2 : B_2 A_2$$
and
$$C_1 N_1 : C_1 A_1 = C_2 N_2 : C_2 A_2,$$

and the ordinal relations of $B_1$, $N_1$ and $C_1$ are in all cases similar to those of $B_2$, $N_2$ and $C_2$ respectively: so that if $N_1$ be distinct from both $B_1$ and $C_1$ we have

$$\angle A_1 B_1 C_1 \equiv \angle A_2 B_2 C_2,$$
and
$$\angle A_1 C_1 B_1 \equiv \angle A_2 C_2 B_2.$$

If $N_1$ coincides with $B_1$ then we know that $N_2$ must coincide with $B_2$, which merely means that $\angle A_1 B_1 C_1$ and $\angle A_2 B_2 C_2$ are both right and therefore are congruent and we still have

$$C_1 N_1 : C_1 A_1 = C_2 N_2 : C_2 A_2,$$
so that
$$\angle A_1 C_1 B_1 \equiv \angle A_2 C_2 B_2.$$

If $N_1$ coincides with $C_1$ we obtain analogous results.

Thus in all cases we have

$$\angle A_1 B_1 C_1 \equiv \angle A_2 B_2 C_2,$$
$$\angle A_1 C_1 B_1 \equiv \angle A_2 C_2 B_2,$$

and by a similar method we can show that

$$\angle B_1 A_1 C_1 \equiv \angle B_2 A_2 C_2.$$

Thus we see that *if $A_1$, $B_1$, $C_1$ be the corners of a triangle in a separation plane $P_1$ and $A_2$, $B_2$, $C_2$ be the corners of a triangle in a separation plane $P_2$, and if further*

$$A_1 B_1 = A_2 B_2, \quad B_1 C_1 = B_2 C_2, \quad C_1 A_1 = C_2 A_2,$$

*then corresponding angle-boundaries in the two triangles will be congruent.*

Another very important result which follows very simply is this:

*If two parallel separation lines a and b lying in a separation plane be intersected by another separation line c in the elements A and B respectively, and if further F, be any element of c such that A is linearly between B and F, and if $A_1$ and $B_1$ be any elements of a and b respectively which both lie on the same side of c, we shall have*

$$\angle A_1 A F \equiv \angle B_1 B F.$$

In the special case where $c$ is normal to $a$, it is also normal to $b$ and the angle-boundaries are both right and therefore are congruent.

If $c$ be not normal to $a$, let a separation line through $F$ in the separation plane be taken normal to $a$ and let it intersect $a$ in $A'$ and $b$ in $B'$. Then, as we saw in the remarks at the end of Theorem 185,

$$AA' : FA = BB' : FB,$$

and so, since $\angle$'s $A'AF$ and $B'BF$ are both acute, we have

$$\angle A'AF \equiv \angle B'BF.$$

Also, since $A$ is linearly between $B$ and $F$, we have $A'$ linearly between $B'$ and $F$, so that $A'$ and $B'$ are both on the same side of $c$.

If this should happen to be the same side of $c$ as $A_1$ and $B_1$ lie on, then $\angle A_1 AF$ is identical with $\angle A'AF$ while $\angle B_1 BF$ is identical with $\angle B'BF$, so that

$$\angle A_1 AF \equiv \angle B_1 BF.$$

If, on the other hand, $A'$ and $B'$ should happen to be on the opposite side of $c$ to that on which $A_1$ and $B_1$ lie, then we should have $\angle A_1 AF$ the supplement of $\angle A'AF$ and $\angle B_1 BF$ the supplement of $\angle B'BF$, so that again we have

$$\angle A_1 AF \equiv \angle B_1 BF.$$

Thus the result holds in all cases.

Again, no matter whether $\angle A_1 AF$ be right, acute or obtuse, let $A_2$ be any element of $a$ such that $A$ is linearly between $A_1$ and $A_2$. Then we already know that:

$$\angle A_1 AF \equiv \angle BAA_2,$$

and so we have $\qquad \angle B_1 BA \equiv \angle BAA_2:$

another important result.

### Theorem 195

*If $A_1$, $B_1$, $C_1$ be the corners of a triangle in a separation plane $P_1$, while $A_2$, $B_2$, $C_2$ are the corners of a triangle in a separation plane $P_2$ and if further*

$$B_1 A_1 = B_2 A_2,$$
$$B_1 C_1 = B_2 C_2,$$
$$\angle A_1 B_1 C_1 \equiv \angle A_2 B_2 C_2,$$

*then we shall also have*

$$A_1C_1 = A_2C_2,$$
$$\angle B_1C_1A_1 \equiv \angle B_2C_2A_2,$$
$$\angle C_1A_1B_1 \equiv \angle C_2A_2B_2.$$

In case $\angle A_1B_1C_1$ and $\angle A_2B_2C_2$ should happen to be right we have already, by Theorem 189, $A_1C_1 = A_2C_2$, and, since the other angle-boundaries are all acute, and since we have

$$A_1B_1 : A_1C_1 = A_2B_2 : A_2C_2,$$
and
$$C_1B_1 : C_1A_1 = C_2B_2 : C_2A_2,$$
it follows that:
$$\angle B_1C_1A_1 \equiv \angle B_2C_2A_2,$$
and
$$\angle C_1A_1B_1 \equiv \angle C_2A_2B_2.$$

Next take the cases where $\angle A_1B_1C_1$ and $\angle A_2B_2C_2$ are both acute or both obtuse and let a separation line through $C_1$, lying in $P_1$ and normal to $A_1B_1$ intersect $A_1B_1$ in $N_1$; while a separation line through $C_2$, lying in $P_2$ and normal to $A_2B_2$ intersects $A_2B_2$ in $N_2$.

Then, since $\angle A_1B_1C_1 \equiv \angle A_2B_2C_2$, we have

$$B_1N_1 : B_1C_1 = B_2N_2 : B_2C_2,$$
and
$$C_1N_1 : B_1C_1 = C_2N_2 : B_2C_2.$$

Thus, since $B_1C_1 = B_2C_2$, we have

$$B_1N_1 = B_2N_2,$$
and
$$C_1N_1 = C_2N_2.$$

If $\angle A_1B_1C_1$ and $\angle A_2B_2C_2$ are both acute $A_1$ and $N_1$ will lie on the same side of $B_1$, while $A_2$ and $N_2$ will lie on the same side of $B_2$.

If $B_1N_1$ should happen to equal $B_1A_1$ then $N_1$ would coincide with $A_1$ while $N_2$ would coincide with $A_2$ and so, in this case, we should have

$$A_1C_1 = A_2C_2.$$

If $B_1N_1$ should happen to be less than $B_1A_1$ we should have $N_1$ linearly between $B_1$ and $A_1$ and also $N_2$ linearly between $B_2$ and $A_2$ and we should also have

$$N_1A_1 = N_2A_2.$$

If $B_1N_1$ should happen to be greater than $B_1A_1$ we should have $A_1$ linearly between $B_1$ and $N_1$ and also $A_2$ linearly between $B_2$ and $N_2$ and again we should have

$$N_1A_1 = N_2A_2.$$

Finally, if $\angle A_1B_1C_1$ and $\angle A_2B_2C_2$ are both obtuse we should have $B_1$ linearly between $A_1$ and $N_1$ and also $B_2$ linearly between $A_2$ and $N_2$ and once more we should have

$$N_1A_1 = N_2A_2.$$

Thus in all three cases, by Theorem 189, we have
$$A_1 C_1 = A_2 C_2.$$

It then follows by the result proved on p. 340 that:
$$\angle B_1 C_1 A_1 \equiv \angle B_2 C_2 A_2,$$
and
$$\angle C_1 A_1 B_1 \equiv \angle C_2 A_2 B_2,$$
and so the theorem is proved.

## THEOREM 196

*If A and B be the extremities of a diameter of a separation circle lying in a separation plane and if C be any other element in the circumference of the separation circle, then AC is normal to BC.*

Let $O$ be the centre of the separation circle. Then $O$ is the mean of $B$ and $A$.

Let $D$ be the mean of $B$ and $C$.

Then we have $\qquad (O, B) \equiv (O, C)$

and so, by Theorem 187, $OD$ is normal to $CB$.

But, since $O$ is the mean of $B$ and $A$, while $D$ is the mean of $B$ and $C$, it follows that $OD$ is parallel to $AC$.

Thus, since $OD$ is normal to $CB$, we must also have $AC$ normal to $CB$, as was to be proved.

*Definition.* If $O$ and $X_0$ be two distinct elements in a separation threefold $W$, then the set of all elements in $W$ such as $X$ where
$$(O, X) \equiv (O, X_0)$$
will be called a *separation sphere*.

The element $O$ will be called the *centre* of the separation sphere.

The terms *radius, diameter, inside, outside,* etc. may be defined in a similar manner to the case of a separation circle.

## REMARKS

As we have already pointed out, any element $B$ in a general line $a$ divides the remaining elements of $a$ into two sets.

In case $a$ is an optical line or an inertia line the two sets consist of those elements of $a$ which are *before* $B$ and those which are *after* $B$; but, in case $a$ is a separation line, the sets are not capable of definition in quite so simple a form.

Confining our attention in the meantime to the cases where $a$ is an optical line or an inertia line, we observe that we may group the element $B$ itself with either of these sets.

Thus, if we divide all the elements of $a$ into all those elements which are *before* $B$ and all those elements which are not *before* $B$; then $B$ itself is grouped with those which are not *before* $B$. If, on the other hand, we divide all the elements of $a$ into all those elements which are *after* $B$ and all those elements which are not *after* $B$; then $B$ itself is grouped with those which are not *after* $B$.

In either case all the elements of $a$ are divided into two sets such that every element of the one (which we may call the lower) set is *before* every element of the other (which we may call the upper) set.

The division is made by means of an element $B$ which is explicitly mentioned.

The question arises as to whether it is possible to have a division of all the elements of $a$ into two sets such that every element of the one set is *before* every element of the other set without the existence of an element making the division in the manner that $B$ does in the above.

Although it seems reasonable to suppose that there should always be an element of this character; yet it appears that there is nothing in the postulates hitherto given which ensures that this must be the case.

These postulates do imply the existence of segments which bear certain incommensurable ratios to one another (as for example, the side and diagonal of a square), but these are only a restricted class of such ratios, and there are other incommensurable ratios conceivable whose existence is not thus implied.

In order to admit of such possibilities, we shall now give the final postulate of our system which is equivalent to the Axiom of Dedekind.

POSTULATE XXI. **If all the elements of an optical line be divided into two sets such that every element of the first set is before every element of the second set, then there is one single element of the optical line which is not before any element of the first set and is not after any element of the second set.**

Since an element is neither *before* nor *after* itself, it is evident that this one single element may belong either to the first or second set.

Again, if $a$ be an optical line in an inertia plane $P$, then through each element of $a$ there passes one single generator of $P$ of the opposite system to that to which $a$ belongs.

Also every such generator intersects $a$.

Thus there is a one-to-one correspondence between the elements of $a$ and the generators of $P$ of the other system and so it follows that: *if either system of generators of an inertia plane be divided into two sets such*

4344    GEOMETRY OF TIME AND SPACE

*that every generator of the first set is a before-parallel of every generator of
the second set, then there is one single generator of the system which is not
a before-parallel of any generator of the first set and is not an after-parallel
of any generator of the second set.*

Again if $b$ be any inertia or separation line and if $P$ be an inertia plane
containing it, then if we select either system of generators of $P$, there
is a one-to-one correspondence between the elements of $b$ and the
generators of the selected system which pass through these elements.

If $b$ be an inertia line and $X$ and $Y$ be any two elements of $b$, then $X$
will be *before* or *after* $Y$ according as the generator through $X$ is a
before- or after-parallel of that through $Y$.

*Thus the property formulated in Post. XXI holds for an inertia line as
well as for an optical line.*

It is also clear that a corresponding result holds in the case of a
separation line, but since here no element is either *before* or *after*
another, the property must be formulated somewhat differently.

In order to state the result when $b$ is a separation line we may make
a perfectly arbitrary convention with regard to the use of the words
*right* and *left*.

Thus if $X$ and $Y$ be any two elements of $b$, we may say that $X$ is *to
the left* or *right* of $Y$ according as the generator of the selected system
which passes through $X$ is a before- or after-parallel of that through $Y$.

We may therefore state the property as follows:

*If all the elements of a separation line be divided into two sets such that
every element of the first set is to the left of every element of the second set,
then there is one single element of the separation line which is not to the left
of any element of the first set and is not to the right of any element of the
second set.*

With the introduction of the equivalent of the Dedekind axiom we
have now reached the stage where we are in a position to set up a one-
to-one correspondence between the elements of a general line $l$ and
the aggregate of *real numbers*.

Thus, if $A_0$ and $A_1$ be two distinct elements in $l$, it may be shown that
there are elements $A_2, A_3, A_4, \ldots A_n \ldots$ in $l$ and on the same side of $A_0$
as is $A_1$ and such that the segment $A_0A_n$ is equal to $n$ times the segment
$A_0A_1$.

Similarly there are elements $A_{-1}, A_{-2}, A_{-3}, \ldots A_n \ldots$ lying in $l$ but
on the opposite side of $A_0$ and such that the segment $A_{-n}A_0$ is equal to
$n$ times the segment $A_0A_1$.

Again, it may easily be shown that corresponding to any positive rational number $r = \dfrac{p}{q}$ there is an element $A_r$ in $l$ and on the same side of $A_0$ as is $A_1$ and such that $q$ times the segment $A_0 A_r$ is equal to $p$ times the segment $A_0 A_1$.

Similarly, corresponding to any negative rational number $-r = -\dfrac{p}{q}$; it may be shown that there is an element $A_{-r}$ in $l$, but on the opposite side of $A_0$ and such that $q$ times the segment $A_{-r} A_0$ is equal to $p$ times the segment $A_0 A_1$.

By making use of our equivalents of the axioms of Archimedes and Dedekind for the elements of $l$ along with the corresponding properties of *real numbers*, it is possible to set up the one-to-one correspondence mentioned above.

The logical steps involved in setting up such a correspondence have been carefully investigated by others and it is unnecessary to go into further details here.

These may be found, for instance, in Pierpont's *Theory of Functions of Real Variables*, vol. I, chapters I and II, and in other works.

The absolute value of the difference of the real numbers corresponding to the two ends of any segment of $l$ gives us a real number which may be called *the numerical value of the length* of the segment in terms of the unit segment $A_0 A_1$.

If $l$ be an inertia or separation line, the length of any segment of a general line of the same kind as $l$ is always expressible in terms of our selected segment; but if $l$ be an optical line we must restrict the meaning of the words "*of the same kind*" to co-directional optical lines.

### THEOREM 197

*If $A$, $B$, $C$ be the corners of a triangle in a separation plane such that $CA$ is normal to $BA$, and if a separation line through $A$ normal to $BC$ intersects $BC$ in $M$, then*

$$\angle BAM \equiv \angle BCA,$$

*and* $\qquad\qquad\qquad\qquad \angle CAM \equiv \angle CBA.$

Since $CA$ is normal to $BA$, therefore, by Theorem 194, $CB$ is greater than $CA$ and so if we take an element $D$ in the half-line $CA$ such that $(C, D) \equiv (C, B)$ we shall have $A$ linearly between $C$ and $D$.

If through $D$ we take a separation line parallel to $AM$ and meeting $CB$ in $N$, then $DN$ must also be normal to $CB$.

Then, as already seen, we shall have

$$(D, N) \equiv (B, A).$$

But, since $DN$ is parallel to $AM$ and $D$ lies in the half-line $CA$, therefore, by Theorem 185,

$$AM : DN = AC : DC,$$

or $$AM : AB = AC : BC.$$

Thus, since $\angle BAM$ and $\angle BCA$ are both acute, we have

$$\angle BAM \equiv \angle BCA.$$

Similarly $$\angle CAM \equiv \angle CBA.$$

THEOREM 198

*If $A$, $B$, $C$ be the corners of a triangle in a separation plane and such that $CA$ is normal to $BA$, then the square of the length of the side $CB$ is equal to the sum of the squares of the lengths of the other two sides.*

With $C$ as centre and $CB$ as radius take a circle in the separation plane and let it intersect the separation line $CA$ in $D$ and $E$.

Now, by Theorem 194, $CA$ must be less than the radius of the circle and so $A$ must be either linearly between $C$ and $E$, or else linearly between $C$ and $D$.

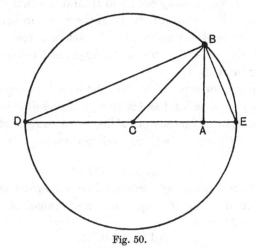

Fig. 50.

Without essential loss of generality we may suppose that $A$ is linearly between $C$ and $E$.

Now, from Theorem 196, we know that $DB$ is normal to $EB$ and so, by Theorem 197,

$$\angle EBA \equiv \angle EDB.$$

Taking the $t$-ratios of these, we have

$$AE : BA = BA : DA.$$

But $$AE = CB - CA$$
and $$DA = CB + CA$$
and therefore $$(CB - CA) : BA = BA : (CB + CA).$$

If then we take any unit of length of a separation line, and let $BA$, $CA$ and $CB$ now represent the numerical values of these lengths in terms of the selected unit we get

$$CB^2 - CA^2 = BA^2,$$
or $$CB^2 = BA^2 + CA^2,$$

as was to be proved.

This result is the equivalent of the "Theorem of Pythagoras": which accordingly holds in a separation plane.

Let us now, as we have previously done, denote the complete separation line of which $x_1$ and $x_2$ are parts by the symbols $\bar{x}_1$ and $\bar{x}_2$ respectively, and let the element $O$ be the common end of $x_1$ and $x_2$.

Let any convenient unit of length be selected and let the element $O$ be associated with the real number $0$ and let elements in $x_1$ be associated with the positive real numbers representing their distances from $O$.

Let elements in $x_1'$ be also associated with real numbers representing their distances from $O$, but having negative signs. We shall suppose the elements in the separation line $\bar{x}_2$ to be treated in an analogous manner.

If $P$ be the element in $x_2$ which is at unit distance from $O$ and if a normal through $P$ on the separation line $\bar{x}_1$, intersects the latter in the element $N$, then the real number associated with $N$ in $\bar{x}_1$ is characteristic of all angle-boundaries which are congruent to that made by $x_2$ and $x_1$.

Thus, when $N$ is distinct from $O$, the absolute value of this real number represents the $c$-ratio $ON : OP$ and its sign is positive for acute angle-boundaries and negative for obtuse angle-boundaries; while, if $N$ coincides with $O$, the real number associated with $N$ is $0$, which is characteristic of the case where we are dealing with right angle-boundaries.

It will be evident that the real number obtained in this way will be the cosine of an angle which $x_2$ makes with $x_1$, but, from the strict logical standpoint, we are not quite in a position to make this identification; since we have not as yet considered angles as distinguished from angle-boundaries. As, however, we require a notation for this function, we shall, in the meantime, denote it by $c(x_2, x_1)$, which from results already obtained is clearly equal to $c(x_1, x_2)$.

If an angle-boundary, which is not null, lies in a separation plane $S$ and has an element $O$ as vertex, it divides all the rays in $S$ having $O$ as end, excluding the sides of the angle-boundary, into two distinct sets.

Let $x_1$ and $x_2$ be the sides of the angle-boundary (which we shall first suppose is not flat) and let $A$ be any element of $x_1$ and $B$ be any element of $x_2$.

Then any ray in $S$ having $O$ as its end which intersects $AB$ in an element $M$ linearly between $A$ and $B$ belongs to the one set; while any ray in $S$ having $O$ as its end, other than $x_1$ and $x_2$, which does not intersect $AB$ in an element linearly between $A$ and $B$, belongs to the other set.

It may easily be proved, by means of the analogues of Peano's axioms 13 and 14, that the property of a ray such as $OM$ is independent of the positions of $A$ and $B$ in $x_1$ and $x_2$ respectively; so that the set of rays of this type is independent of the positions of $A$ and $B$ in the sides of the angle-boundary, and accordingly, the other set of rays is also independent of these positions.

Next, taking the case where $x_1$ and $x_2$ form a flat angle-boundary; any ray in $S$ having $O$ as its end which lies on one side of the complete separation line formed by $x_1$, $x_2$ and the element $O$, belongs to the one set; while any ray in $S$ having $O$ as its end which lies on the other side of this separation line belongs to the other set.

Thus in all these cases, the angle-boundary together with any one ray of a set determines that set and distinguishes it from the other set of rays having the same boundary.

In the case of a null angle-boundary, since $x_1$ coincides with $x_2$, there is no separation of the remaining rays into two sets; but, instead of this, there is only one set comprising all such rays.

We may now introduce the following:

*Definitions:* If $x_1$ and $x_2$ be the sides of an angle-boundary in a separation plane $S$ having an element $O$ as vertex, then either of the sets of rays in $S$ having $O$ as end separated off in the above manner by the angle-boundary (or, in the case of a null angle-boundary, the single set) will be called an *angular segment*.

The rays $x_1$ and $x_2$ will be called the *sides* of the angular segment, but are not included in it.

An angle-boundary which is not a null one together with either of the angular segments which it separates will be called an *angular interval*.

A null angle-boundary without any angular segment will be called a *null angular interval*; while a null angle-boundary together with the single angular segment associated with it will be called a *circuit angular interval*: the segment itself being called a *circuit angular segment*.

Reverting to the case of an angle-boundary which is neither null nor flat: an angular segment or interval which contains a ray such as $OM$ will be said to be of the *first type*; while one which does not contain a ray such as $OM$ will be said to be of the *second type*.

An angular segment or interval of the first type will be said to be *acute, right* or *obtuse* according as the angle-boundary is acute, right or obtuse; but these terms do not apply to angular segments or intervals of the second type.

The two angular segments or intervals associated with a flat angle-boundary will be called flat angular segments or intervals and will also be regarded as *obtuse*.

Again, if $x_1$ and $x_2$ form an angle-boundary which is neither null nor flat, the angular segment or interval of the first type formed by $x_1$ and $x_2'$ or by $x_2$ and $x_1'$ will be said to be a *supplement* of the angular segment or interval of the first type formed by $x_1$ and $x_2$; but this term does not apply to angular segments or intervals of the second type.

When an angle-boundary is not null the two angular segments or intervals associated with it will be said to be *conjugate* to one another.

Also a null angular interval and a circuit angular interval will be said to be *conjugate* to one another; but there is no angular segment conjugate to a circuit angular segment.

Consider now the case of an angular segment which is not a complete circuit one, but has $x_1$ and $x_2$ as its sides and $O$ as its vertex. We have to make a distinction between the two sides of the separation line $\bar{x}_1$ with respect to the angular segment, and this is done as follows:

If (1) all the rays of the angular segment lie on one side of $\bar{x}_1$;

or (2) all the rays in $S$ having $O$ as end and lying on one side of $x_1$ are rays of the angular segment;

then such side will be called the *positive side of $\bar{x}_1$ with respect to that angular segment* (or the corresponding angular interval), and the other will be called the negative side.

This gives a unique determination of the positive side of $\bar{x}_1$ with respect to a given angular segment when this is not a circuit segment; but fails to do this if it be such.

It will be found however that, for our present purpose, this does not matter.

If we consider the case where the angle-boundary formed by $x_1$ and $x_2$ is neither flat nor null; then of the two conjugate angular segments into which it divides the other rays of $S$ which have $O$ as end, one will be of what we have called the first type and the other will be of the second type.

It is clear that the angular segment of the second type will contain the two rays $x_1'$ and $x_2'$; since a separation line such as $AB$ which intersects $x_1$ and $x_2$ cannot intersect either $x_1'$ or $x_2'$.

Thus $x_2$ and all the rays belonging to the angular segment of the first type lie on one side of $\bar{x}_1$, and this is the positive side of $\bar{x}_1$ with respect to the angular segment of the first type.

On the other hand, all the rays in $S$ which have $O$ as end and which lie on the opposite side of $\bar{x}_1$ belong to the angular segment of the second type, and accordingly, this will be the positive side of $\bar{x}_1$ with respect to the angular segment of the second type.

Thus $x_2$ lies on the negative side of $\bar{x}_1$ with respect to the angular segment of the second type; but it lies on the positive side of $\bar{x}_1$ with respect to the angular segment of the first type.

In the case of a flat angular segment, all the rays belonging to it lie on one side of $\bar{x}_1$ and also all rays in $S$ having $O$ as end and lying on this same side of $\bar{x}_1$ are rays of the angular segment; so that, for a double reason in this case, this will be the positive side of $\bar{x}_1$ with respect to this flat angular segment.

The ray $x_2$ of course in this case actually lies in $\bar{x}_1$.

Now let $\bar{y}_1$ be the separation line passing through $O$ and lying in $S$ which is normal to $\bar{x}_1$ and let the half-lines into which $\bar{y}_1$ is divided by $O$ be denoted by $y_1$ and $y_1'$; of which $y_1$ is taken to be the one which is on the positive side of $\bar{x}_1$ with respect to the particular angular segment we are considering.

Then the angle-boundary which $x_2$ and $y_1$ make is characterised by the function $c\,(x_2, y_1)$, which may be positive, zero or negative according as the angle-boundary is acute, right or obtuse.

If the angular segment we are considering be of the first type, then, as we have seen, $x_2$ will be on the positive side of $\bar{x}_1$, so that $c\,(x_2, y_1)$ will be positive.

If the angular segment be of the second type, then, as we have also seen, $x_2$ will be on the negative side of $\bar{x}_1$ and so $c\,(x_2,\,y_1)$ will be negative.

If the angular segment be flat, then $x_2$ will lie in $\bar{x}_1$ so that $c\,(x_2,\,y_1)$ will be zero.

If the angular segment be a circuit one, then $x_2$ coincides with $x_1$, so that again we have $c\,(x_2,\,y_1)$ zero, no matter which side of $\bar{x}_1$ be taken as the positive one.

This case however differs from that of a flat angular segment in that for the latter, $c\,(x_2,\,x_1) = -1$; while for a circuit segment, $c\,(x_2,\,x_1) = +1$.

In fact it follows directly from Theorem 198, that in all cases

$$\{c\,(x_2,\,x_1)\}^2 + \{c\,(x_2,\,y_1)\}^2 = 1.$$

The angular segment, as here defined, will be characterised by the two functions $c\,(x_2,\,x_1)$ and $c\,(x_2,\,y_1)$ taken in conjunction: the one being determinate from the other except as regards sign.

It will hereafter be found convenient to denote such a pair of functions taken in conjunction by the symbol

$$c\,(x_2,\,x_1) + ic\,(x_2,\,y_1),$$

and we shall call this the De Moivre function of the angular segment.

Any angular segment similar to this will have the same De Moivre function.

*Definition.* Angular segments will be said to be *congruent* when their De Moivre functions are equal.

Such complex functions are regarded as equal when the corresponding component functions are separately equal each to each.

As regards angular intervals which are neither null nor circuit intervals, these will also be said to be *congruent* when their De Moivre functions are equal, but a null angular interval has the same De Moivre function as a circuit one.

They are however distinguished from one another in that a null interval has no corresponding angular segment, while a circuit interval has.

### ADDITION OF ANGLES

We have now to consider a series of half-lines, $x_0, x_1, x_2, \ldots$ all having a common end $O$ and lying in a separation plane $S$, along with a second series $y_0, y_1, y_2, \ldots$ also having the common end $O$ and lying in $S$ and such that $\bar{y}_0, \bar{y}_1, \bar{y}_2, \ldots$ are respectively normal to $\bar{x}_0, \bar{x}_1, \bar{x}_2, \ldots$.

As regards $\bar{y}_0$, it is perfectly arbitrary which of the component half lines into which it is divided by $O$ we denote by $y_0$ and which by $y_0'$, but, a selection being made in the case of $\bar{y}_0$, we are able to assign a definite systematic nomenclature in the case of $\bar{y}_1, \bar{y}_2, \ldots.$

If now we take $x_0$ and $y_0$ as standards we shall make the following conventions:

If $x_1$ is either identical with $x_0$ or else makes acute angle-boundaries with both $x_0$ and $y_0$, it will be said to lie in the first quadrant.

If $x_1$ is either identical with $y_0$ or else makes acute angle-boundaries with both $y_0$ and $x_0'$, it will be said to lie in the second quadrant.

If $x_1$ is either identical with $x_0'$ or else makes acute angle-boundaries with both $x_0'$ and $y_0'$, it will be said to lie in the third quadrant.

If $x_1$ is either identical with $y_0'$ or else makes acute angle-boundaries with both $y_0'$ and $x_0$, it will be said to lie in the fourth quadrant.

These cover all the possibilities which are open with regard to $x_1$; as is readily seen.

Omitting for the present the cases where $x_1$ is identical with one of the half-lines $x_0, y_0, x_0', y_0'$, we shall consider the other possibilities in succession.

(1) Let us suppose that $x_1$ makes acute angle-boundaries with both $x_0$ and $y_0$ and let $A$ be any element in $x_0$. Let the normal through $A$ to $\bar{x}_1$ intersect it in $M$.

Then, since $x_1$ makes an acute angle-boundary with $x_0$, it follows that $M$ must lie in $x_1$.

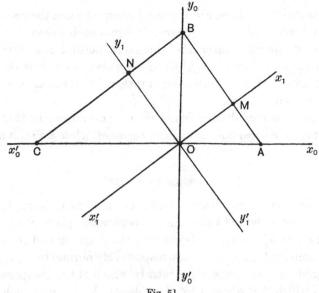

Fig. 51.

Also, since $\bar{x}_1$ does not coincide with $\bar{x}_0$, it follows that $AM$ cannot be parallel to $\bar{y}_0$ and therefore must intersect $\bar{y}_0$ in some element $B$. Further, since $x_1$ makes an acute angle-boundary with $y_0$, it follows that $B$ must lie in $y_0$, and, by the remarks at the end of Theorem 194, $M$ must be linearly between $A$ and $B$.

Now, since $\bar{x}_0$ is supposed to be distinct from $\bar{x}_1$, a separation line through $B$ parallel to $\bar{x}_1$ must intersect $\bar{x}_0$ in some element $C$.

Further, since $M$ is linearly between $A$ and $B$ we must have $O$ linearly between $A$ and $C$, so that $C$ must lie in $x_0'$.

Now let the separation line $\bar{y}_1$, which passes through $O$ and is normal to $\bar{x}_1$, intersect $BC$ in $N$.

Then, since $\angle BOC$ is a right angle-boundary, it follows that $N$ must be linearly between $B$ and $C$; so that $\angle NOB$ and $\angle NOC$ are both acute.

We shall select the part of $\bar{y}_1$ which contains $N$ as the one to be called $y_1$.

Thus, if $x_1$ makes acute angle-boundaries with $x_0$ and $y_0$, then our notation is so chosen that $y_1$ makes acute angle-boundaries with $y_0$ and $x_0'$.

Now, since $ON$ is parallel to $AB$, it follows that:

$$ON:OC = AB:AC,$$

and, since both are acute angle-boundaries, it follows that:

$$\angle CON \equiv \angle OAB.$$

But, since $BO$ is normal to $AO$, and $OM$ is normal to $AB$, it follows, by Theorem 197, that:

$$\angle OAB \equiv \angle BOM.$$

Thus $\qquad\qquad\qquad \angle CON \equiv \angle BOM.$

Similarly we can show that:

$$\angle AOM \equiv \angle BON.$$

These last two congruences may be expressed thus:

$$\angle(y_1, x_0') \equiv \angle(x_1, y_0) \qquad \dots\dots(a)$$
and $\qquad\quad \angle(y_1, y_0) \equiv \angle(x_1, x_0) \qquad \dots\dots(b).$

(2) If we carry out the above investigation making the substitution:

$$\begin{pmatrix} x_0, y_0, x_0', y_0' \\ y_0, x_0', y_0', x_0 \end{pmatrix}, \text{ implying that } \begin{pmatrix} \bar{x}_0, \bar{y}_0 \\ \bar{y}_0, \bar{x}_0 \end{pmatrix},$$

we get $\qquad\qquad \angle(y_1, y_0') \equiv \angle(x_1, x_0') \qquad \dots\dots(a)$
and $\qquad\qquad \angle(y_1, x_0') \equiv \angle(x_1, y_0) \qquad \dots\dots(b).$

R

(3) If instead, we make the substitution:

$$\begin{pmatrix} x_0, y_0, x_0', y_0' \\ x_0', y_0', x_0, y_0 \end{pmatrix}, \text{ implying that } \begin{pmatrix} \bar{x}_0, \bar{y}_0 \\ \bar{x}_0, \bar{y}_0 \end{pmatrix},$$

we get $\qquad \angle(y_1, x_0) \equiv \angle(x_1, y_0')$ ......(a)

and $\qquad \angle(y_1, y_0') \equiv \angle(x_1, x_0')$ ......(b).

(4) Finally, if we carry out the original investigation making the substitution:

$$\begin{pmatrix} x_0, y_0, x_0', y_0' \\ y_0', x_0, y_0, x_0' \end{pmatrix}, \text{ implying that } \begin{pmatrix} \bar{x}_0, \bar{y}_0 \\ \bar{y}_0, \bar{x}_0 \end{pmatrix},$$

we get $\qquad \angle(y_1, y_0) \equiv \angle(x_1, x_0)$ ......(a)

and $\qquad \angle(y_1, x_0) \equiv \angle(x_1, y_0')$ ......(b).

The above results may be transformed thus:

(1) $\begin{cases}(a) \\ (b)\end{cases}$ $\quad$ supplement$\angle(x_1, y_0) \equiv \angle(y_1, x_0)$,
$\qquad\qquad\quad \angle(x_1, x_0) \equiv \angle(y_1, y_0)$,

(2) $\begin{cases}(a) \\ (b)\end{cases}$ $\qquad\qquad\quad \angle(x_1, x_0) \equiv \angle(y_1, y_0)$,
$\qquad\quad$ supplement$\angle(x_1, y_0) \equiv \angle(y_1, x_0)$,

(3) $\begin{cases}(a) \\ (b)\end{cases}$ $\quad$ supplement$\angle(x_1, y_0) \equiv \angle(y_1, x_0)$,
$\qquad\qquad\quad \angle(x_1, x_0) \equiv \angle(y_1, y_0)$,

(4) $\begin{cases}(a) \\ (b)\end{cases}$ $\qquad\qquad\quad \angle(x_1, x_0) \equiv \angle(y_1, y_0)$,
$\qquad\quad$ supplement$\angle(x_1, y_0) \equiv \angle(y_1, x_0)$.

Let us now complete the conventions of notation as regards the part of $\bar{y}_1$ which we shall call $y_1$ in the following manner:

If $x_1$ coincides with $x_0$, then $y_1$ coincides with $y_0$.
If $x_1$ coincides with $y_0$, then $y_1$ coincides with $x_0'$.
If $x_1$ coincides with $x_0'$, then $y_1$ coincides with $y_0'$.
If $x_1$ coincides with $y_0'$, then $y_1$ coincides with $x_0$.

Then, since the supplement of a right angle-boundary is a right angle-boundary, while the supplement of a null angle-boundary is a flat angle-boundary and *vice versa*, it follows at once that, with these conventions, we still have the two congruences:

$$\text{supplement}\angle(x_1, y_0) \equiv \angle(y_1, x_0)$$

and $\qquad\qquad \angle(x_1, x_0) \equiv \angle(y_1, y_0)$.

The conventions of notation may now be summed up as follows:
If $x_1$ lies in the first quadrant $\quad y_1$ lies in the second quadrant.
If $x_1$ lies in the second quadrant $y_1$ lies in the third quadrant.
If $x_1$ lies in the third quadrant $\quad y_1$ lies in the fourth quadrant.
If $x_1$ lies in the fourth quadrant $y_1$ lies in the first quadrant.

With these conventions, whatever quadrant $x_1$ lies in we have the above two congruences; so that, if we employ the numerical form of the $c$-ratios, we have in all cases:

$$-c(x_1, y_0) = c(y_1, x_0)$$
and
$$c(x_1, x_0) = c(y_1, y_0).$$

It will be observed that these results have been obtained without making use of the "addition of angles".

If now we make the substitution $\begin{pmatrix} x_0, y_0, x_1, \bar{y}_1 \\ x_1, y_1, x_2, \bar{y}_2 \end{pmatrix}$, we may carry out a similar argument fixing the part of $\bar{y}_2$ to be called $y_1$ and proving that:

$$-c(x_2, y_1) = c(y_2, x_1)$$
and
$$c(x_2, x_1) = c(y_2, y_1)$$

and so on in succession any number of times.

Now let $A$ be an element in the half-line $x_2$ and let the length $OA$ be denoted by $r$.

We may treat the pair of separation lines $\bar{x}_0$, $\bar{y}_0$ as a pair of Cartesian coordinate axes having $O$ as origin and the half-lines $x_0$, $y_0$ as the positive parts of the axes.

Let $\xi_0$, $\eta_0$ be the coordinates of the element $A$ in this system.

Similarly we may treat $\bar{x}_1$, $\bar{y}_1$ as another pair of Cartesian coordinate axes having the same origin $O$ and the half-lines $x_1$, $y_1$ as the positive parts of the axes.

Let $\xi_1$, $\eta_1$ be the coordinates of the element $A$ in this second system and let the normal through $A$ to $\bar{x}_1$ intersect $\bar{x}_1$ in the element $M$.

Now the principle that the algebraic sum of the projections of the parts $A_0 A_1$, $A_1 A_2$, ... $A_{n-1} A_n$ of a broken line upon a given line is equal algebraically to the projection of the linear interval $A_0 A_n$ joining the extremities of the broken line upon the given line clearly holds in our geometry just as it does in ordinary Euclidean geometry: the proof being exactly analogous.

Accordingly the projection of $OA$ upon $\bar{x}_0$ or $\bar{y}_0$ is equal to the algebraic sum of the projections of $OM$ and $MA$ upon the same separation lines.

Taking first the projections upon $\bar{x}_0$, we have

$$OM = \xi_1 \quad \text{and} \quad MA = \eta_1$$

while $MA$ is co-directional with $\bar{y}_1$.

Thus
$$\xi_0 = \xi_1 c(x_1, x_0) + \eta_1 c(y_1, x_0),$$
$$= \xi_1 c(x_1, x_0) - \eta_1 c(x_1, y_0).$$
But
$$\xi_0 = rc(x_2, x_0), \quad \xi_1 = rc(x_2, x_1), \quad \eta_1 = rc(x_2, y_1).$$

Thus we get

$$c(x_2, x_0) = c(x_2, x_1) c(x_1, x_0) - c(x_2, y_1) c(x_1, y_0) \quad \dots\dots(1).$$

Taking the projections upon $\bar{y}_0$ we get

$$\eta_0 = \xi_1 c(x_1, y_0) + \eta_1 c(y_1, y_0),$$
$$= \xi_1 c(x_1, y_0) + \eta_1 c(x_1, x_0);$$

while

$$\eta_0 = rc(x_2, y_0).$$

Thus

$$c(x_2, y_0) = c(x_2, x_1) c(x_1, y_0) + c(x_2, y_1) c(x_1, x_0) \quad \dots\dots(2).$$

The formulae (1) and (2) may be combined into a single one by means of the symbol $i = \sqrt{-1}$; thus:

$$c(x_2, x_0) + ic(x_2, y_0)$$
$$= \{c(x_2, x_1) + ic(x_2, y_1)\}\{c(x_1, x_0) + ic(x_1, y_0)\} \quad \dots\dots(3).$$

It will be observed that formulae (1) and (2) are the equivalents of the addition formulae for cosine and sine respectively; while (3) is equivalent to the formula of De Moivre.

Now we have already seen that any angular segment is characterised by one definite De Moivre function and any De Moivre function is characteristic of all angular segments which are congruent to one another.

The same is true with regard to angular intervals except that a null interval and a circuit interval have the same De Moivre function.

We also remarked that it would be found convenient to denote a De Moivre function in a certain manner involving a symbol $i$.

We now see that by taking $i$ to stand for $\sqrt{-1}$ we have got an interpretation for the product of the De Moivre functions of two angular segments as the De Moivre function of an angular segment bearing a simple relation to the first two taken in conjunction.

We have now to make a diversion on the exponential function.

The exponential function of an argument $z$ or exp $(z)$ is defined as the limit of the infinite series

$$1 + \frac{z}{1!} + \frac{z^2}{2!} + \frac{z^3}{3!} + \frac{z^4}{4!} + \dots$$

and, as is well known, it has the property that

$$\exp(u) . \exp(v) = \exp(u + v)$$

for all values of $u$ and $v$, real and imaginary.

If we put $z = i\theta$ we get

$$\exp(i\theta) = 1 + \frac{i\theta}{1!} - \frac{\theta^2}{2!} - \frac{i\theta^3}{3!} + \frac{\theta^4}{4!} + \dots$$

and it is well known that the real and imaginary parts of this, namely:

$$1 - \frac{\theta^2}{2!} + \frac{\theta^4}{4!} - \frac{\theta^6}{6!} + \cdots$$

and

$$\theta - \frac{\theta^3}{3!} + \frac{\theta^5}{5!} - \frac{\theta^7}{7!} + \cdots$$

called $\cos\theta$ and $\sin\theta$ respectively, are such that the sum of their squares is equal to unity.

Also it is known that corresponding to any value of $\exp(i\theta)$ there are an infinite number of values of $\theta$ of the general form $\theta_0 + 2n\pi$, where $n$ is any integer positive, zero or negative.

We can thus write

$$c(x_1, x_0) + ic(x_1, y_0) = \exp(i\theta)$$

and

$$c(x_2, x_1) + ic(x_2, y_1) = \exp(i\phi)$$

and then we have

$$c(x_2, x_0) + ic(x_2, y_0) = \exp(i\overline{\theta + \phi}).$$

The quantities $\theta$ and $\phi$ so introduced may be either positive or negative according to the signs of the $n$'s.

The different positive values of $\theta$; that is to say $\theta_0 + 2n\pi$, where $n$ is zero or positive and $\theta_0$ is the smallest positive value of $\theta$, will be called the *angles* (in natural measure) *corresponding congruently to* an angular interval whose De Moivre function is

$$c(x_1, x_0) + ic(x_1, y_0).$$

The different values of $\theta_0 + 2n\pi$, where $n$ has the negative values $-1, -2, -3, \ldots$ with their signs reversed, will be the angles corresponding congruently to the conjugate angular interval.

In the special case of a null interval we have the set of angles

$$0, \ 2\pi, \ 4\pi, \ 6\pi, \ \ldots$$

corresponding congruently to it, and, giving negative values to $n$, and reversing the signs, we have the set of angles

$$2\pi, \ 4\pi, \ 6\pi, \ \ldots$$

corresponding congruently to the conjugate angular interval, which is here a circuit interval.

Thus, for a null interval, $\theta_0$ is taken as zero, while for a circuit interval $\theta_0$ is taken as $2\pi$, though the De Moivre function is the same for both.

The value $\theta_0$ will then represent the magnitude of the angular interval or segment (when such exists); while $\theta_0 + 2n\pi$ for positive integral values of $n$ will represent that magnitude $+n$ times the magnitude of a circuit interval or segment.

It is to be observed that in a separation plane there exists no entity corresponding to an angular segment or interval greater than a circuit one, any more than any linear segment or interval can exist inside a circle which is greater than the diameter of the circle; although the sum of the magnitudes of several such segments or intervals may be as great as we please.

It is unnecessary to go into this subject in further detail, since it is obvious that we should merely be covering ground which is already familiar.

We have employed the above method in dealing with the measures of angles in order to avoid making use of conceptions alien to our subject: such, for instance, as the "rotation of a half-line about its end".

This mode of speech, although familiar, is appropriate to Kinematics rather than to Pure Geometry, and would be quite out of place in treating of a separation plane in which no element is either *before* or *after* any other one and in which no motion can occur.

This is particularly important in a work like the present, where we are concerned with showing how a system of geometry may be built up from certain fundamental concepts, and not merely with seeing, more or less intuitively that certain things are the case.

Once a firm basis is laid down one may proceed with the development of a subject with less circumspection, and in the present case we have reached a stage where we are safely entitled to say that the geometry of a separation plane is formally identical with that of a Euclidean plane, and, in consequence, *the geometry of a separation threefold is formally identical with the ordinary (Euclidean) geometry of three dimensions*.

We do not propose to consider the theory of areas, volumes, etc. since these are formally identical with the corresponding theories in ordinary geometry.

### Theorem 199

*If B and C be two distinct elements in a separation line and O be their mean, and if A be any element in an optical line a which passes through O and is normal to BC, then*

$$(A, B) \equiv (A, C).$$

Since $a$ is an optical line which is normal to the separation line $BC$, it follows that $a$ and $BC$ lie in an optical plane, say $P$.

If the element $A$ should happen to coincide with $O$ then, since $BC$ is a separation line, the theorem obviously holds.

Suppose next that $A$ does not coincide with $O$, and let $d$ be a separation line passing through $O$ and normal to $P$.

Then $a$ and $d$ determine an optical plane, say $Q$, which is completely normal to $P$; and, since $BC$ and $d$ are both separation lines and are normal to one another, it follows that they lie in a separation plane, say $S$.

Let $D$ be any element of $d$ distinct from $O$.

Then $DO$ is normal to $BC$ and so, by Theorem 186, we have

$$(D, B) \equiv (D, C).$$

But, since $Q$ is completely normal to $P$, it follows that $DA$ is normal to $P$ and so $DA$ is normal to both $AB$ and $AC$.

Also, since $D$ is not an element of $a$, and $a$ is a generator of the optical plane $Q$, it follows that $DA$ must be a separation line.

Similarly, since $B$ and $C$ are not elements of $a$, and $a$ is a generator of the optical plane $P$, it follows that both $BA$ and $CA$ are separation lines.

Thus $DA$ and $BA$ must lie in a separation plane, say $R_1$, and $DA$ and $CA$ must lie in a separation plane, say $R_2$.

Thus, by Theorem 190, since

$$(A, D) \equiv (A, D)$$
and
$$(D, B) \equiv (D, C),$$
it follows that:
$$(A, B) \equiv (A, C)$$

as was to be proved.

### THEOREM 200

*If $O$ and $X_0$ be two distinct elements in a separation line lying in an optical plane $P$, then the set of all elements in $P$ such as $X$ where $OX$ is a separation line and*

$$(O, X) \equiv (O, X_0)$$

*consists of a pair of parallel optical lines.*

Let $X_0'$ be an element in $OX_0$ and on the opposite side of $O$ to that on which $X_0$ lies, and such that:

$$(O, X_0') \equiv (O, X_0).$$

Then $X_0'$ is an element of the set we are considering, and it is evident that it is the only one besides $X_0$ lying in the separation line $OX_0$.

Further it is evident that $O$ is the mean of $X_0$ and $X_0'$.

Let $a$, $b$ and $c$ be three generators of the optical plane $P$ passing through $X_0$, $X_0'$ and $O$ respectively.

Let $X_1$ be any element in $a$ distinct from $X_0$, and let $OX_1$ intersect $b$ in $X_1'$.

Further, let $c$ intersect $X_0'X_1$ in the element $M$.

Then $OX_1$ and $X_0'X_1$ are both separation lines, since they have each got elements in two distinct generators of the optical plane.

Now, since $c$ must be parallel to $a$, and since $O$ is the mean of $X_0$ and $X_0'$, it follows, by Theorem 94, that $M$ is the mean of $X_1$ and $X_0'$.

But, since $OM$ is an optical line and $X_0'X_1$ is a separation line in the same optical plane $P$ with it, it follows that $OM$ is normal to $X_0'X_1$.

Thus, by Theorem 199, we must have
$$(O, X_1) \equiv (O, X_0').$$
But, since $\quad (O, X_0') \equiv (O, X_0),$

it follows that: $\quad (O, X_1) \equiv (O, X_0).$

Similarly $\quad (O, X_1') \equiv (O, X_0'),$

and so $\quad (O, X_1') \equiv (O, X_0).$

Thus $X_1$ and $X_1'$ are evidently elements of the set we are considering, and are clearly the only ones lying in the separation line $OX_1$.

Similarly any other separation line passing through $O$ and lying in $P$ will intersect $a$ and $b$ in elements belonging to the set considered, and these will be the only ones lying in that separation line.

Thus the parallel optical lines $a$ and $b$ together constitute the set of elements in $P$, such as $X$, where $OX$ is a separation line and
$$(O, X) \equiv (O, X_0),$$
and so the theorem is proved.

### REMARKS

Certain interesting results follow directly from the last theorem.

Thus if we consider any triangle in an optical plane whose corners are $A$, $B$ and $C$, then not more than one of the general lines $AB$, $BC$, $CA$ can be an optical line, since no two optical lines in an optical plane can intersect.

If $BC$ be an optical line, then $AB$ and $CA$ must be separation lines, and from the last theorem it follows that:
$$(A, B) \equiv (A, C).$$

If, on the other hand, neither $AB$, $BC$ nor $CA$ be an optical line, they must all be separation lines.

In this case, let $a$, $b$ and $c$ be generators of the optical plane passing through $A$, $B$ and $C$ respectively, and intersecting $BC$, $CA$ and $AB$ in $A'$, $B'$ and $C'$ respectively.

Then, since neither $AB$, $BC$ nor $CA$ are optical lines, it follows that neither $A'$, $B'$ nor $C'$ can coincide with a corner of the triangle.

Thus we must either have

(1)  $A'$ linearly between $B$ and $C$,

or   (2)  $C$ linearly between $A'$ and $B$,

or   (3)  $B$ linearly between $C$ and $A'$.

In the first case, we shall also have

$\qquad\qquad\qquad$ $A$ linearly between $B'$ and $C$,

and $\qquad\qquad\quad$ $A$ linearly between $B$ and $C'$.

In the second case, we shall also have

$\qquad\qquad\qquad$ $C$ linearly between $A$ and $B'$,

and $\qquad\qquad\quad$ $C'$ linearly between $A$ and $B$.

In the third case, we shall also have

$\qquad\qquad\qquad$ $B$ linearly between $C'$ and $A$,

and $\qquad\qquad\quad$ $B'$ linearly between $C$ and $A$.

Thus in all cases one of the three elements $A'$, $B'$, $C'$, and only one, lies linearly between a pair of the corners $A$, $B$, $C$.

Now let us consider the case, for instance, where $A'$ is linearly between $B$ and $C$.

It follows directly from the last theorem that:

$$(B, A) \equiv (B, A'),$$

and $\qquad\qquad\quad$ $(C, A) \equiv (C, A').$

This remarkable result may be expressed as follows:

*If all three sides of a triangle in an optical plane be separation segments, then the sum of the lengths of a certain two of the sides is equal to that of the third side.*

Again, if $a$ and $b$ be a pair of neutral-parallel optical lines and if $A_1$ and $A_2$ be any elements in $a$, while $B_1$ and $B_2$ are any elements in $b$, we have

$$(A_1, B_1) \equiv (A_1, B_2),$$

and $\qquad\qquad\quad$ $(B_2, A_1) \equiv (B_2, A_2).$

Thus we see that we must have

$$(A_1, B_1) \equiv (A_2, B_2).$$

It will be observed that, in the case of an optical plane, a pair of parallel optical lines is the analogue of a circle, in so far as any analogue exists.

Again, if $W$ be an optical threefold and $O$ be any element in it, while $c$ is the generator of $W$ which passes through $O$, then any general plane in $W$ which contains $c$ is an optical plane, while any one which passes through $O$, but does not contain $c$, is a separation plane.

If then $S$ be any separation plane lying in $W$ and passing through $O$ and $X_0$ be any element in it distinct from $O$, the set of elements in $S$, such as $X$, where

$$(O, X) \equiv (O, X_0),$$

constitutes a separation circle.

If through each element of the separation circle a generator of $W$ be taken, then any element $X$ on any such generator will also satisfy the relation

$$(O, X) \equiv (O, X_0).$$

Further, it is clear that no other element of $W$ does satisfy it.

The set of elements thus obtained lie on a sort of cylinder which, in the case of an optical threefold, takes the place of a sphere.

We shall call this an *optical circular cylinder*.

## THEOREM 201

*If $A_1$, $B_1$, $C_1$ be the corners of a triangle in an inertia plane $P_1$ and $A_2$, $B_2$, $C_2$ be the corners of a triangle in an inertia plane $P_2$, and if further $B_1 C_1$ be a separation line which is normal to the inertia line $A_1 C_1$, while $B_2 C_2$ is a separation line which is normal to the inertia line $A_2 C_2$, then:*

(1) *If* $\qquad (C_1, A_1) \equiv (C_2, A_2)$

*and* $\qquad (C_1, B_1) \equiv (C_2, B_2),$

*we shall either have* $\qquad (A_1, B_1) \equiv (A_2, B_2),$

*or else both $A_1 B_1$ and $A_2 B_2$ will be optical lines.*

(2) *If* $\qquad (A_1, C_1) \equiv (C_2, A_2)$

*and* $\qquad (C_1, B_1) \equiv (C_2, B_2),$

*we shall either have* $\qquad (A_1, B_1) \equiv (B_2, A_2),$

*or else both $A_1 B_1$ and $B_2 A_2$ will be optical lines.*

Consider first part (1) of the theorem.

Since $(C_1, A_1) \equiv (C_2, A_2)$, and since these are inertia pairs, we must have either $A_1$ before $C_1$ and $A_2$ before $C_2$, or else have $A_1$ *after* $C_1$ and $A_2$ *after* $C_2$.

We shall only consider the case where $A_1$ is before $C_1$ and $A_2$ before $C_2$, since the other case is quite analogous.

If $A_1 B_1$ were an optical line we should have $(C_1,\ A_1)$ a before-conjugate to $(C_1,\ B_1)$ and if $A_2'$ were an element in $C_2 A_2$, such that

$(C_2, A_2')$ were a before-conjugate to $(C_2, B_2)$, then it would follow, by Theorem 179, that we must have

$$(C_1, A_1) \equiv (C_2, A_2'),$$

and so we should have

$$(C_2, A_2) \equiv (C_2, A_2').$$

Thus $A_2'$ would be identical with $A_2$, and so $A_2 B_2$ would also be an optical line.

We are not, however, at liberty to assert in this case that:

$$(A_1, B_1) \equiv (A_2, B_2),$$

but only that they are both optical pairs.

We shall suppose next that $A_1 B_1$ is not an optical line, and that accordingly $A_2 B_2$ is not an optical line.

Let $D_1$ and $D_2$ be elements in $C_1 A_1$ and $C_2 A_2$ respectively, such that $(C_1, D_1)$ is a before-conjugate to $(C_1, B_1)$, and $(C_2, D_2)$ is a before-conjugate to $(C_2, B_2)$.

Then, by Theorem 179, we must have

$$(C_1, D_1) \equiv (C_2, D_2).$$

Now two cases occur; we may have

(1) $A_1$ linearly between $D_1$ and $C_1$,

or (2) $D_1$ linearly between $A_1$ and $C_1$.

In the first case, since we also have

$$(C_1, A_1) \equiv (C_2, A_2),$$

it follows that we must also have $A_2$ linearly between $D_2$ and $C_2$, as was shown in the remarks at the end of Theorem 184.

Similarly, in the second case we must also have $D_2$ linearly between $A_2$ and $C_2$.

Again, in the first case we have $D_1$ *before* $C_1$, and must therefore have $A_1$ *after* $D_1$ and *before* $C_1$.

But $A_1$ could not be *before* $B_1$, for then $A_1$ would require to lie in the optical line $D_1 B_1$, which we know is not the case.

Further, $A_1$ could not be *after* $B_1$, for then, since $C_1$ is *after* $A_1$, we should have $C_1$ *after* $B_1$ contrary to the hypothesis that $B_1 C_1$ is a separation line.

It follows that in case (1) $A_1 B_1$ is a separation line, and similarly $A_2 B_2$ is a separation line.

In case (2), on the other hand, we must have $A_1$ *before* $D_1$ and so, since $D_1$ is *before* $B_1$, we must have $A_1$ *before* $B_1$.

Thus, since $D_1 A_1$ is an inertia line, and since $D_1$ is the only element common to it and the $\beta$ sub-set of $B_1$, it follows in this case that $A_1 B_1$ is an inertia line, and similarly $A_2 B_2$ is an inertia line.

We shall consider cases (1) and (2) separately.

*Case* (1).

We have here got $A_1 B_1$ and $A_2 B_2$, both separation lines.

Now let $W_1$ and $W_2$ be inertia threefolds containing $P_1$ and $P_2$ respectively, and let $S_1$ and $S_2$ be the separation planes in $W_1$ and $W_2$ which pass through $C_1$ and $C_2$, and are normal to the inertia lines $A_1 C_1$ and $A_2 C_2$ respectively.

Then, since $B_1 C_1$ is normal to $A_1 C_1$, it follows that $B_1 C_1$ must lie in $S_1$ and similarly $B_2 C_2$ must lie in $S_2$.

Now, since $A_1 B_1$ is a separation line, there is an inertia plane which passes through $A_1$, lies in $W_1$ and is normal to $A_1 B_1$.

This inertia plane contains two optical lines which pass through $A_1$ and must be normal to $A_1 B_1$ and which must intersect $S_1$, since $S_1$ is a separation plane in the same inertia threefold along with these optical lines.

Let one of these optical lines intersect $S_1$ in the element $E_1$.

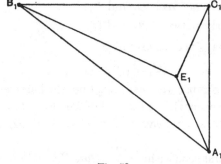

Fig. 52.

Similarly we can show that there are two optical lines passing through $A_2$ and lying in $W_2$, and which are normal to $A_2 B_2$.

These optical lines may be shown in a similar manner to intersect $S_2$, and we shall suppose that one of them intersects $S_2$ in the element $E_2$.

Now, since the optical line $A_1 E_1$ is normal to the separation line $A_1 B_1$, it follows that $A_1 E_1$ and $A_1 B_1$ lie in an optical plane.

Similarly $A_2 E_2$ and $A_2 B_2$ lie in an optical plane.

But, since an optical line in an optical plane is normal to every general line in the optical plane, it follows that $A_1 E_1$ is normal to $E_1 B_1$, and similarly $A_2 E_2$ is normal to $E_2 B_2$.

Again, since $E_1 B_1$ lies in $S_1$ and since $S_1$ is normal to $A_1 C_1$, it follows that $A_1 C_1$ is normal to $E_1 B_1$.

Thus $E_1 B_1$ is normal to the two intersecting general lines $A_1 E_1$ and $A_1 C_1$, and is therefore normal to the general plane containing them.

It follows that $E_1 B_1$ is normal to $E_1 C_1$, and similarly $E_2 B_2$ is normal to $E_2 C_2$.

Again, since $S_1$ is normal to $A_1 C_1$, it follows that $E_1 C_1$ is normal to $A_1 C_1$, and similarly it follows that $E_2 C_2$ is normal to $A_2 C_2$.

Thus, since $A_1 C_1$ and $A_2 C_2$ are inertia lines while $A_1 E_1$ and $A_2 E_2$ are optical lines, it follows that $(C_1, E_1)$ and $(C_2, E_2)$ are after-conjugates to $(C_1, A_1)$ and $(C_2, A_2)$ respectively.

But, since $$(C_1, A_1) \equiv (C_2, A_2),$$ it follows by Theorem 179 that:
$$(C_1, E_1) \equiv (C_2, E_2).$$

Thus $C_1$, $B_1$, $E_1$ are the corners of a triangle in the separation plane $S_1$ and $C_2$, $B_2$, $E_2$ are the corners of a triangle in the separation plane $S_2$, while further
$$(E_1, C_1) \equiv (E_2, C_2),$$
$$(C_1, B_1) \equiv (C_2, B_2),$$
and also $B_1 E_1$ is normal to $C_1 E_1$ and $B_2 E_2$ is normal to $C_2 E_2$, and so, by Theorem 190,
$$(E_1, B_1) \equiv (E_2, B_2).$$

But since $E_1 B_1$ and $A_1 B_1$ are separation lines lying in an optical plane, of which $A_1 E_1$ is a generator, it follows from the remarks at the end of Theorem 200 that:
$$(E_1, B_1) \equiv (A_1, B_1).$$
Similarly $\qquad (E_2, B_2) \equiv (A_2, B_2).$
Thus we get finally $\qquad (A_1, B_1) \equiv (A_2, B_2),$
and so the theorem is proved in case (1).

*Case* (2).

We have here got $A_1 B_1$ and $A_2 B_2$, both inertia lines.

As before, let $W_1$ and $W_2$ be inertia threefolds containing $P_1$ and $P_2$ respectively, and let $S_1$ and $S_2$ be the separation planes in $W_1$ and $W_2$ which pass through $C_1$ and $C_2$ and are normal to the inertia lines $A_1 C_1$ and $A_2 C_2$ respectively.

Then, as in the first case, $B_1 C_1$ lies in $S_1$ and $B_2 C_2$ lies in $S_2$.

Let $b_1$ be the separation line in $S_1$ which passes through $B_1$ and is normal to $B_1 C_1$, and similarly let $b_2$ be the separation line in $S_2$ which passes through $B_2$ and is normal to $B_2 C_2$.

Then, since $A_1 B_1$ is an inertia line, it follows that $A_1 B_1$ and $b_1$ lie in an inertia plane, say $Q_1$, and similarly $A_2 B_2$ and $b_2$ lie in an inertia plane, say $Q_2$.

Let one of the generators of $Q_1$ which pass through $A_1$ intersect $b_1$ in the element $F_1$, and let one of the generators of $Q_2$ which pass through $A_2$ intersect $b_2$ in the element $F_2$.

Now, since $A_1 C_1$ is an inertia line, it follows that $A_1 C_1$ and $A_1 F_1$ determine an inertia plane, and similarly $A_2 C_2$ and $A_2 F_2$ determine an inertia plane.

Since $C_1 F_1$ lies in $S_1$, it must be normal to $A_1 C_1$, and since $C_2 F_2$ lies in $S_2$, it must be normal to $A_2 C_2$.

Thus, since $A_1 F_1$ and $A_2 F_2$ are optical lines, it follows that $(C_1, F_1)$ $(C_2, F_2)$ are after-conjugates to $(C_1, A_1)$ and $(C_2, A_2)$ respectively, and so, since

$$(C_1, A_1) \equiv (C_2, A_2),$$

it follows, by Theorem 179, that:

$$(C_1, F_1) \equiv (C_2, F_2).$$

But now $C_1$, $F_1$, $B_1$ are the corners of a triangle in the separation

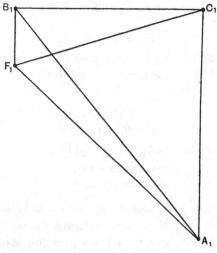

Fig. 53.

plane $S_1$ and $C_2$, $F_2$, $B_2$ are the corners of a triangle in the separation plane $S_2$, while further

$$(C_1, B_1) \equiv (C_2, B_2),$$
$$(C_1, F_1) \equiv (C_2, F_2),$$

and also $F_1 B_1$ is normal to $C_1 B_1$ and $F_2 B_2$ is normal to $C_2 B_2$ and so, by Theorem 190,
$$(B_1, F_1) \equiv (B_2, F_2).$$

Since $F_1 B_1$ lies in $S_1$, it is normal to $A_1 C_1$, and by hypothesis it is also normal to $B_1 C_1$ and so, since $A_1 C_1$ and $B_1 C_1$ are intersecting general lines in $P_1$, it follows that $F_1 B_1$ is normal to $P_1$.

Thus $F_1 B_1$ must be normal to $A_1 B_1$ and similarly $F_2 B_2$ must be normal to $A_2 B_2$.

But $A_1 F_1$ and $A_2 F_2$ are optical lines while $A_1 B_1$ and $A_2 B_2$ are inertia lines and so $(B_1, F_1)$ and $(B_2, F_2)$ are after-conjugates to $(B_1, A_1)$ and $(B_2, A_2)$ respectively.

Thus since          $$(B_1, F_1) \equiv (B_2, F_2),$$
it follows, by Theorem 179, that:
$$(B_1, A_1) \equiv (B_2, A_2),$$
that is          $$(A_1, B_1) \equiv (A_2, B_2),$$
as was to be proved.

Consider now part (2) of the theorem.

Since $(A_1, C_1) \equiv (C_2, A_2)$ and since these are inertia pairs we must either have $A_1$ *before* $C_1$ and $A_2$ *after* $C_2$ or else have $A_1$ *after* $C_1$ and $A_2$ *before* $C_2$.

There is then no difficulty in showing that:
$$(A_1, B_1) \equiv (B_2, A_2),$$
provided that $A_1 B_1$ be not an optical line.

The proof is quite analogous to that of the first part of the theorem except that we make use of the result given at the end of Theorem 179 in place of Theorem 179 itself.

It is also evident that if $A_1 B_1$ be an optical line, then $B_2 A_2$ must also be an optical line.

Thus both parts of the theorem hold.

It will be observed that the two parts of Theorem 201 are the analogue for inertia planes of Theorem 189.

### Analogues of the Theorem of Pythagoras
#### in inertia and optical planes

In Theorem 198 we proved that the equivalent of the theorem of Pythagoras holds in a separation plane and we now propose to make use of the constructions of the two cases of Theorem 201 in order to obtain the analogue for the case of an inertia plane.

It is only necessary to consider the construction in connexion with one of the two triangles: say that whose corners are $A_1$, $B_1$, $C_1$.

In case (1) $B_1 C_1$ is a separation line, $A_1 C_1$ is an inertia line normal to $B_1 C_1$, and $A_1 B_1$ is a separation line.

But now we obtained a triangle whose corners were $B_1$, $C_1$ and $E_1$ which lay in the separation plane $S_1$ and such that $E_1 B_1$ was normal to $E_1 C_1$ and in which accordingly we must have the segment relation:

$$(B_1 C_1)^2 = (E_1 B_1)^2 + (E_1 C_1)^2.$$

This triangle was related to the one whose corners are $A_1$, $B_1$, $C_1$ in such a way that

$$(E_1, B_1) \equiv (B_1, A_1),$$

while $(C_1, E_1)$ was a before- or after-conjugate to $(C_1, A_1)$.

Thus taking segments instead of pairs we get

$$(B_1 C_1)^2 = (B_1 A_1)^2 + (\text{conjugate } C_1 A_1)^2.$$

Thus the analogue of the theorem of Pythagoras is in this case

$$(B_1 A_1)^2 = (B_1 C_1)^2 - (\text{conjugate } C_1 A_1)^2 \qquad \ldots\ldots(\text{i}).$$

Again if we consider case (2) we have $B_1 C_1$ is a separation line, $A_1 C_1$ is an inertia line normal to $B_1 C_1$, and $A_1 B_1$ is also an inertia line.

In this case we obtained a triangle whose corners were $C_1$, $B_1$ and $F_1$ which lay in the separation plane $S_1$ and such that $B_1 F_1$ was normal to $B_1 C_1$.

Thus we must have the segment relation:

$$(C_1 F_1)^2 = (B_1 C_1)^2 + (B_1 F_1)^2.$$

This triangle was related to the one whose corners are $A_1$, $B_1$, $C_1$ in such a way that $(C_1, F_1)$ was a before- or after-conjugate to $(C_1, A_1)$ while $(B_1, F_1)$ was a before- or after-conjugate to $(B_1, A_1)$, and so, taking segments instead of pairs, we get

$$(\text{conjugate } C_1 A_1)^2 = (B_1 C_1)^2 + (\text{conjugate } B_1 A_1)^2.$$

Thus the analogue of the theorem of Pythagoras is in this case

$$-(\text{conjugate } B_1 A_1)^2 = (B_1 C_1)^2 - (\text{conjugate } C_1 A_1)^2 \ \ldots(\text{ii}).$$

In the case where $A_1 B_1$ is an optical line we obviously have

$$0 = (B_1 C_1)^2 - (\text{conjugate } C_1 A_1)^2 \qquad \ldots\ldots(\text{iii}).$$

Thus (i), (ii) and (iii) constitute the complete analogue of the Pythagorean theorem in an inertia plane provided that $A_1$, $B_1$ and $C_1$ form the corners of a triangle.*

If we consider a triangle whose corners are $A_1$, $B_1$, $C_1$ and which lies in an optical plane, then if $B_1 C_1$ be a separation line and $A_1 C_1$ be normal

* Cf. footnote, p. 369.

to $B_1 C_1$ we know that $A_1 C_1$ must be an optical line, while $A_1 B_1$ must be another separation line.

Now we have shown that:

$$(B_1, A_1) \equiv (B_1, C_1),$$

and so taking segments instead of pairs we see that:

$$(B_1 A_1)^2 = (B_1 C_1)^2 \qquad \ldots\ldots\text{(iv)}.$$

This is the analogue of the Pythagorean theorem in an optical plane provided that $A_1$, $B_1$ and $C_1$ form the corners of a triangle.*

Considering now equations (i), (ii), (iii) and (iv) we observe that *the modifications which take place in the theorem of Pythagoras are such that when any side of the triangle becomes an inertia segment the corresponding square is replaced by the negative square of the conjugate of this inertia segment, while if any side becomes an optical segment, the corresponding square is replaced by zero.*

If we consider equation (i) we see that:

$$(B_1 A_1)^2 < (B_1 C_1)^2$$

and accordingly $\qquad B_1 A_1 < B_1 C_1.$

Again, if we consider equation (ii) we see that:

$$(\text{conjugate } B_1 A_1)^2 < (\text{conjugate } C_1 A_1)^2$$

and accordingly $\qquad B_1 A_1 < C_1 A_1.$

Thus, *provided that the hypotenuse is not an optical line, its length is less than the length of that side which is of the same kind as itself.*

We shall now make use of this result in order to prove another important theorem concerning triangles in an inertia plane.

Let $A$, $B$, $C$ be the corners of a triangle in an inertia plane $P$, and let $AB$, $BC$ and $CA$ be all separation lines or all inertia lines.

It is easy to see that triangles of both these kinds exist, although, as Theorem 14 shows, it is not possible for $AB$, $BC$ and $CA$ to be all optical lines.

Let $a_1$, $b_1$, $c_1$ be generators of $P$ of one set, which pass through $A$, $B$, $C$ respectively and intersect $BC$, $CA$, $AB$ in $A_1$, $B_1$, $C_1$ respectively.

---

* There is a limiting case of analogue to the theorem of Pythagoras in which the elements $A_1$, $B_1$ and $C_1$ do not form the corners of a triangle but all lie in one optical line. It may be stated in the following form:

If $A_1$, $B_1$ and $C_1$ be three distinct elements and if $B_1 C_1$ and $C_1 A_1$ be segments of optical lines which are normal to one another, then $B_1 A_1$ is also a segment of an optical line. This follows directly since $B_1 C_1$ and $C_1 A_1$ being normal to one another, must be segments of the same optical line, as must also $B_1 A_1$.

Then we may show by a method similar to that employed in the remarks at the end of Theorem 200, that one and only one of the elements $A_1$, $B_1$, $C_1$ is linearly between a pair of the corners $A$, $B$, $C$.

Similarly we may show that if $a_2$, $b_2$, $c_2$ be generators of $P$ of the opposite set passing through the elements $A$, $B$, $C$ respectively and intersecting $BC$, $CA$, $AB$ in $A_2$, $B_2$, $C_2$ respectively, then one and only one of the elements $A_2$, $B_2$, $C_2$ is linearly between a pair of the corners $A$, $B$, $C$.

Now consider the case, for instance, where $A_1$ is linearly between $B$ and $C$, and suppose first that $AB$, $BC$, $CA$ are all separation lines.

Then $B$ cannot be linearly between $A_1$ and $A_2$ for then, by Theorem 73 (a) and (b), $AB$ would require to be an inertia line, contrary to hypothesis.

Similarly $C$ cannot be linearly between $A_1$ and $A_2$.

Thus, since obviously $A_2$ cannot be identical with either $B$ or $C$, it follows that $A_2$ must be also linearly between $B$ and $C$.

Now let $O$ be the mean of $A_1$ and $A_2$.

Then $O$ is linearly between $A_1$ and $A_2$, and therefore clearly it must lie linearly between $B$ and $C$.

But now $A_1$, $A$, $A_2$ are three corners of an optical parallelogram of which $O$ is the centre, and so $AO$ must be normal to $A_1A_2$: that is to $BC$.

Again, if instead of $AB$, $BC$, $CA$ being all separation lines they are all inertia lines, a similar result holds.

Let us take the case where $A_1$ is linearly between $B$ and $C$.

Then clearly $B$ cannot be linearly between $A_1$ and $A_2$, for then $AB$ would require to be a separation line, and, for a similar reason, $C$ cannot be linearly between $A_1$ and $A_2$.

Thus, since $A_2$ cannot coincide with either $B$ or $C$, it follows that $A_2$ must also be linearly between $B$ and $C$.

As in the former case, if $O$ be the mean of $A_1$ and $A_2$, then $O$ must be linearly between $B$ and $C$, and $AO$ must be normal to $BC$.

Making use of the result proved above we see that, whether the sides be all separation lines, or all inertia lines we have

segment $BA$ is less than segment $BO$,

and        segment $AC$ is less than segment $OC$.

Thus it follows that the sum of the lengths of the segments $BA$ and $AC$ is less than that of the segment $BC$.

Now we know that in a separation plane the sum of the lengths of any two sides of a triangle is greater than that of the third, and thus, remembering what was proved at the end of Theorem 199, we have the following interesting results:

*If A, B, C be the corners of a general triangle all whose sides are segments of one kind, then:*

(1) *If the triangle lies in a separation plane, the sum of the lengths of any two sides is greater than that of the third side.*

(2) *If the triangle lies in an optical plane, the sum of the lengths of a certain two sides is equal to that of the third side.*

(3) *If the triangle lies in an inertia plane, the sum of the lengths of a certain two sides is less than that of the third side.*

## THEOREM 202

*If A, B, C be three distinct elements in an inertia plane P which do not all lie in one general line and if*

$$(A, B) \equiv (A, C),$$

*or if*
$$(B, A) \equiv (A, C),$$

*then BC cannot be an optical line.*

Since the only congruence of optical pairs is co-directional, it is evident that neither $AB$ nor $AC$ can be optical lines and must therefore be either inertia or separation lines.

Consider first the case where they are inertia lines and

$$(A, B) \equiv (A, C).$$

It is evident that we must either have $A$ *before* both $B$ and $C$ or *after* both $B$ and $C$.

Suppose $A$ is *before* both $B$ and $C$ and let $a$ be a separation line passing through $A$ and normal to the inertia plane containing $AB$ and $AC$.

Let $D$ be an element in $a$ such that $(A, D)$ is a before-conjugate to $(A, B)$.

Then $(A, D)$ will also be a before-conjugate to $(A, C)$ since

$$(A, B) \equiv (A, C).$$

Thus $DB$ and $DC$ will both be optical lines, and so $BC$ cannot be an optical line.*

If $A$ be *after* both $B$ and $C$ the result follows in a similar manner.

Next consider the case where $AB$ and $AC$ are inertia lines but where

$$(B, A) \equiv (A, C).$$

---

* It is also to be noted that $B$ is neither *before* nor *after* $C$ in this case.

We must then either have $A$ *after* $B$ and $C$ *after* $A$ or else $A$ *before* $B$ and $C$ *before* $A$.

In either case it is evident that $BC$ could not be an optical line, for otherwise $A$ would be *after* one element of it and *before* another and yet not lie in the optical line; which we know to be impossible.

Consider next the case where $AB$ and $AC$ are separation lines and where accordingly

$$(A, B) \equiv (A, C)$$

implies $\qquad\qquad (B, A) \equiv (A, C),$

and conversely.

Now we know that there is one single optical parallelogram in $P$ having $A$ as centre and $B$ as one of its corners.

Suppose, if possible, that $BC$ is an optical line which we shall denote shortly by $b$.

Then $b$ would be one of the side lines of this optical parallelogram, and we shall denote the opposite side line by $b'$.

Let $B'$ be the corner opposite to $B$ and let $D$ and $D'$ be the remaining two corners: $D$ lying in $b$ and $D'$ lying in $b'$.

Let $CA$ intersect $b'$ in the element $C'$ and let optical lines passing through $C$ and $C'$ respectively and parallel to $BD'$ intersect $b'$ and $b$ in $E'$ and $E$ respectively.

Then $E'$, $C$, $E$, $C'$ would form the corners of an optical parallelogram having also $b$ and $b'$ as a pair of opposite side lines.

Thus, since the diagonal line $CC'$ passes through $A$, it follows, by Theorem 64, that these two optical parallelograms would have a common centre $A$.

But now either $(A, D)$ or $(A, D')$ would be an after-conjugate to $(A, B)$ while $(A, E)$ or $(A, E')$ would be an after-conjugate to $(A, C)$ and $DE$ and $D'E'$ would both be optical lines.

Thus by the first case of the theorem it is impossible that we should have

$$(A, D) \equiv (A, E),$$

or $\qquad\qquad (A, D') \equiv (A, E').$

If however we had $\qquad (A, B) \equiv (A, C),$

these other congruences would require to hold and so it is impossible to have $BC$ an optical line if

$$(A, B) \equiv (A, C).$$

Thus the theorem holds in all cases.

It is important to note that while this result holds for an inertia plane, it does not, as we have already shown, hold for an optical plane.

Thus since an optical line can only lie in an inertia or optical plane, it follows that:

*If B and C be two distinct elements in an optical line while A is an element which does not lie in BC, then if*

$$(A, B) \equiv (A, C)$$

*the elements A, B, C must lie in an optical plane.*

In the remaining theorems to be considered dealing with triangles in inertia planes, we propose to treat of *equality of segments* rather than of *congruence of pairs*; as was done in connexion with triangles in separation planes.

In the latter case it is a matter of indifference whether we consider the congruence of pairs or the equality of segments; since the two subjects run parallel; but in inertia planes things are somewhat different.

In order to deal with the congruence of pairs in inertia planes, it is often necessary to make the enunciation of theorems very complicated in order to cover the various possibilities which occur as regards *before* and *after* relations of the pairs in inertia or optical lines.

If we deal with the equality of segments, on the other hand, it is generally easy, in any particular case, to express the result in the notation of pairs if so required.

### Theorem 203

*If A, B and C be three distinct elements which lie in an inertia plane P, but do not all lie in one general line and if O be the mean of B and C, then if*

$$AB = AC,$$

*or if AB and AC be both optical lines, we must have AO normal to BC.*

The conditions of this theorem could not be satisfied if $BC$ were an optical line for, in the first place, since $A$, $B$ and $C$ lie in an inertia plane, it follows from the last theorem that we could not have $AB = AC$.

In the second place, Theorem 14 shows that, if $BC$ were an optical line, we could not have $AB$ and $AC$ also optical lines.

It follows that $BC$ must either be a separation line or an inertia line.

In either of these cases, if $AB$ and $AC$ were both optical lines, then $A$, $B$ and $C$ would be three corners of an optical parallelogram of which $O$ would be the centre and so $AO$ would be normal to $BC$ by definition.

Next suppose that $BC$ is a separation or inertia line and let the normal to $BC$ through $A$ intersect $BC$ in $O'$.

Then in all cases, considering the two triangles $AO'B$ and $AO'C$, we have

$$AB = AC$$

and $AO'$ common, so that, applying the analogue of the theorem of Pythagoras, we get either

$$(BO')^2 = (CO')^2,$$

or          $$(\text{conjugate } BO')^2 = (\text{conjugate } CO')^2.$$

Thus we have          $$BO' = CO',$$

so that $O'$ must be the mean of $B$ and $C$ and therefore must be identical with $O$.

Thus since, by hypothesis $AO'$ is normal to $BC$, we must have $AO$ normal to $BC$, as was to be proved.

We have incidentally proved that:

*If $A$, $B$ and $C$ be three distinct elements which lie in an inertia plane $P$, but do not all lie in one general line and if $O$ be an element in $BC$ such that $AO$ is normal to $BC$, then if*

$$AB = AC,$$

*or if $AB$ and $AC$ be both optical lines, the element $O$ must be the mean of $B$ and $C$.*

## THEOREM 204

*If $A_1$, $B_1$, $C_1$ be the corners of a triangle in an inertia plane $P_1$, and $A_2$, $B_2$, $C_2$ be the corners of a triangle in an inertia plane $P_2$, and if further $B_1 C_1$ is normal to $A_1 C_1$ and $B_2 C_2$ is normal to $A_2 C_2$, then if the segments*

$$C_1 A_1 = C_2 A_2,$$

*and*          $$A_1 B_1 = A_2 B_2,$$

*or, alternately to the latter equality, if $A_1 B_1$ and $A_2 B_2$ be both optical lines, we shall also have*

$$C_1 B_1 = C_2 B_2.$$

It will be observed that, except that we have taken segments instead of pairs, this is the analogue of Theorem 190 and may be proved in a similar manner, using Theorem 201 in place of Theorem 189 and Theorem 203 in place of Theorem 187.

It should be noticed that neither $C_1A_1$ nor $C_1B_1$ can be optical lines for, since they are normal, they would then coincide and $A_1$, $B_1$, $C_1$ could not be corners of a triangle. Similarly, neither $C_2A_2$ nor $C_2B_2$ can be optical lines.

This point is required in applying Theorem 203 but otherwise the proof is quite analogous.

## THEOREM 205

*If $A_1$, $B_1$, $C_1$ be the corners of a triangle in an inertia plane $P_1$, and $A_2$, $B_2$, $C_2$ be the corners of a triangle in an inertia plane $P_2$, and if $A_1C_1$ be a separation line which is normal to the inertia line $B_1C_1$, then if the segments*

$$A_1C_1 = A_2C_2,$$
$$B_1C_1 = B_2C_2,$$

*and if* $\qquad\qquad A_1B_1 = A_2B_2,$

*or, alternately to the last equality, if $A_1B_1$ and $A_2B_2$ be both optical lines, we must also have $A_2C_2$ normal to $B_2C_2$.*

From the equalities it follows that, since $A_1C_1$ is a separation line, $A_2C_2$ must be a separation line, and, since $B_1C_1$ is an inertia line, $B_2C_2$ must be an inertia line.

Thus any general line normal to $B_2C_2$ must be a separation line.

Except that we have taken segments instead of pairs, this is the analogue of Theorem 191 and may be proved in a similar manner with some slight modifications.

Using the same notation employed in Theorem 191, the only point to be noticed in the case where $A_1B_1$ and $A_2B_2$ are not optical lines is that, since

$$A_2C_2 = A_2'C_2,$$

the hypothetical general line $A_2A_2'$ could not be an optical line, by Theorem 202, and so the normal to it through $O$ could not coincide with itself.

Apart from this the proof is similar to that of Theorem 191 using Theorem 201 in place of Theorem 189 and Theorem 203 in place of Theorem 187.

In the case where $A_1B_1$ and $A_2B_2$ are both optical lines the following point is to be noted: If $A_2'$ lies in $A_2B_2$ it must coincide with $A_2$: not because we are entitled to assert the equality of $A_2B_2$ and $A_2',B_2$, but because, if it did not do so, we should have

$$A_2C_2 = A_2'C_2$$

and $A_2A_2'$ an optical line which we know is impossible.

Apart from this the proof is similar to that in the case where $A_1 B_1$ and $A_2 B_2$ are not optical lines.

## REMARKS

This theorem may be used to prove a converse to the analogue of the theorem of Pythagoras in an inertia plane.

Let $A_1$, $B_1$ and $C_1$ be the corners of a triangle in an inertia plane and let $B_1 C_1$ be a segment of a separation line while $C_1 A_1$ is a segment of an inertia line and suppose that one of the three following conditions holds:

(i) $B_1 A_1$ is a segment of a separation line and
$$(B_1 A_1)^2 = (B_1 C_1)^2 - (\text{conjugate } C_1 A_1)^2;$$

(ii) $B_1 A_1$ is a segment of an inertia line and
$$- (\text{conjugate } B_1 A_1)^2 = (B_1 C_1)^2 - (\text{conjugate } C_1 A_1)^2;$$

(iii) $B_1 A_1$ is a segment of an optical line and
$$O = (B_1 C_1)^2 - (\text{conjugate } C_1 A_1)^2;$$

we shall prove that $A_1 C_1$ is normal to $B_1 C_1$.

Consider a second triangle in an inertia plane and let its corners be $A_2$, $B_2$, $C_2$.

Let $B_2 C_2$ be a separation line and let $A_2 C_2$ be an inertia line which is normal to $B_2 C_2$ and let the segments
$$B_2 C_2 = B_1 C_1$$
and
$$C_2 A_2 = C_1 A_1.$$

Then we shall have one of the three conditions:

(1) $B_2 A_2$ is a segment of a separation line and
$$(B_2 A_2)^2 = (B_2 C_2)^2 - (\text{conjugate } C_2 A_2)^2;$$

(2) $B_2 A_2$ is a segment of an inertia line and
$$- (\text{conjugate } B_2 A_2)^2 = (B_2 C_2)^2 - (\text{conjugate } C_2 A_2)^2;$$

(3) $B_2 A_2$ is a segment of an optical line and
$$O = (B_2 C_2)^2 - (\text{conjugate } C_2 A_2)^2.$$

Thus we must either have
$$B_2 A_2 = B_1 A_1,$$
or else both $B_2 A_2$ and $B_1 A_1$ must be segments of optical lines.

Thus the conditions of Theorem 205 hold between the two triangles and so $A_1 C_1$ must be normal to $B_1 C_1$.

It is obvious that the conditions (i), (ii) or (iii) could only hold in an inertia plane since inertia segments are involved in each of them.

Consider now a triangle whose corners are $A_1$, $B_1$ and $C_1$ and suppose the condition holds:

(iv) $A_1 C_1$ is a segment of an optical line and

$$(B_1 A_1)^2 = (B_1 C_1)^2.$$

Then $$B_1 A_1 = B_1 C_1,$$

and so, as was shown at the end of Theorem 202, the triangle must lie in an optical plane and, since $A_1 C_1$ is a segment of an optical line, we must have $A_1 C_1$ normal to $B_1 C_1$ and also normal to $B_1 A_1$.

Consider now a triangle whose corners are $A_1$, $B_1$ and $C_1$ and in which the segment relation holds:

(v) $$(B_1 A_1)^2 = (B_1 C_1)^2 + (C_1 A_1)^2.$$

It is obvious that in this triangle the sides are all segments of one kind and the sum of the lengths of any two sides is greater than that of the third side, and accordingly it can only lie in a separation plane and the sides must be segments of separation lines.

By a method similar to that which we employed in dealing with conditions (i), (ii) and (iii), but using Theorem 191 in place of Theorem 205, we can prove that $A_1 C_1$ must be normal to $B_1 C_1$.

There is a limiting case of analogue to the theorem of Pythagoras in which the segments do not form a triangle and which was mentioned in the footnote on p. 369.

The converse of this also holds and may be stated as follows:

(vi) If $A_1$, $B_1$ and $C_1$ be three distinct elements and if $B_1 C_1$, $C_1 A_1$ and $B_1 A_1$ be all segments of optical lines, then $A_1 C_1$ must be normal to $B_1 C_1$.

For, by Theorem 14, $A_1$, $B_1$ and $C_1$ cannot lie in pairs in three distinct optical lines and must therefore all lie in one optical line.

Thus $A_1 C_1$ must be normal to $B_1 C_1$.

We have thus got the complete converse to the various forms of analogue to the theorem of Pythagoras.

### THEOREM 206

*If $A_1$, $B_1$, $C_1$ be the corners of a triangle in an inertia plane $P_1$, and $A_2$, $B_2$, $C_2$ be the corners of a triangle in an inertia plane $P_2$: the triangles being such that no side of either is an optical line; and if the segments*

$$A_1 B_1 = A_2 B_2,$$
$$A_1 C_1 = A_2 C_2,$$
$$B_1 C_1 = B_2 C_2,$$

*while $N_1$ is an element in $B_1 C_1$ such that $A_1 N_1$ is normal to $B_1 C_1$, and*

$N_2$ is an element in $B_2 C_2$ such that $A_2 N_2$ is normal to $B_2 C_2$; and if $N_1$ is distinct from both $B_1$ and $C_1$, then $N_2$ will be distinct from both $B_2$ and $C_2$, and we shall also have

$$A_1 N_1 = A_2 N_2,$$
$$B_1 N_1 = B_2 N_2,$$
$$C_1 N_1 = C_2 N_2.$$

1ST EXTENSION. *The same results hold if, instead of the segments $A_1 B_1$ and $A_2 B_2$ being equal, the general lines $A_1 B_1$ and $A_2 B_2$ are both optical lines.*

2ND EXTENSION. *The same results hold, if in addition to $A_1 B_1$ and $A_2 B_2$ being optical lines, $A_1 C_1$ and $A_2 C_2$ are optical lines instead of the segments $A_1 C_1$ and $A_2 C_2$ being equal.*

It will be observed that, except that we have taken segments instead of pairs, this is the analogue of Theorem 192 and, when none of the sides of the triangles is an optical line, it may be proved in a similar manner using Theorem 181 instead of Theorem 182 in those cases where $B_1 C_1$ is an inertia line and using Theorem 201 instead of Theorem 189. We also must take note of the consequences of Theorem 202 as was done in proving Theorem 205.

As regards the 1st extension it may be proved in a similar manner again taking note of the consequences of Theorem 202 as was done in proving Theorem 205 in the case where $A_1 B_1$ and $A_2 B_2$ are both optical lines.

As regards the 2nd extension it is clear in this case that $B_1$, $A_1$ and $C_1$ are the corners of an optical parallelogram of which $B_1 C_1$ is one diagonal line and $N_1$ is the centre. Similarly $B_2$, $A_2$ and $C_2$ are the corners of an optical parallelogram of which $B_2 C_2$ is a diagonal line of the same kind as $B_1 C_1$, while $N_2$ is the centre.

Thus        $$B_1 N_1 = B_2 N_2 = C_1 N_1 = C_2 N_2,$$

while $A_1 N_1$ is a conjugate to $B_1 N_1$ and $A_2 N_2$ is a conjugate to $B_2 N_2$, so that we must have

$$A_1 N_1 = A_2 N_2.$$

It is also evident, as in Theorem 192, that we must have

|                | $N_2$ linearly between $B_2$ and $C_2$, |
| -------------- | --------------------------------------- |
| or             | $C_2$ linearly between $B_2$ and $N_2$, |
| or             | $B_2$ linearly between $C_2$ and $N_2$, |
| according as   | $N_1$ is linearly between $B_1$ and $C_1$, |
| or             | $C_1$ is linearly between $B_1$ and $N_1$, |
| or             | $B_1$ is linearly between $C_1$ and $N_1$. |

## REMARKS

If $B$, $C_1$ and $C_2$ be three distinct elements in an inertia plane which do not all lie in one general line, but such that $BC_1$ and $BC_2$ are both inertia lines or both separation lines, and if the segments

$$BC_1 = BC_2,$$

while $N_1$ and $N_2$ are elements in the general lines $BC_1$ and $BC_2$ respectively, such that $C_2N_1$ is normal to $BC_1$ and $C_1N_2$ is normal to $BC_2$, then we may take the triangle whose corners are $B$, $C_1$ and $C_2$ and apply the results of the last theorem to the one triangle taken in two aspects, as was done in the case of Theorem 192, and so prove that

$$C_2N_1 = C_1N_2,$$
$$BN_1 = BN_2,$$
$$C_1N_1 = C_2N_2.$$

Also the linearly between relations of $B$, $N_2$ and $C_2$ will be similar to those of $B$, $N_1$ and $C_1$ respectively.

## PROPER HYPERBOLIC ANGLES

While any optical line in a given inertia plane intersects every inertia line and every separation line lying in the inertia plane, this is clearly not the case for every inertia half-line or for every separation half-line in it; since the half-line under consideration may be a part of the complete inertia or separation line which does not contain the element of intersection.

If we have two inertia or two separation half-lines with a common end lying in an inertia plane, such half-lines may be such that they may both be intersected by the same optical lines or such that cannot both be intersected by any optical line.

It should be remembered that, by definition, the end of a half-line is not included in it.

If we revert to the remarks at the end of Theorem 14, employing the notation there used we see that the general half-line $OF$ (where $F$ is any element of $b$ which is *after* $E$) must be an inertia half-line having $F$ *after* the end element $O$. Taking any number of positions for $F$, we get any number of inertia half-lines having the common end $O$ and which are all intersected by the optical line $b$.

Similarly taking $F'$ as any element of $b'$ which is *before* $E'$ we get an inertia half-line $OF'$ having $F'$ *before* the end element $O$ and, by taking any number of positions for $F'$ we get any number of inertia half-lines having the common end $O$ and which are all intersected by the optical line $b'$; which we shall here suppose to be in the same inertia plane as $b$.

An inertia half-line such as $OF$ and one such as $OF'$ cannot be intersected by the same optical line, for, if $F_1$ be any element of the half-line $OF$ and $F_1'$ be any element of the half-line $OF'$, then we should have $O$ *after* $F_1'$ and also $O$ *before* $F_1$ so that if $O$ does not lie in $F_1 F_1'$ we know that $F_1 F_1'$ could not be an optical line, while if $O$ does lie in $F_1 F_1'$ then $F_1 F_1'$ must be an inertia line of which the half-lines $OF$ and $OF'$ are parts.

Thus in all cases $F_1 F_1'$ must be an inertia line, since $F_1$ must be *after* $F_1'$.

Again, since every optical line in the inertia plane intersects every inertia line in it, it follows that every optical line in the inertia plane which intersects one such half-line as $OF$ must intersect all such half-lines; while every optical line in the inertia plane which intersects one such half-line as $OF'$ must intersect all such half-lines.

Again, with the same notation as in the remarks at the end of Theorem 14 we see that any such half-line as $OD$ (where $D$ is any element of $b$ which is *before* $E$) must be a separation half-line and, taking any number of positions for $D$ we get any number of separation half-lines having the common end $O$ and which are all intersected by the optical line $b$.

Similarly, any such half-line as $OD'$ (where $D'$ is any element of $b'$ which is *after* $E'$) must be a separation half-line and, taking any number

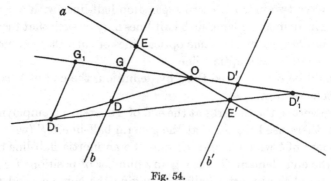

Fig. 54.

of positions for $D'$ we get any number of separation half-lines having the common end $O$ and which are all intersected by the optical line $b'$.

We shall now prove that if $D_1$ be any element of the separation half-line $OD$ and $D_1'$ be any element of the separation half-line $OD'$, then

$D_1D_1'$ is a separation line. This is obvious if $OD$ and $OD'$ form parts of the same separation line, so we shall suppose that this is not the case.

Now the complete separation line of which $OD'$ is a part must intersect $b$ in some element, say $G$.

Also, since $O$ is between the parallel optical lines $b$ and $b'$ and, since $D'$ is *after* $E'$, it follows, by Theorem 69, that $E$ is *after* $G$.

Since we have supposed that the half-lines $OD$ and $OD'$ are not parts of the same separation line, it follows that $G$ must be distinct from $D$ and may be either *before* or *after* it. The method of proof is similar in the two cases, so that we shall merely consider the case where $G$ is *after* $D$.

Then, since $D_1$ and $D$ both lie in the half-line $OD$ which has the same end $O$ as the half-line $OG$, it follows, by Theorem 67, that, since $G$ is *after* $D$, an optical line through $D_1$ co-directional with $DG$ will intersect the half-line $OG$ in an element, say $G_1$, such that $G_1$ is *after* $D_1$.

Now $D_1$ could not be *after* $D_1'$, for then we should have $G_1$ *after* $D_1'$; which is impossible, since $G_1D_1'$ is a separation line.

Again $D_1$ could not be *before* $D_1'$, for, since $O$ is linearly between $D_1'$ and $G_1$, it would follow, by Theorem 73, that $D_1O$ must be an inertia line, which it is not.

Thus $D_1D_1'$ must in all cases be a separation line, and no optical line can intersect two such half-lines as $OD$ and $OD'$.

It follows, as in the case of inertia half-lines, that any optical line in the inertia plane which intersects one such separation half-line as $OD$ must intersect all such separation half-lines, while any optical line in the inertia plane which intersects one such separation half-line as $OD'$ must intersect all such separation half-lines.

Suppose now that we have any two distinct separation half-lines $OA$ and $OB$ having a common end $O$ and lying in an inertia plane, and suppose further that the pair of half-lines are such as may both be intersected by the same optical lines.

Let $A$ and $B$ be any elements of the half-lines and let $C$ be any element which is linearly between $A$ and $B$.

We shall show that $CO$ must be a separation line.

Let a general line be taken through $A$ parallel to $CO$. Then this general line must intersect the separation line $BO$ in some element, say $H$, and since $C$ is linearly between $A$ and $B$, the element $O$ must be linearly between $H$ and $B$.

Then the separation half-lines $OA$ and $OH$ having the common end $O$ are such as cannot both be intersected by any optical line and accordingly $AH$ must be a separation line. Thus, since $CO$ is parallel to $AH$, it follows that $CO$ must also be a separation line.

Now any general line lying in the inertia plane and normal to $CO$ must be an inertia line. Thus in the particular case where the segment $OA =$ the segment $OB$ and where $C$ is taken to be the mean of $A$ and $B$, it follows, by Theorem 203, that $AB$ must be normal to $CO$ and therefore $AB$ must be an inertia line.

Thus we get finally that: if $OA$ and $OB$ be two distinct separation lines having a common end $O$ and lying in an inertia plane, and if they are such as can both be intersected by the same optical lines, then if segment $OA =$ segment $OB$ the general line $AB$ must be an inertia line.

If instead of being separation half-lines $OA$ and $OB$ be inertia half-lines such that both $A$ and $B$ are *after* $O$ or else both $A$ and $B$ are *before* $O$, and if segment $OA =$ segment $OB$, it follows from the footnote to Theorem 202 that $AB$ must be a separation line.

These are the cases in which the half-lines $OA$ and $OB$ may both be intersected by the same optical lines.

Again, let $OA$ and $OA'$ be two inertia half-lines or two separation half-lines lying in an inertia plane and having a common end $O$, and such that both half-lines may be intersected by the same optical lines.

Let the elements $A$ and $A'$ be so selected that $OA = OA'$.

Let the two optical lines which pass through $A'$ and lie in the inertia plane intersect the half-line $OA$ in $F$ and $G$, and let the notation be such that $F$ is linearly between $O$ and $G$.

Similarly, let the two optical lines which pass through $A$ and lie in the inertia plane intersect the half-line $OA'$ in $F'$ and $G'$, and let the notation be such that $F'$ is linearly between $O$ and $G'$.

Let $N$ be the mean of $F$ and $G$ while $N'$ is the mean of $F'$ and $G'$.

Then $A'N$ is normal to $AO$ and $AN'$ is normal to $A'O$.

Also $N$ must lie in the half-line $OA$, while $N'$ must lie in the half-line $OA'$.

Further, from the remarks at the end of Theorem 106, it follows that
$$ON = ON'$$
and
$$A'N = AN'.$$

But $GN$ is a conjugate to $A'N$, while $G'N'$ is a conjugate to $AN'$ and so
$$GN = G'N'.$$

Thus, since $N$ must be linearly between $O$ and $G$, while $N'$ is linearly between $O$ and $G'$, it follows that we must have
$$OG = OG'.$$

Also, since $$OA = OA'$$
and since $OA$ is the hypotenuse of the triangle whose corners are $O$, $N'$, $A$, while the side $ON'$ is the same kind of general line as $OA$, it follows that:
$$OA' < ON'.$$
But $ON' < OG'$ and so $A'$ is linearly between $O$ and $G'$.

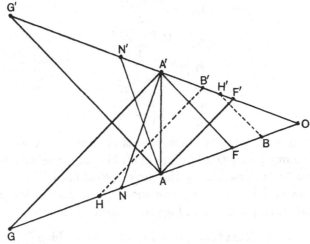

Fig. 55.

Similarly $A$ is linearly between $O$ and $G$.

Thus, by Theorem 76, the optical lines $AG'$ and $A'G$ intersect and therefore are generators of opposite sets of the inertia plane.

Also, since $OG = OG'$ and $OA = OA'$, it follows that:
$$OG : OA' = OG' : OA.$$

If now $BH'$ be any optical line parallel to $AG'$ and intersecting the half-line $OA$ in $B$ and the half-line $OA'$ in $H'$, while $HB'$ is any optical line parallel to $GA'$ and intersecting the half-line $OA$ in $H$ and the half-line $OA'$ in $B'$; it follows from the remarks at the end of Theorem 183 that:
$$OH' : OB = OG' : OA$$
and
$$OH : OB' = OG : OA'.$$

It follows that if any generator intersects the pair of half-lines and cuts off from them a pair of segments in a particular ratio, then any generator of the same set which intersects the half-lines will cut off segments in the same ratio, while any generator of the opposite set which intersects the half-lines will cut off segments in the reciprocal ratio.

We see from Theorem 202 that these ratios can never be ratios of equality so long as the half-lines $OA$ and $OA'$ are distinct and thus, if expressed numerically, one ratio must be greater than unity, while the other is less than unity.

Let the greater of these reciprocal ratios be $z$.

Then, since $N'$ is the mean of $F'$ and $G'$, we have

$$ON' = \frac{OG' + OF'}{2},$$

and

$$N'G' = \frac{OG' - OF'}{2}.$$

Thus

$$\frac{ON'}{OA} = \tfrac{1}{2}\left(z + \frac{1}{z}\right) \qquad \ldots\ldots(1),$$

while

$$\frac{N'G'}{OA} = \tfrac{1}{2}\left(z - \frac{1}{z}\right) \qquad \ldots\ldots(2).$$

Suppose now that we have three inertia or three separation half-lines having a common end $O$ and lying in an inertia plane, and such that they may all be intersected by the same optical lines.

Let them be intersected by one such optical line in the elements $A_0$, $A_1$ and $A_2$ respectively and let $A_1$ be linearly between $A_0$ and $A_2$.

Further suppose that this optical line be one for which $\dfrac{OA_1}{OA_0}$ is greater than unity.

Let the second optical line passing through $A_0$ and lying in the inertia plane intersect $OA_1$ in $B_1$ and $OA_2$ in $B_2$.

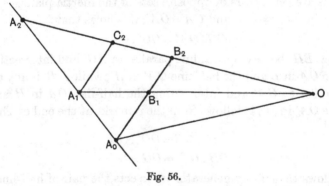

Fig. 56.

Then $OB_1 < OA_1$ so that $B_1$ is linearly between $O$ and $A_1$.

But, since $A_1$ is linearly between $A_0$ and $A_2$, it follows, from Theorem 77, that $B_2$ is linearly between $O$ and $A_2$ and therefore $OB_2 < OA_2$.

It follows that $\dfrac{OA_2}{OA_0}$ is greater than unity.

Again let the optical line through $A_1$ parallel to $A_0 B_2$ intersect $OA_2$ in $C_2$.

Then, since $A_1$ is linearly between $A_0$ and $A_2$, it follows that $C_2$ is linearly between $B_2$ and $A_2$ and therefore $C_2$ is linearly between $O$ and $A_2$, so that $OC_2 < OA_2$.

Thus $\dfrac{OA_2}{OA_1}$ is greater than unity.

Then
$$\frac{OA_2}{OA_0} = \frac{OA_1}{OA_0} \cdot \frac{OA_2}{OA_1};$$

so that taking the logarithms of these ratios we have
$$\log \frac{OA_2}{OA_0} = \log \frac{OA_1}{OA_0} + \log \frac{OA_2}{OA_1}.$$

If, instead of three inertia half-lines or three separation half-lines having the common end $O$ and intersecting the optical line $A_0 A_1$, we have any further number intersecting it in the elements $A_3, A_4, \ldots A_n$ and such that:

$\qquad A_2$ is linearly between $A_1$ and $A_3$,

$\qquad A_3$ is linearly between $A_2$ and $A_4$,

$\qquad \cdots\cdots\cdots\cdots\cdots\cdots\cdots\cdots\cdots\cdots\cdots\cdots$

$\qquad \cdots\cdots\cdots\cdots\cdots\cdots\cdots\cdots\cdots\cdots\cdots\cdots$

$\qquad A_{n-1}$ is linearly between $A_{n-2}$ and $A_n$;

we have $\qquad \log \dfrac{OA_n}{OA_0} = \log \dfrac{OA_1}{OA_0} + \log \dfrac{OA_2}{OA_1} + \ldots + \log \dfrac{OA_n}{OA_{n-1}}.$

Reverting now to formulae (1) and (2) and putting $u = \log_e z$ we have $z = e^u$ so that these formulae become

$$\frac{ON'}{OA} = \frac{e^u + e^{-u}}{2} = \cosh u \qquad \ldots\ldots(3),$$

and
$$\frac{N'G'}{OA} = \frac{(\text{conjugate } AN')}{OA} = \frac{e^u - e^{-u}}{2} = \sinh u \qquad \ldots\ldots(4),$$

where
$$u = \log_e \frac{OG'}{OA} = \log_e \frac{OA}{OF'}.$$

If $x_0$ and $x_1$ be two inertia half-lines or two separation half-lines having a common end $O$ and lying in an inertia plane and such that they may both be intersected by the same optical lines, then, together with the element $O$, they will be said to form a *proper hyperbolic angle-boundary* and the element $O$ will be called its *vertex* while $x_0$ and $x_1$ will be called its *sides*.

If $A_0$ be any element in $x_0$ and an optical line through $A_0$ intersects $x_1$ in $A_1$, then any general half-line having $O$ as its end and intersecting

$A_0A_1$ in an element $M$ linearly between $A_0$ and $A_1$ must be the same type of half-line as $x_0$ and $x_1$.

The set of all such inertia or separation half-lines as $OM$ will be called a *proper hyperbolic angular segment*.

A proper hyperbolic angular segment together with the angle-boundary will be called a *proper hyperbolic angular interval*.

Taking the case of the optical line which makes $OA_1 > OA_0$, then $\log_e \dfrac{OA_1}{OA_0}$ will be the magnitude of the angular interval in natural measure and will be called a *proper hyperbolic angle*.

Unlike the case of angular intervals in a separation plane there exist proper hyperbolic angular intervals of any magnitude however great.

Other types of angular intervals exist in an inertia plane besides those which we have designated "proper". In these the rays are not all of one kind and it will be found most convenient when we have to deal with them to describe them in terms of proper hyperbolic angular intervals to which they are related. Thus, for example, we may have the supplement or the conjugate of a proper hyperbolic angular interval; or, when one side is an inertia half-line and the other a separation half-line we can construct an auxiliary proper hyperbolic angular interval whose one side is normal to one side of the given one.

Various results may be deduced by means of formulae (3) and (4) analogous to theorems in ordinary geometry. Thus if $A$, $B$ and $C$ be the corners of a triangle in an inertia plane and if we denote the sides by $a$, $b$ and $c$ as in ordinary geometry; then if the sides $b$ and $c$ form a proper hyperbolic angle-boundary with one another which we denote by $A$, and we form the expression

$$b^2 + c^2 - 2bc \cosh A,$$

then, if this expression be positive, the side $a$ will be a separation segment if $b$ and $c$ are separation segments and will be an inertia segment if $b$ and $c$ are inertia segments and in either case we shall have

$$b^2 + c^2 - 2bc \cosh A = a^2.$$

If the expression be zero, then $a$ will be a segment of an optical line.

If the expression be negative, then the side $a$ will be an inertia segment if $b$ and $c$ are separation segments and will be a separation segment if $b$ and $c$ are inertia segments and in either case we shall have

$$2bc \cosh A - b^2 - c^2 = (\text{conjugate } a)^2.$$

These results may easily be deduced and we shall not trouble to prove them.

It should be noted however that when the side $a$ is an optical segment, nothing is said as to its length and we may even have two triangles with two sides and the included proper hyperbolic angle of the one respectively equal to the two sides and the included proper hyperbolic angle of the other while the third side of the one forms a part of the third side of the other; but this can only happen if these third sides are optical segments.

On the other hand, no comparison whatever can be made in the lengths of the third sides if they be optical segments but not co-directional.

### INTRODUCTION OF COORDINATES

If we take any element $O$ of the set as origin, we have already seen that we may obtain systems of four general lines through $O$, say $OX$, $OY$, $OZ$, $OT$, which are mutually normal to one another.

Three of these, say $OX$, $OY$, $OZ$, will be separation lines, while the fourth, $OT$, will be an inertia line.

The three separation lines $OX$, $OY$, $OZ$ will determine a separation threefold, say $W$, and $OT$ will be normal to it.

If we select any arbitrary separation segment as a unit of length and associate the number zero with the element $O$, we may associate every other element of $OX$, $OY$, $OZ$ with a real number, positive or negative, corresponding to the length of the segment of which that element is one end and the origin $O$ is the other.

In this way we set up a coordinate system in $W$ which will be quite similar to that with which we are familiar.

Since all the theorems of ordinary Euclidean geometry hold for a separation threefold, the length of a segment in $W$ will be given by the ordinary Cartesian formula.

Again, nor confining our attention merely to the elements of $W$, let $A$ be any element of the whole set.

Then $A$ must either lie in $OT$, or else there is an inertia line through $A$ parallel to $OT$, and, as has already been proved, this inertia line will intersect $W$ in some element, say $N$.

Further, $AN$ must be normal to $W$.

Now if $A$ does not lie in $W$ there will be a separation threefold, say $W'$, passing through $A$ and parallel to $W$, and the inertia line $OT$ must intersect $W'$ in some element, say $M$.

Further, since $W'$ is parallel to $W$, both $OT$ and $AN$ must be normal to $W'$.

Thus, if $OM$ and $NA$ are distinct, $MA$ and $ON$ must both be separation lines normal to $OM$, and so, since $OM$ and $NA$ lie in an inertia plane, we must have $MA$ parallel to $ON$.

Now we may select a unit inertia segment, just as we selected a unit separation segment, and with each element of $OT$ distinct from $O$ we may associate a real number positive or negative corresponding to the length of the segment of which that element is one end and the origin $O$ is the other.

*We shall suppose this correspondence to be set up in such a way that a positive real number corresponds to any element which is after $O$ and a negative real number to any element which is before $O$.*

As regards the relationship between the unit separation segment and the unit inertia segment, the simplest convention to make is to take the unit inertia segment such that its conjugate is equal to the unit separation segment.

More generally, we may take the unit inertia segment such that:

(conjugate of unit inertia segment) = $v$ (unit separation segment),

where $v$ is a constant afterwards to be identified with what we call the "*velocity of light*".

Now the element $N$ lies in $W$ and is determined by three coordinates, say $x_1$, $y_1$, $z_1$, taken parallel to $OX$, $OY$, $OZ$ respectively in the usual manner.

Further        segment $NA =$ segment $OM$,

and so if $t_1$ be the length of $OM$ in terms of the unit inertia segment, then the element $A$ will be determined by the four coordinates $x_1$, $y_1$, $z_1$, $t_1$.

Let the length of the segment $ON$ be denoted by $w_1$.

Then as in ordinary coordinate geometry

$$w_1{}^2 = x_1{}^2 + y_1{}^2 + z_1{}^2.$$

Thus if $OA$ should be an optical line, we must have

$$w_1{}^2 = v^2 t_1{}^2,$$

or $\qquad\qquad x_1{}^2 + y_1{}^2 + z_1{}^2 - v^2 t_1{}^2 = 0 \qquad\qquad \ldots\ldots(1).$

Again, if $OA$ should be a separation segment and if $r_1$ be its length, it follows from the analogue of the theorem of Pythagoras for this case that:

$$w_1{}^2 - v^2 t_1{}^2 = r_1{}^2,$$

or $\qquad\qquad x_1{}^2 + y_1{}^2 + z_1{}^2 - v^2 t_1{}^2 = r_1{}^2 \qquad\qquad \ldots\ldots(2).$

Finally, if $OA$ should be an inertia segment and $\bar{r}_1$ its length, it follows from the corresponding analogue of the Pythagoras theorem that:

$$w_1{}^2 - v^2 t_1{}^2 = -v^2 \bar{r}_1{}^2,$$

or $$x_1{}^2 + y_1{}^2 + z_1{}^2 - v^2 t_1{}^2 = -v^2 \bar{r}_1{}^2 \qquad \text{......(3)}.$$

*Thus from (1), (2) and (3) it follows that the expression*

$$x_1{}^2 + y_1{}^2 + z_1{}^2 - v^2 t_1{}^2$$

*is positive, zero, or negative according as OA is a separation line, an optical line, or an inertia line.*

If $A$ be *after* $O$, it is clear from the convention which we have made that $t_1$ must be positive, and so the conditions that $A$ should be *after* $O$ are:

(1) $\qquad x_1{}^2 + y_1{}^2 + z_1{}^2 - v^2 t_1{}^2$ is zero or negative$\Big\}$.
(2) $\qquad\qquad\qquad t_1$ is positive

The conditions that $A$ should be *before* $O$ are similarly:

(1) $\qquad x_1{}^2 + y_1{}^2 + z_1{}^2 - v^2 t_1{}^2$ is zero or negative$\Big\}$.
(2) $\qquad\qquad\qquad t_1$ is negative

The conditions that $A$ should be neither *before* nor *after* $O$ are either that:

$$A \text{ is identical with } O,$$

in which case $\qquad x_1 = y_1 = z_1 = t_1 = 0$ $\Big\}$.
or else $\qquad x_1{}^2 + y_1{}^2 + z_1{}^2 - v^2 t_1{}^2$ is positive

More generally, it is clear that: if $(x_0, y_0, z_0, t_0)$ and $(x_1, y_1, z_1, t)_1$ be the coordinates of two elements which we call $A_0$ and $A_1$ respectively, then if $A_0$ and $A_1$ lie in an optical line we must have

$$(x_1 - x_0)^2 + (y_1 - y_0)^2 + (z_1 - z_0)^2 - v^2 (t_1 - t_0)^2 = 0 \quad \text{......(4)}.$$

If $A_0 A_1$ be a separation segment and $r_1$ be its length we must have

$$(x_1 - x_0)^2 + (y_1 - y_0)^2 + (z_1 - z_0)^2 - v^2 (t_1 - t_0)^2 = r_1{}^2 \quad \text{......(5)}.$$

While if $A_0 A_1$ be an inertia segment and $\bar{r}_1$ be its length we must have

$$(x_1 - x_0)^2 + (y_1 - y_0)^2 + (z_1 - z_0)^2 - v^2 (t_1 - t_0)^2 = -v^2 \bar{r}_1{}^2 \quad \text{...(6)}.$$

*Thus the expression*

$$(x_1 - x_0)^2 + (y_1 - y_0)^2 + (z_1 - z_0)^2 - v^2 (t_1 - t_0)^2$$

*is positive, zero, or negative according as $A_0 A_1$ is a separation line, an optical line, or an inertia line.*

Accordingly if $A_0$ and $A_1$ be any elements of the set, the conditions that $A_1$ should be *after* $A_0$ are:

(1) $\quad (x_1 - x_0)^2 + (y_1 - y_0)^2 + (z_1 - z_0)^2 - v^2 (t_1 - t_0)$

$$\text{is zero or negative}$$

and (2) $\qquad\qquad t_1 - t_0$ is positive

The conditions that $A_1$ should be *before* $A_0$ are:

(1)  $(x_1 - x_0)^2 + (y_1 - y_0)^2 + (z_1 - z_0)^2 - v^2 (t_1 - t_0)^2$

$\qquad\qquad\qquad\qquad\qquad$ is zero or negative $\Big\}$.

and  (2)  $\qquad\qquad\qquad t_1 - t_0$ is negative

The conditions that $A_1$ should be neither *before* nor *after* $A_0$ are (if we include the case where $A_0$ and $A_1$ are identical):

$$x_1 - x_0 = y_1 - y_0 = z_1 - z_0 = t_1 - t_0 = 0$$

or else $\qquad\qquad\qquad\qquad\qquad\qquad\qquad\qquad\qquad\Big\}$.

$$(x_1 - x_0)^2 + (y_1 - y_0)^2 + (z_1 - z_0)^2 - v^2 (t_1 - t_0)^2 \text{ is positive}$$

Now the condition that two distinct elements lie in an optical line gives us also the condition that the one should lie in the $\alpha$ sub-set of the other.

Thus if $(x_0, y_0, z_0, t_0)$ be the coordinates of an element $A_0$, the equation of the combined $\alpha$ and $\beta$ sub-sets of $A_0$ is

$$(x - x_0)^2 + (y - y_0)^2 + (z - z_0)^2 - v^2 (t - t_0)^2 = 0 \qquad \ldots\ldots(7).$$

*The $\alpha$ sub-set will then consist of all elements $(x, y, z, t)$ for which this equation is satisfied and for which $t - t_0$ is zero or positive; while the $\beta$ sub-set of $A_0$ will consist of all elements for which the equation is satisfied and for which $t - t_0$ is zero or negative.*

*Definition.* The set of all elements whose coordinates satisfy equation (7) will be called the *standard cone* with respect to the element whose coordinates are $(x_0, y_0, z_0, t_0)$.

Taking $v$ equal to unity, for the sake of simplicity, it is evident that the equation

$$x^2 + y^2 + z^2 - t^2 = c^2$$

represents the set of elements such as $A$, where $OA$ is a separation segment whose length is $c$.

Similarly, the equation

$$x^2 + y^2 + z^2 - t^2 = -c^2$$

represents the set of elements such as $A$, where $OA$ is an inertia segment whose length is $c$.

If we put $y = 0$ and $z = 0$ in the first of these we obtain

$$x^2 - t^2 = c^2,$$

which gives us the relation between $x$ and $t$ for the portion of the corresponding set which lies in the inertia plane containing the axes of $x$ and $t$.

This then represents the analogue of a circle in the inertia plane.

Similarly for the case of inertia segments, putting $y=0$ and $z=0$ we get

$$x^2 - t^2 = -c^2.$$

The two equations:

$$x^2 - t^2 = c^2 \quad \text{and} \quad x^2 - t^2 = -c^2$$

are of the same forms as the equations of a hyperbola and its conjugate in ordinary plane geometry.

The equation $\qquad x^2 - t^2 = 0$

along with $y=0$ and $z=0$ represents the two optical lines through the origin in the same inertia plane, and these correspond to the common asymptotes of the hyperbolas.

## NORMALITY OF GENERAL LINES

Let $A$, $B$ and $C$ be three distinct elements whose coordinates are $(x_0, y_0, z_0, t_0)$, $(x_1, y_1, z_1, t_1)$ and $(x_2, y_2, z_2, t_2)$ respectively and such that $AB$ is normal to $AC$.

For the sake of simplicity we shall take $v$ equal to unity.

For brevity let us write

$$(x_2 - x_1)^2 + (y_2 - y_1)^2 + (z_2 - z_1)^2 - (t_2 - t_1)^2 = H,$$
$$(x_1 - x_0)^2 + (y_1 - y_0)^2 + (z_1 - z_0)^2 - (t_1 - t_0)^2 = S_1,$$
$$(x_2 - x_0)^2 + (y_2 - y_0)^2 + (z_2 - z_0)^2 - (t_2 - t_0)^2 = S_2.$$

Considering all the six cases of analogue to the theorem of Pythagoras (including the limiting case mentioned in the footnote on p. 369), we see that they are all included in the formula:

$$H = S_1 + S_2;$$

or expanding, rearranging and omitting a factor 2, in the formula:

$$(x_2 - x_0)(x_1 - x_0) + (y_2 - y_0)(y_1 - y_0)$$
$$+ (z_2 - z_0)(z_1 - z_0) - (t_2 - t_0)(t_1 - t_0) = 0. \quad (1)$$

We have to show that this is not merely a necessary, but also a sufficient condition of the normality of $AB$ to $AC$.

Now let us consider the various possibilities which are conceivable with regard to the types of general line which $AB$, $AC$ and $BC$ might be.

It is obvious that the two sides of the equation

$$H = S_1 + S_2$$

must either be both positive, both negative, or both zero.

For $H$ positive, it is clear that we could only conceivably have $S_1$ and $S_2$ either (1) both positive, or (2) one positive and other negative, or (3) one positive and other zero.

For $H$ negative, we could only conceivably have $S_1$ and $S_2$ either (4) both negative, or (5) one negative and other positive, or (6) one negative and other zero.

For $H$ zero, we could only conceivably have $S_1$ and $S_2$ either (7) both zero, or (8) one positive and other negative.

However, cases (4) and (6) can be shown to be impossible from other considerations.

We showed geometrically that this was so, but it is desirable to show the reasons analytically.

In order to do so we shall first investigate a certain lemma.

Suppose that we have two series of four corresponding quantities,

say
$$Q_1, \quad Q_2, \quad Q_3, \quad Q_4,$$
and
$$R_1, \quad R_2, \quad R_3, \quad R_4.$$

Then the following identity may easily be verified:

$$(R_1 \pm R_4)^2 \{Q_1{}^2 + Q_2{}^2 + Q_3{}^2 - Q_4{}^2\} + (Q_1 \pm Q_4)^2 \{R_1{}^2 + R_2{}^2 + R_3{}^2 - R_4{}^2\}$$
$$- 2(Q_1 \pm Q_4)(R_1 \pm R_4)\{Q_1 R_1 + Q_2 R_2 + Q_3 R_3 - Q_4 R_4\}$$
$$\equiv \{Q_2 R_1 - Q_1 R_2 \pm (Q_2 R_4 - Q_4 R_2)\}^2 + \{Q_3 R_1 - Q_1 R_3 \pm (Q_3 R_4 - Q_4 R_3)\}^2;$$

where, in the ambiguities, either the positive sign is to be used throughout, or else the negative sign throughout.

In this identity we may obviously interchange $R_1$ with either $R_2$ or $R_3$, while at the same time we interchange $Q_1$ with $Q_2$ or $Q_3$ respectively.

We shall now prove that if

$$Q_1 R_1 + Q_2 R_2 + Q_3 R_3 - Q_4 R_4 = 0,$$

while $Q_1{}^2 + Q_2{}^2 + Q_3{}^2 - Q_4{}^2$ is negative and equal to $-k^2$, then

$$R_1{}^2 + R_2{}^2 + R_3{}^2 - R_4{}^2$$

must be positive unless

$$R_1 = R_2 = R_3 = R_4 = 0;$$

when it is zero.

Let
$$R_1{}^2 + R_2{}^2 + R_3{}^2 - R_4{}^2 = \theta,$$

and our identity gives us

$$(Q_1 \pm Q_4)^2 \theta = (R_1 \pm R_4)^2 k^2 + \{Q_2 R_1 - Q_1 R_2 \pm (Q_2 R_4 - Q_4 R_2)\}^2$$
$$+ \{Q_3 R_1 - Q_1 R_3 \pm (Q_3 R_4 - Q_4 R_3)\}^2.$$

If $R_1 \pm R_4$ be not zero, the right-hand side of this equation must be positive and therefore neither $(Q_1 \pm Q_4)^2$ nor $\theta$ can be zero, and, since $(Q_1 \pm Q_4)^2$ must be positive, therefore $\theta$ must be positive.

Thus $\theta$ must be positive if $R_1{}^2$ is not equal to $R_4{}^2$.

By the use of similar identities we may prove that $\theta$ must be positive if $R_2{}^2$ is not equal to $R_4{}^2$ and also if $R_3{}^2$ is not equal to $R_4{}^2$. Thus $\theta$ must be positive unless

$$R_1{}^2 = R_2{}^2 = R_3{}^2 = R_4{}^2.$$

But in this case we should have

$$\theta = R_1{}^2 + R_2{}^2 + R_3{}^2 - R_4{}^2 = 2R_4{}^2,$$

which is positive unless $R_4 = 0$; when it is zero.

Thus $R_1{}^2 + R_2{}^2 + R_3{}^2 - R_4{}^2$ must always be positive unless

$$R_1 = R_2 = R_3 = R_4 = 0.$$

If now we substitute

$$(x_1 - x_0), \quad (y_1 - y_0), \quad (z_1 - z_0), \quad (t_1 - t_0)$$

for $Q_1, Q_2, Q_3, Q_4$ respectively, and

$$(x_2 - x_0), \quad (y_2 - y_0), \quad (z_2 - z_0), \quad (t_2 - t_0)$$

for $R_1, R_2, R_3, R_4$ respectively, and suppose that equation (1) (which is equivalent to $H = S_1 + S_2$) holds and that

$$(x_1 - x_0)^2 + (y_1 - y_0)^2 + (z_1 - z_0)^2 - (t_1 - t_0)^2$$

is negative; then

$$(x_2 - x_0)^2 + (y_2 - y_0)^2 + (z_2 - z_0)^2 - (t_2 - t_0)^2$$

must be positive or zero and, in the latter case, we must have

$$(x_2 - x_0) = (y_2 - y_0) = (z_2 - z_0) = (t_2 - t_0) = 0,$$

or $C$ coincident with $A$, contrary to the hypothesis that $C$ and $A$ are distinct.

Thus cases (4) and (6) are both excluded as possibilities, and we are left only with cases (1), (2), (3), (5), (7), (8), which are precisely the six cases considered in the remarks at the end of Theorem 205.

Thus, provided that equation (1) holds, the general line $AB$ must be normal to the general line $AC$.

We may also make use of our identity in order to obtain the equations of an optical line from the definition give on p. 30 and incidentally to give an analytical demonstration that, in case (7) above, the three elements, $A$, $B$ and $C$, lie in one optical line.

Thus putting

$$Q_1 R_1 + Q_2 R_2 + Q_3 R_3 - Q_4 R_4 = 0,$$
$$Q_1{}^2 + Q_2{}^2 + Q_3{}^2 - Q_4{}^2 = 0,$$
$$R_1{}^2 + R_2{}^2 + R_3{}^2 - R_4{}^2 = 0,$$

in our identity, we get

$$0 = \{Q_2 R_1 - Q_1 R_2 \pm (Q_2 R_4 - Q_4 R_2)\}^2 + \{Q_3 R_1 - Q_1 R_3 \pm (Q_3 R_4 - Q_4 R_3)\}^2.$$

Since the right-hand side is a sum of two squares equated to zero, they must be each separately zero, and so

$$Q_2 R_1 - Q_1 R_2 \pm (Q_2 R_4 - Q_4 R_2) = 0,$$
$$Q_3 R_1 - Q_1 R_3 \pm (Q_3 R_4 - Q_4 R_3) = 0,$$

and, since either the $+$ or $-$ sign may be taken in the ambiguities, we see that:

$$Q_2 R_1 - Q_1 R_2 = 0, \quad Q_2 R_4 - Q_4 R_2 = 0, \quad Q_3 R_1 - Q_1 R_3 = 0, \quad Q_3 R_4 - Q_4 R_3 = 0$$

or
$$Q_1 : Q_2 : Q_3 : Q_4 = R_1 : R_2 : R_3 : R_4.$$

Now let $A$ and $B$ be two distinct elements such that the one lies in the $\alpha$ sub-set of the other and let $(x_0, y_0, z_0, t_0)$ and $(x_1, y_1, z_1, t_1)$ be the coordinates of $A$ and $B$ respectively.

Then
$$(x_1 - x_0)^2 + (y_1 - y_0)^2 + (z_1 - z_0)^2 - (t_1 - t_0)^2 = 0.$$

If $(x, y, z, t)$ be any element which lies in the standard cones with respect to both $A$ and $B$, we know from definition that such element lies in the optical line containing $A$ and $B$.

But in this case we must have

$$(x - x_0)^2 + (y - y_0)^2 + (z - z_0)^2 - (t - t_0)^2 = 0,$$
and
$$(x - x_1)^2 + (y - y_1)^2 + (z - z_1)^2 - (t - t_1)^2 = 0.$$

It follows that we must have

$$(x - x_0)(x_1 - x_0) + (y - y_0)(y_1 - y_0) + (z - z_0)(z_1 - z_0) - (t - t_0)(t_1 - t_0) = 0.$$

Thus substituting

$$(x_1 - x_0), \quad (y_1 - y_0), \quad (z_1 - z_0), \quad (t_1 - t_0)$$

for $Q_1, Q_2, Q_3, Q_4$ respectively and

$$(x - x_0), \quad (y - y_0), \quad (z - z_0), \quad (t - t_0)$$

for $R_1, R_2, R_3, R_4$ respectively, the result above obtained enables us to write

$$\frac{x - x_0}{x_1 - x_0} = \frac{y - y_0}{y_1 - y_0} = \frac{z - z_0}{z_1 - z_0} = \frac{t - t_0}{t_1 - t_0};$$

which are the equations of the optical line containing $A$ and $B$.

If $C$ be any element distinct from both $A$ and $B$ and such that both $AC$ and $BC$ are optical segments and if $(x_2, y_2, z_2, t_2)$ be the coordinates of $C$, it is evident that $(x_2, y_2, z_2, t_2)$ satisfies the above conditions and accordingly, $C$ must lie in the optical line through $A$ and $B$. This is equivalent to case (7).

The above identity may be generalised to $n$ dimensions and will be found very useful in the further analytical development of this subject. See paper by the author " On the Connexion of a Certain Identity with the Extension of Conical Order to $n$ Dimensions", *Camb. Phil. Soc.* vol. XXIV, pp. 357–74, 1928.

## EQUATIONS OF GENERAL LINES, PLANES
### AND THREEFOLDS

Making use of the notation employed in the preceding section let us take the elements $A$ and $B$ as fixed while $C$ is variable and substituting the running coordinates $(x, y, z, t)$ for $(x_2, y_2, z_2, t_2)$ in equation (1) we get

$$(x-x_0)(x_1-x_0)+(y-y_0)(y_1-y_0)+(z-z_0)(z_1-z_0)-(t-t_0)(t_1-t_0)=0,$$
$$\ldots\ldots(2)$$

as the equation of the general threefold passing through $A$ and normal to $AB$.

This will be an inertia, an optical, or a separation threefold according as $AB$ is a separation, an optical or an inertia line. That is to say, equation (2) will represent an inertia, an optical, or a separation threefold according as the expression

$$(x_1-x_0)^2+(y_1-y_0)^2+(z_1-z_0)^2-(t_1-t_0)^2$$

is positive, zero, or negative.

If $l, m, n, p$ be any four quantities such that:

$$\frac{x_1-x_0}{l}=\frac{y_1-y_0}{m}=\frac{z_1-z_0}{n}=\frac{t_1-t_0}{p},$$

then $\qquad l(x-x_0)+m(y-y_0)+n(z-z_0)-p(t-t_0)=0 \qquad \ldots\ldots(3)$

will obviously represent the same set of elements as equation (2) and is therefore the equation of a threefold passing through the element whose coordinates are $(x_0, y_0, z_0, t_0)$, and its type will be, inertia, optical, or separation, according as the expression

$$l^2+m^2+n^2-p^2$$

is positive, zero or negative.

If $(x', y', z', t')$ be any element such that:

$$\frac{x'-x_0}{x_1-x_0}=\frac{y'-y_0}{y_1-y_0}=\frac{z'-z_0}{z_1-z_0}=\frac{t'-t_0}{t_1-t_0},$$

then we could substitute

$$(x'-x_0), (y'-y_0), (z'-z_0), (t'-t_0)$$

for $\qquad (x_1-x_0), (y_1-y_0), (z_1-z_0), (t_1-t_0)$

respectively in equation (2) and the resultant equation would still represent the same threefold while the general line through $(x_0, y_0, z_0, t_0)$ and $(x', y', z', t')$ would still be the normal to the threefold through the element $A$ which, as we know, is unique.

Thus $(x', y', z', t')$ would always lie in the general line $AB$.

Thus, removing the accents we get

$$\frac{x-x_0}{x_1-x_0} = \frac{y-y_0}{y_1-y_0} = \frac{z-z_0}{z_1-z_0} = \frac{t-t_0}{t_1-t_0} \qquad \ldots\ldots(4)$$

as the equations of the general line passing through the two elements $(x_0, y_0, z_0, t_0)$ and $(x_1, y_1, z_1, t_1)$; and these will represent a separation, an optical, or an inertia line according as the expression

$$(x_1-x_0)^2 + (y_1-y_0)^2 + (z_1-z_0)^2 - (t_1-t_0)^2$$

is positive, zero, or negative.

As before, we may substitute $l, m, n, p$ for

$$(x_1-x_0), (y_1-y_0), (z_1-z_0), (t_1-t_0)$$

respectively and the equations

$$\frac{x-x_0}{l} = \frac{y-y_0}{m} = \frac{z-z_0}{n} = \frac{t-t_0}{p} \qquad \ldots\ldots(5)$$

will represent the same general line, which will be separation, optical, or inertia according as the expression

$$l^2 + m^2 + n^2 - p^2$$

is positive, zero, or negative.

Again, if $(x_2, y_2, z_2, t_2)$ be any element of the general threefold (2) distinct from the element $(x_0, y_0, z_0, t_0)$, then the general line joining these elements will be normal to the general line $AB$ and

$$(x_2-x_0)(x_1-x_0) + (y_2-y_0)(y_1-y_0) + (z_2-z_0)(z_1-z_0) - (t_2-t_0)(t_1-t_0)$$
$$= 0.$$

This will be the condition that the lines are normal to one another.

If the general lines be expressed in the forms:

$$\frac{x-x_0}{l} = \frac{y-y_0}{m} = \frac{z-z_0}{n} = \frac{t-t_0}{p}$$

and

$$\frac{x-x_0}{l'} = \frac{y-y_0}{m'} = \frac{z-z_0}{n'} = \frac{t-t_0}{p'}$$

the condition of normality may be expressed in the form

$$ll' + mm' + nn' - pp' = 0 \qquad \ldots\ldots(6).$$

It can readily be seen that this is still the condition of normality if the lines do not both pass through the same element.

Since any two general threefolds which are not parallel have a general plane in common, it follows that any two equations of the forms:

$$l_1 x + m_1 y + n_1 z - p_1 t = c_1 \qquad \ldots\ldots(7)$$

and

$$l_2 x + m_2 y + n_2 z - p_2 t = c_2 \qquad \ldots\ldots(8)$$

where $\qquad l_1 : m_1 : n_1 : p_1 \neq l_2 : m_2 : n_2 : p_2,$

will represent some form of general plane $P$.

In order to determine to which type this belongs consider the two general lines:

$$\frac{x}{l_1} = \frac{y}{m_1} = \frac{z}{n_1} = \frac{t}{p_1} = r \qquad \ldots\ldots(9),$$

$$\frac{x}{l_2} = \frac{y}{m_2} = \frac{z}{n_2} = \frac{t}{p_2} = r \qquad \ldots\ldots(10).$$

These general lines both pass through the origin and therefore lie in some general plane $Q$.

Further, since the general line (9) must be normal to the general threefold (7), while (10) is normal to (8), it follows that (9) and (10) are both normal to $P$ and consequently $Q$ must be completely normal to $P$.

We shall determine the type of $Q$ and thence deduce the type of $P$.

Let $(x_1, y_1, z_1, t_1)$ be any element in (9) distinct from the origin.

Then the equation of a standard cone having this element as vertex will be

$$(x - l_1 r_1)^2 + (y - m_1 r_1)^2 + (z - n_1 r_1)^2 - (t - p_1 r_1)^2 = 0.$$

If this cone intersects the general line (10) in an element $(x_2, y_2, z_2, t_2)$, we shall have

$$(l_2 r_2 - l_1 r_1)^2 + (m_2 r_2 - m_1 r_1)^2 + (n_2 r_2 - n_1 r_1)^2 - (p_2 r_2 - p_1 r_1)^2 = 0$$

or

$$(l_2{}^2 + m_2{}^2 + n_2{}^2 - p_2{}^2) r_2{}^2 - 2 (l_2 l_1 + m_2 m_1 + n_2 n_1 - p_2 p_1) r_1 r_2$$
$$+ (l_1{}^2 + m_1{}^2 + n_1{}^2 - p_1{}^2) r_1{}^2 = 0.$$

Let us first suppose that neither of the expressions:

$$(l_2{}^2 + m_2{}^2 + n_2{}^2 - p_2{}^2)$$
$$(l_1{}^2 + m_1{}^2 + n_1{}^2 - p_1{}^2)$$

is zero.

Regarded as an equation for $r_2$ in terms of $r_1$ and the direction ratios, the condition that the roots should be real and distinct is that

$$(l_2 l_1 + m_2 m_1 + n_2 n_1 - p_2 p_1)^2$$
$$- (l_2{}^2 + m_2{}^2 + n_2{}^2 - p_2{}^2)(l_1{}^2 + m_1{}^2 + n_1{}^2 - p_1{}^2) > 0.$$

If this be the case the general plane $Q$ will be such that it contains two optical lines passing through an element of it.

It follows that $Q$ will be an inertia plane.

If the above expression be zero there will be only one optical line in $Q$ which passes through the given element and accordingly, in this case, $Q$ will be an optical plane.

If the above expression be negative there will be no optical line in $Q$ which passes through the given element and accordingly, in this case, $Q$ will be a separation plane.

Let us next consider the case where one of the general lines (9) and (10) is an optical line. It will be sufficient to suppose that

$$l_2^2 + m_2^2 + n_2^2 - p_2^2 = 0$$

while

$$l_1^2 + m_1^2 + n_1^2 - p_1^2 \neq 0.$$

Then provided that the expression

$$l_2 l_1 + m_2 m_1 + n_2 n_1 - p_2 p_1$$

be not zero, $r_2$ may be determined and $Q$ contains a second optical line which intersects the optical line $(l_2, m_2, n_2, p_2)$ and accordingly $Q$ will be an inertia plane.

It is obvious however that in this case also

$$(l_2 l_1 + m_2 m_1 + n_2 n_1 - p_2 p_1)^2$$
$$- (l_2^2 + m_2^2 + n_2^2 - p_2^2)(l_1^2 + m_1^2 + n_1^2 - p_1^2) > 0.$$

In case $l_2 l_1 + m_2 m_1 + n_2 n_1 - p_2 p_1$ be zero, the above expression is zero and no value of $r_2$ can be determined. Thus in this case there is no optical line in $Q$ which will intersect the optical line $(l_2, m_2, n_2, p_2)$; so that $Q$ must be an optical plane.

Finally if

$$l_2^2 + m_2^2 + n_2^2 - p_2^2 = 0$$

and

$$l_1^2 + m_1^2 + n_1^2 - p_1^2 = 0,$$

there are evidently two intersecting optical lines in $Q$, which must therefore be an inertia plane.

Since these optical lines intersect it is not possible to have

$$l_2 l_1 + m_2 m_1 + n_2 n_1 - p_2 p_1$$

zero: for this would be the condition of their normality which would imply coincidence.

Thus in this case also we should have

$$(l_2 l_1 + m_2 m_1 + n_2 n_1 - p_2 p_1)^2$$
$$- (l_2^2 + m_2^2 + n_2^2 - p_2^2)(l_1^2 + m_1^2 + n_1^2 - p_1^2) > 0.$$

Accordingly in all cases this expression will be positive, zero or negative, according as $Q$ is an inertia plane, an optical plane or a separation plane.

But since $P$ is completely normal to $Q$, it follows that when:

  (i) $Q$ is an inertia plane, $P$ is a separation plane.
 (ii) $Q$ is an optical plane, $P$ is an optical plane.
(iii) $Q$ is a separation plane, $P$ is an inertia plane.

Thus $P$ is a separation plane, an optical plane or an inertia plane according as the expression

$$(l_2 l_1 + m_2 m_1 + n_2 n_1 - p_2 p_1)^2$$
$$- (l_2^2 + m_2^2 + n_2^2 - p_2^2)(l_1^2 + m_1^2 + n_1^2 - p_1^2)$$

is positive, zero, or negative.

Having thus obtained the equations of general lines, planes and threefolds and the conditions that they should be of any one of the three different types and having also obtained the condition that general lines should be normal to one another; the analytical development of the subject may be carried forward in the usual manner.

SYMMETRICAL COORDINATES

The systems of coordinates which we have considered are those in which we have four coordinate axes which are normal to one another, and such systems are those which are most generally useful; but they give an expression for the square of the distance between two elements in the form of a sum of squares, in which one square is of different sign from the remaining ones.

This want of symmetry takes away from the analytical attractiveness of the subject, although the fact that it can be built up entirely from *before* and *after* relations gives it a special importance.

The writer has shown that it is possible to introduce symmetrical systems of coordinates for Conical Order in four or any larger number of dimensions; which, however, are not orthogonal.*

Thus if we put:

$$(X_1 + X_2 - X_3 - X_4)/\sqrt{6} = x,$$
$$(X_1 - X_2 + X_3 - X_4)/\sqrt{6} = y,$$
$$(X_1 - X_2 - X_3 + X_4)/\sqrt{6} = z,$$
$$\sqrt{3}\,(X_1 + X_2 + X_3 + X_4)/\sqrt{6} = t,$$

we can easily verify that:

$$t^2 - x^2 - y^2 - z^2 = \tfrac{4}{3}\{X_1 X_2 + X_2 X_3 + X_3 X_1 + X_1 X_4 + X_2 X_4 + X_3 X_4\};$$

which is symmetrical in the four coordinates: $X_1, X_2, X_3, X_4$.

---

* In order to construct a conical order of five dimensions it is only necessary to omit Post. XX, and substitute for it a postulate of the form:

*If W be any optical threefold there is at least one element which is neither* before *nor* after *any element of W.*

The subject can then very easily be developed, since the main difficulties have already been overcome in treating of four dimensions. We may limit the geometry to five dimensions, if so desired, by means of a postulate analogous to Post. XX; or extend it to six dimensions or any larger number in an analogous manner.

To go into any further details on this subject would be outside the scope of the present work, which is concerned with the development in four dimensions.

The transformation has been made of such a form as to introduce a coefficient $\frac{4}{3}$ for reasons connected with the interpretation of the system.

We do not purpose going into this in the present work, but refer the reader to a paper by the author: "On a Symmetrical Analysis of Conical Order and its Relation to Time-Space Theory in the *Proc. Roy. Soc.* A (1930), vol. CXXIX, pp. 549–79.

## INTERPRETATION OF RESULTS

It is evident that any element whose coordinates are $(a, b, c, 0)$ must lie in the separation threefold $W$ and accordingly the three equations

$$x=a, \quad y=b, \quad z=c$$

must represent an inertia line normal to $W$ and therefore co-directional with the axis of $t$.

Again, any equation of the first degree in $x, y, z$, together with the equation $t = 0$, will represent a separation plane in $W$, while any two independent but consistent equations of the first degree in $x, y, z$, together with the equation $t = 0$, will represent a separation line in $W$.

Thus any equation of the first degree in $x, y, z$ (leaving out the equation $t = 0$) will represent an inertia threefold containing inertia lines parallel to the axis of $t$; while any two independent but consistent equations of the first degree in $x, y, z$ will represent an inertia plane containing inertia lines parallel to the axis of $t$.

Thus corresponding to any theorem concerning the elements of $W$ there will be a theorem concerning inertia lines normal to $W$ and passing through these elements.

Conversely, if we consider the system consisting of any selected inertia line together with all others parallel to it, then any two such inertia lines will determine an inertia plane, while any three which do not lie in one inertia plane will determine an inertia threefold.

Since these inertia lines must all intersect any separation threefold to which they are normal, it follows that they have a geometry similar to that of the separation threefold and therefore of the ordinary Euclidean type.

*If then we call any element of the entire set an "instant"; any inertia line of the selected system a "point"; any inertia plane of the selected system a "straight line"; and any inertia threefold of the selected system a "plane"; we can speak of succeeding instants at any given point, and have thus obtained a representation of the space and time of our experience in so far as their geometrical relations are concerned.*

The distance between two parallel inertia lines of the system will naturally be taken as the length of the segment intercepted by them in a separation line which intersects them both normally.

This, then, will be the meaning to be attached to the *distance between two points*.

Time intervals in the usual sense will be measured by the lengths of segments of the corresponding inertia lines: that is to say, by differences of the *t* coordinates.

Since we have defined the equality of separation and inertia segments in terms of the relations of *after* and *before* and have assigned an interpretation of these, it follows that the equality of length and time intervals in the ordinary sense is rendered precise.

It is to be observed that the particular system of parallel inertia lines which we may select is quite arbitrary although the set of elements or instants contained in the entire system is in all cases identical.

The distinction between different systems is that while two parallel inertia lines represent the time paths of unaccelerated particles which are at rest relative to one another; two non-parallel inertia lines represent the time paths of unaccelerated particles which are in motion with uniform velocity with respect to one another.

Thus we are able to give a definition of absence of acceleration, but, since all inertia lines are on a par with one another, we can attach no meaning to a particle or system being at "absolute rest".

The definition of absence of acceleration based upon the relations of *after* and *before* and as regards a finite interval of time, may be thus expressed:

*Definition.* If *A* and *B* be two distinct elements of any inertia line (*B* being *after A*), then a particle will be said to be *unaccelerated from the instant A to the instant B* provided it lies in the inertia line *AB* throughout that interval.

The physical signification of an optical line is: that a flash of light or other instantaneous electromagnetic disturbance in going directly from one particle to another would follow this time path.

As regards a separation line; since no element of it is either *before* or *after* another, then if our view be correct, no single particle could occupy more than one element, and so particles which occupy distinct elements of any separation line must be separate particles.

The above considerations indicate the reasons for adopting the names we have assigned to the three types of general line.

The names inertia, optical and separation, as applied to general planes and general threefolds, have been given on account of certain analogies with the corresponding types of general lines.

In the first edition of this work the names: "acceleration plane" and "rotation threefold" were used instead of inertia plane and inertia threefold respectively; but the present nomenclature is more systematic and permits of systematic extension to Conical Order in $n$ dimensions.

Results involving only three coordinates $x$, $y$ and $t$ may be visualised by means of the three-dimensional conical order described in the introduction, but a certain amount of distortion appears in a model of this kind, since equal lengths in the model do not in general represent equal lengths as we have defined them.

The optical significations of Posts. I to XVIII are however made clear by such models, and it is easily seen that the assertions made in these postulates, when interpreted in the manner described, are in accordance with the ordinarily accepted ideas.

Post. XXI also finds an interpretation in such a model, but its significance is concerned rather with the logic of continuity than with any observable physical phenomenon.

Since it is possible to define equality of lengths in terms of *after* and *before* it seems superfluous to introduce any other conception of length, since the effect of this would merely be to destroy the symmetry which otherwise exists.

It is again to be emphasised that the application of the theory of conical order does not in itself require that the $\alpha$ and $\beta$ sub-sets should be determined by optical phenomena, but merely that there should exist some influence having the properties which we have ascribed to light.

Accordingly if it should be found hereafter that some other influence than light possessed these properties we should merely require to substitute this influence for light and interpret our results in terms of it.

## CONCLUSION

Our task now approaches completion.

We have shown how from some twenty-one postulates involving the ideas of *after* and *before* it is possible to set up a system of geometry in which any element may be represented by four coordinates $x, y, z, t$.

Three of these, $x, y, z$, correspond to what we ordinarily call space coordinates, while the fourth corresponds to time as generally understood.

Since however an element in this geometry corresponds to an instant, and bears the relations of *after* and *before* to certain other instants, it appears that the theory of space is really a part of the theory of time.

Of the postulates used: nineteen, namely I to XVIII and Post. XXI, may easily be seen to have an interpretation in three-dimensional geometry by making use of cones as described in the introduction.

It follows that if ordinary geometry be consistent with itself, these nineteen postulates must be consistent with one another.

Of the remaining two postulates, Post. XIX has the effect of introducing one more dimension, while Post. XX limits the number of dimensions to four.

Since by means of these we have been enabled to set up a coordinate system in the four variables $x, y, z, t$, the question of the consistency of the whole twenty-one postulates is reduced to analysis.

It is not proposed to go further into this matter in the present volume, having said sufficient to leave little doubt that they are all consistent with one another.

The question as to whether the postulates are all independent is mainly a matter of logical nicety and is of comparatively little importance provided that the number of redundant postulates be not large.

In the course of development of the present work the writer succeeded in eliminating a considerable number of postulates which he had provisionally laid down: the redundancies being generally indicated by the possibility of proving some particular result from several sets of postulates.

One known redundancy has been permitted to remain: namely Post. II (*a*) and (*b*), which might have been deduced directly from Post. V and Post. VI (*a*) and (*b*).

By retaining Post. II, however, our first four postulates will be seen to hold for the set of instants of which any one individual is directly conscious, and the subject is thus better exhibited as an extension of the commonly accepted ideas of time.

A still further diminution of the number of postulates might have been made by combining Posts. VI and XI in the way mentioned on p. 42, but to have done so would have complicated still further the initial part of the subject, since Post. VI implies merely a two-dimensional conical order, while its combination with Post. XI makes the set of elements at least three-dimensional from the very beginning.

Apart from the above-mentioned, no further definite indications of redundancy have been observed, and, although some redundant postulates may still remain, it seems unlikely that there can be many.

This opinion is confirmed by a comparison with the number of fundamental assumptions given by various writers on the foundations of ordinary geometry.

We have now concluded the exposition of the argument by which we have been led to the view expressed in the introduction: that *spacial relations are to be regarded as the manifestation of the fact that the elements of time form a system in conical order: a conception which may be analysed in terms of the relations of* after *and* before.

This view would appear to have important bearings on general philosophy, but into these we do not purpose here to enter.

One point may however be mentioned:

The fundamental properties of time must, on any theory, be regarded as possessing a character which is not transitory, but in some sense persistent; since otherwise, statements about the past or future would be meaningless.

We here touch on the difficult problem as to the nature of "universals": a problem which has been much discussed by philosophers, but appears to be still far from a satisfactory solution.

Though space may be analysable in terms of time relations, yet these remain in their ultimate nature as mysterious as ever; and though events occur in time, yet any logical theory of time itself must always imply the Unchangeable.

## APPENDIX

It is worthy of note that, just as by treating a system of co-directional inertia lines as points, we may represent ordinary Euclidean geometry in our time-space continuum: so by means of a system of inertia lines having a common element, we may represent the geometry of Lobatschewski in a very analogous manner.

We shall first prove a certain theorem with regard to three such inertia lines.

Let these be denoted by $l_1$, $l_2$ and $l_3$ and let them have the common element $O$.

Let $A$ be any element of $l_1$ which is *after* $O$ and let a separation line be taken through $A$ normal to $l_1$ and lying in the inertia plane containing $l_1$ and $l_2$ and let it intersect $l_2$ in $B'$.

Similarly let a separation line be taken through $A$ normal to $l_1$ and lying in the inertia plane containing $l_1$ and $l_3$ and let it intersect $l_3$ in $C'$.

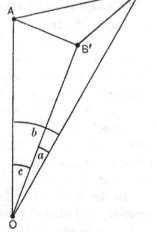

Then, since $OA$ is normal to both $AB'$ and $AC'$, it follows that $AB'$ and $AC'$ lie in a separation plane, so that $B'C'$ is a separation line and the relations of the sides and angles of the triangle whose corners are $A$, $B'$, $C'$ are the same as in ordinary Euclidean geometry, and we shall denote $\angle B'AC'$ by $A$.

Also, since both $B'$ and $C'$ are neither *before* nor *after* $A$, while $A$ is *after* $O$, it follows that both $B'$ and $C'$ are also *after* $O$, and accordingly, the inertia half-lines $OA$, $OB'$ and $OC'$ having the common end $O$ must make proper hyperbolic angle-boundaries with one another.

Fig. 57.

We shall denote the proper hyperbolic angles $\angle C'OB'$, $\angle AOC'$ and $\angle B'OA$ by $a$, $b$ and $c$ respectively, and shall use the abbreviation *conj* for the word conjugate.

Then, as we have seen

$$(\operatorname{conj} B'C')^2 = 2OB' . OC' \cosh a - OB'^2 - OC'^2.$$

Also $$B'C'^2 = AB'^2 + AC'^2 - 2AB' . AC' \cos A;$$

so that

$$(\operatorname{conj} B'C')^2 = (\operatorname{conj} AB')^2 + (\operatorname{conj} AC')^2 - 2 (\operatorname{conj} AB')(\operatorname{conj} AC') \cos A.$$

Thus

$$2OB' \cdot OC' \cosh a = OB'^2 + (\text{conj } AB')^2 + OC'^2 + (\text{conj } AC')^2$$
$$- 2 (\text{conj } AB') (\text{conj } AC') \cos A,$$

or $\quad 2OB' \cdot OC' \cosh a = 2OA^2 - 2 (\text{conj } AB') (\text{conj } AC') \cos A.$

This may be written in the form

$$\cosh a = \frac{OA}{OC'} \cdot \frac{OA}{OB'} - \frac{(\text{conj } AC')}{OC'} \cdot \frac{(\text{conj } AB')}{OB'} \cos A,$$

or $\qquad \cosh a = \cosh b \cosh c - \sinh b \sinh c \cos A \qquad \ldots\ldots(1);$

where $\cos A$ is equal to the cosine of the di-hedral angle which the two inertia planes containing $l_1$ make with one another.

By similar constructions taken with respect to $l_2$ and $l_3$, denoting the corresponding di-hedral angles by $B$ and $C$ respectively, we may deduce the equations

$$\cosh b = \cosh c \cosh a - \sinh c \sinh a \cos B \qquad \ldots\ldots(2),$$

and $\qquad \cosh c = \cosh a \cosh b - \sinh a \sinh b \cos C \qquad \ldots\ldots(3).$

These equations (1), (2) and (3) are the relations connecting the sides and angles of a triangle in the geometry of Lobatschewski.

Now let $R$ be the set of elements which are *after* $O$ exclusive of those which lie in the $\alpha$ sub-set of $O$, and consider the portions lying in the region $R$ of all inertia lines, inertia planes and inertia threefolds which pass through the element $O$.

If then, for the purpose of this representation, we call such a portion of an inertia line a *point*; such a portion of an inertia plane a *line*, and such a portion of an inertia threefold a *plane*; we see that: any two points determine a line; while any three points which do not lie in one line determine a plane.

Also the sides and angles of any triangle satisfy the relations of the geometry of Lobatschewski.

It is to be observed that, since optical and separation lines which pass through $O$ do not lie in the region $R$, two inertia planes passing through $O$ and intersecting in an optical or separation line through $O$ correspond to lines in a plane which have no point in common. In the case where the inertia planes intersect in an optical line they correspond to Lobatschewski parallels; since an optical half-line does not make a finite hyperbolic angle with any inertia half-line.

It is obvious that similar results hold if, instead of the region $R$, we take a region $R'$ consisting of all elements which are *before* $O$ exclusive of those which lie in the $\beta$ sub-set of $O$.

# INDEX

*References in* **dark** *type are to definitions*

Addition of angles, 351 *et seq.*
*After*, 6 *et seq.*
α sub-set, **27**
Angle-boundary in inertia plane, proper hyperbolic, **385**
  vertex, sides, **385**
Angle-boundary in separation plane, **337**
  vertex, sides, acute, obtuse, etc., **337**
  supplement of, congruence of, 338
Angle in inertia plane, natural measure of proper hyperbolic, **386**
Angle in separation plane, natural measure of, **357**
Angular interval in inertia plane, proper hyperbolic, **386**
Angular interval in separation plane, **348**
  null, circuit, etc., supplement of, conjugate, **349**
Angular segment in inertia plane, proper hyperbolic, **386**
Angular segment in separation plane, **348**
  acute, obtuse, right, flat, etc., **349**
  congruence of, **351**
Archimedes, 104, 308, 345

*Before* in terms of *after*, 27
β sub-set, **28**
*Between* a pair of optical lines in an inertia plane, 67
Between, linearly, **119**

Circle, separation, **328**
  centre, radius, diameter, **328**
  inside, outside, **329**
Co-directional congruence of pairs, **283**
Co-directional general lines, **125**
Complete normality of general planes, **221**
Cone, standard, **390**
Congruence, co-directional of pairs, **283**
  general, of inertia pairs, **295**
  general, of separation pairs, **295**
  of inertia pairs having a self-corresponding element, 276
  of separation pairs having a self-corresponding element, 279
Conjugate angular segments and intervals, **349**
Conjugate inertia and separation lines, **163**
Conjugate pairs, **278**
Co-ordinates, introduction of, 387 *et seq.*
  symmetrical, 399
Cylinder, optical circular, **362**

Dedekind, 343, 344, 345
De Moivre, 351, 356, 357

Einstein, 11, 12, 13, 21
Equations of general lines, planes and threefolds, 395 *et seq.*
Euclid, 2, 5, 313, 314, 358
Eudoxos, 313

First element of an inertia line which is *after* an element, **167**
FitzGerald, 10
Fizeau, 8, 9, 21

General line, **63**
General plane, **192**
General threefold, **229**
  sets of elements which determine different types of, 242–245
Generator of inertia plane, **57**
  of optical plane, **146**
  of optical or inertia threefold, **246**

Half-line, general, **309**
  end of, **309**
Half-plane, general, **310**
  boundary of, **310**
Hill, 313, 314
Hyperbolic angles, proper, 379 *et seq.*
Hypotenuse, **333**

Inertia line, **63**
Inertia plane, **52**
  sets of elements which determine, 70–72
Inertia threefold, **245**
Interpretation of results, 400 *et seq.*
Intersection, of optical lines, **42**
  of general lines, **77**
  of general line and general plane, **229**
  of general line and general threefold, **260**
Interval, linear, **308**

James, 22

Kelvin, Lord, 13

Lagrange, 13
Larmor, Sir J., 11
Last element of an inertia line which is *before* an element, **167**
Length of segment, numerical value of in terms of unit segment, 345
Line, general, **63**
  inertia, **63**
  optical, **30**
  separation, **63**
Linear interval, **308**
*Linearly between*, **119**
Lobatschewski, 405, 406
Lodge, Sir O., 11
Lorentz, 10, 11

Mean, of two elements in an inertia or separation line, **98**
  of two elements in an optical line, **98**
Michelson, 9, 11
Minkowski, 1, 6, 13, 17, 18, 21, 22
Morley, 9, 11

Printed in the United States
By Bookmasters